DYNAMICS OF THE UPPER ATMOSPHERE

DEVELOPMENTS IN EARTH AND PLANETARY SCIENCES

DYNAMICS OF THE UPPER ATMOSPHERE

Susumu Kato

Developments in Earth and Planetary Sciences

01

Center for Academic Publications Japan / Tokyo
D. Reidel Publishing Company / Dordrecht · Boston · London

Library of Congress Cataloging in Publication Data

DATA-APPEARS ON SEPARATE CARD

ISBN 90-277-1132-1

Published by Center for Academic Publications Japan, Tokyo, in co-publication with D. Reidel Publishing Company, P. O. Box 17, 3300 AA Dordrecht.

Sold and distributed in Japan, China, Korea, Taiwan, Indonesia, Cambodia, Laos, Malaysia, Philippines, Thailand, Vietnam, Burma, India, Bangla Desh, Sri Lanka by Center for Academic Publications Japan, 4-16, Yayoi 2-chome, Bunkyo-ku, Tokyo 113, Japan.

Sold and distributed in the U.S.A. and Canada by Kluwer Boston Inc., Lincoln Building, 160 Old Derby Street, Hingham, MA 02043, U.S.A.

Sold and distributed in all other countries by Kluwer Academic Publishers Group, P. O. Box 322, 3300 AH Dordrecht Holland.

Printed in Japan

PREFACE

Studies related to the earth and planets along with their surroundings are of great concern for modern scientists. Global geodynamics as represented by the plate tectonics has now become one of the most powerful tools by which we can study the causes of earthquakes, volcanic eruptions, mountain formation and the like. Various missions sent out to the space, manned or unmanned, brought out geoscientific features of the moon, Mars, Venus and other planets. Earthquake prediction that was the business of astrologers and fortune-tellers some twenty years ago, has now grown up to be an important science. A number of destructive earthquakes were successfully forecast in the People's Republic of China.

In the light of the above-mentioned and other accomplishments in geosciences, we feel that it is a good thing to publish a series of monographs which review selected topics of earth and planetary sciences. We are of course well aware of the fact that similar monographs have been and will be published from overseas publishers. The series, which we plan to publish, will therefore put stress on Japanese work. But we hope that the series will also include review articles by distinguished overseas authors.

The series, which is named the "Developments in Earth and Planetary Sciences" will be published by the Center for Academic Publications Japan and the D. Reidel Publishing Company. It is my great pleasure to work as the Editor of the series. I should like to have comments on what subjects we shall choose in the future publications. I shall be greatly obliged if anyone would suggest suitable subjects and potential authors for the series to me.

Tsuneji Rikitake
Editor

PREFACE TO VOLUME 1

This book is a mathematically oriented introduction to the dynamics of the earth's upper atmosphere, with special emphasis on acoustic-gravity and tidal waves and their ionospheric effects. The basic equations governing atmospheric motions and the equilibrium state of the atmosphere are presented in Chapter 1. In Chapter 2 the perturbation equations about the basic state are examined for an isothermal plane atmosphere. The dispersion equation relating wave frequency and horizontal and vertical wavelengths is then examined in detail to determine the propagation characteristics of the waves. Several examples of the excitation of these waves by simple sources are then considered. Finally the ionospheric perturbations due to these atmospheric waves will be considered. Chapter 3 is devoted to the development of the theory of atmospheric tides, i.e. global oscillations on a non-isothermal spherical atmosphere with periods which are an integral fraction of a day. Classical tidal theory will be presented in considerable mathematical detail, due both to its continuing relevance today and its value as a mathematical training for students. In addition, recent developments in tidal theory and observations will be presented. Chapter 4 is a brief introduction to the electrodynamics of the ionosphere. Basic plasma concepts are introduced, with an emphasis on anisotropic electrical conductivities and ionospheric currents. Chapter 5 is then devoted to a thorough discussion of the dynamo theory, which explains quiet time magnetic perturbations as the results of current flows in the ionosphere driven by tidal winds.

The author feels that much of the progress that has been made in understanding the dynamics of the upper atmosphere in the last two decades is the direct result of three technological innovations; rockets and the satellites they launch, computers, and high-powered radars. Space vehicles have made it possible to make in-situ measurements in the upper armosphere. Computers have made possible the construction of increasingly realistic mathematical models of upper atmosphere dynamics.

The large incoherent scatter radars were built in the early 1960's in order to investigate the earth's ionosphere. In the succeeding years these radars have proved to be one of the most powerful tools for investigating the dynamics of the neutral atmosphere as well. These ground-based radars are able to measure continuously in time and simultaneously over a wide range of altitudes various ionosphere parameters, tremendous advantages in the study of wave phenomena. Recently, it has been shown that these radars can make important measurements in the lower and middle atmosphere. Observations by these radars will be discussed in many parts of this book.

Much of the material in this book was first presented in a graduate course taught by the author, while he was a visiting professor in the Meteorology Department at the University of California at Los Angeles in 1973–74.

The author was fortunate to have Dr. R. M. Harper of Rice University, Houston, Texas, as a visiting scientist at the Ionosphere Research Laboratory, Kyoto University, during the period when this book was written. The author is indebted to Dr. Harper for carefully reading the original manuscript and making several helpful comments.

The author's thanks also go to the many colleagues who permitted the use of tables and diagrams from their works. The author is indebted to Dr. T. Aso, Ionosphere Research Laboratory, Kyoto University, for his carefully reading the manuscript. The author acknowledges with thanks that Mr. T. Tsuda, a research assistant, Ionosphere Research Laboratory, has done numerical calculations for Table 3. 1 and Figs. 3. 1–3. 3.

The author is very thankful to Professor T. Rikitake for supervising the editing of this book. The author's sincere thanks also go to Mrs. Kato, who typed the manuscript, and to Miss M. Tatsukawa, who assisted in checking references and diagrams.

Publication of this book is partly subsidized by the Ministry of Education, Science, and Culture of Japan.

Susumu Kato

Contents

LIST OF SYMBOLS

Only those symbols which appear often are shown. Other symbols will be defined wherever necessary. The numbering of mathematical formulas is shown as (i, j) where i is the chapter number and j the formula number in the chapter. (A, i, j) is used for formulas in Appendix.

Symbols	Definition
a	Planet radius, 6,370 km for the earth.
f	Frequency normalized by 2Ω as $f=(\omega/2\Omega)$.
$\boldsymbol{g}$	Acceleration of gravity on unit mass, 9.81 m sec^{-2} for the earth on the surface.
e	Electron charge, 1.60×10^{-19} Coulomb.
$\boldsymbol{k}$	Wave number.
$m_{e,i}$	Electron and ion mass, respectively.
m	Mach number, $m=1$ for sonic speed, $m>1$ super sonic, $m<1$ subsonic.
m_0	(ω_g/ω_a) which is about 0.9 for diatomic gas.
$n_{e,i}$	Electron and ion number density, respectively.
p	Atmospheric pressure, $p_{0,1}$ static and perturbed parts, respectively, as $p=p_0+p_1$.
r	Radial distance measured from the planet center.
$s_g(x)$	Sign function $s_g(x)=\pm1$ for $x\gtrless0$, respectively.
t	Time
x, y, z	A right-handed Cartesian coordinate; usually x is taken positive southward, y eastward and z upward. z is measured from the ground, $r=a+z$.
$\boldsymbol{B}$	Magnetic flux density. $\boldsymbol{B}_0$ is $\boldsymbol{B}$ of the geomagnetic field.

$$B_0 = \frac{1}{2} B_{00} \sqrt{1+3\cos^2\theta}$$

$$B_{0\theta} = -\frac{1}{2} B_{00} \sin\theta$$

$$B_{0\phi} = 0$$

Symbols	Definition
	$B_{0z} = -B_{00}\cos\theta$ where $B_{00}=0.63$ gauss $=6.3\times10^{-5}$ Weber m^{-2} for Centered dipole parallel to the earth's rotation axis.
C	Sonic speed, $\sqrt{\kappa RT}$
$\boldsymbol{E}$	Electric field.
H	Atmospheric scale height, RT/g.
$H(x)$	Step function, $H(x)=0$ for $x<0$, $H(x)=1$ for $x>0$.
$\Delta\boldsymbol{H}$	Geomagnetic variation (in Section 5.1), $\Delta\boldsymbol{H}(\Delta H^x, \Delta H^y, \Delta H^z)$
I	Geomagnetic dip angle.
$\boldsymbol{I}$	Integrated horizontal electric current, $\boldsymbol{I}=\int\boldsymbol{J}_{x,y}\,dz$.
$\boldsymbol{J}$	Electric current density.
R	Gas constant per unit mass per unit time.
Q	Heat input per unit mass per unit time.
T	Temperature, $T_{0,1}$ as in $p_{0,1}$, $\rho_{0,1}$ and $W_{0,1}$.
$\boldsymbol{W}$	Atmospheric velocity, $\boldsymbol{W}_{0,1}$ as in $p_{0,1}$, $T_{0,1}$ and $\rho_{0,1}$.
α, β, γ	x, y and z components of wave number $\boldsymbol{k}$, respectively.
$\delta(x)$	Delta function, $\delta(x)=0$ for $x\neq0$ and $\int_{-\infty}^{\infty}\delta(x)=1$.
θ	Colatitude, $90°-$(latitude), $0\leq\theta\leq180°$.
σ	Boltzmann constant, 1.38×10^{-23} J °K^{-1}.
$\mu_{e,i}{}^{(1),(2),(0)}$	Mobilities of electrons and ions denoted by subscripts e and i, respectively. Superscripts (1), (2) and (0) denote Pedersen, Hall and parallel mobilities, respectively.
μ_0	Viscosity coefficient.
μ_v	Kinematic viscosity, (μ_0/ρ_0).
$\nu_{e,i}$	Collision frequencies of electrons and ions, respectively.
ρ	Atmospheric mass density, $\rho_{0,1}$ as in $p_{0,1}$, $T_{0,1}$ and $\boldsymbol{W}_{0,1}$.
κ	Ratio of heat capacities at constant pressure and at constant volume.
$\sigma_{0,1,2}$	Parallel, Pedersen and Hall conductivities, respectively.
ϕ	Longitude, positive eastward.
ω	Angular frequency.
ω_g	Brünt-Väisälä frequency, $\sqrt{(\kappa-1)}\,g/C$ for isothermal atmosphere.
ω_a	Acoustic cutoff frequency, $\kappa\, g/(2C)$ for isothermal atmosphere.
∇_H	Gradient in the horizontal plane.
Λ	Gravitational tidal potential.
χ	Divergence of velocity,

$$\frac{\partial W_x}{\partial x}+\frac{\partial W_y}{\partial y}+\frac{\partial W_z}{\partial z}=\frac{1}{r^2}\frac{\partial}{\partial r}(r^2W_r)+\frac{1}{r\sin\theta}\frac{\partial(W_\theta\sin\theta)}{\partial\theta}+\frac{1}{r\sin\theta}\frac{\partial W_\phi}{\partial\phi}$$

Symbols	Definition
	$\simeq \frac{\partial W_z}{\partial z} + \frac{1}{a \sin\theta} \frac{\partial (W_\theta \sin\theta)}{\partial \theta} + \frac{1}{a \sin\theta} \frac{\partial W_\phi}{\partial \phi}.$
Ω	Planetary angular velocity.
Ω_c	$(\lvert z\rvert / \sqrt{y^2+z^2})\, \omega_a.$
$\Omega_{e,i}$	Gyrofrequencies of electrons and ions, respectively; $eB_0/m_{e,i}$.
Ω_0	$\sqrt{\lvert (m^2-m_0^2)/(m^2-1)\rvert}\; \omega_a.$

CHAPTER 1

GENERAL INTRODUCTION

An atmosphere is defined as a " spheroidal gaseous envelope surrounding a heavenly body " (Oxford Dictionary, 1923). The atmosphere is generally held around the planet in approximate hydrostatic equilibrium by the gravitational force of the planet. Small oscillations can take place about this equilibrium state. This book will deal with these atmospheric waves and their geophysical effects in the upper atmosphere of the earth. The bulk of the earth's atmosphere is electrically neutral; however above about 50 km a small fraction of the atmosphere is ionized. The regions where ionization is important have been called the ionosphere, since their discovery by Appleton in 1924. A nomenclature for regions of the ionosphere with different properties has since developed, with the D (70–80 km), E (80–160 km), F1 (160–250 km) and F2 (250–500 km) regions. However, no clear boundaries exist for any of the regions. The ionization is mainly produced by solar extreme-ultra- and ultra-violet radiation (for the E and F1, 2 regions) and solar soft X-rays (for the D and E regions). Precipitating energetic particles also ionize the neutral atmosphere.

There is another system of nomenclature for the earth's atmosphere based on the temperature distribution with height (Fig. 1. 1): the troposphere (ground –15 km), stratosphere (15–50 km), mesosphere (50–80 km) and thermosphere (80 km–), regions of decreasing, increasing, decreasing and increasing temperature, respectively. Recently, the stratosphere and mesosphere have been called the middle atmosphere.

The uppermost part of the atmosphere is called the exosphere. Here a neutral particle moves along a ballistic trajectory with little probability of collision. The exosphere begins near 400 km height.

1.1 Fundamental Equations for Atmospheric Motion

In this book we shall adopt a fluid dynamic approach. The atmospheric

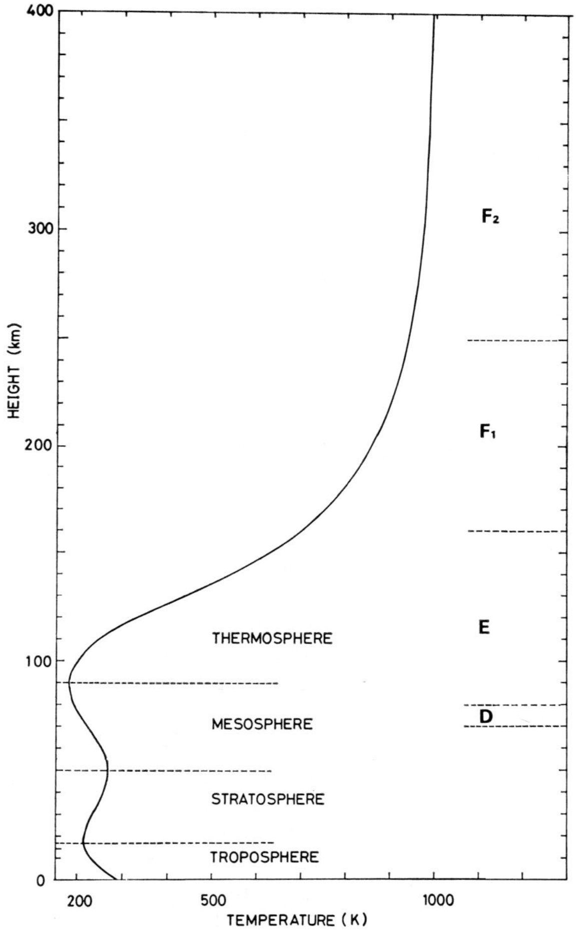

FIG. 1. 1. The earth's atmosphere; temperature distribution (CIRA, 1972) and nomenclature.

density varies very widely between the troposphere and the thermosphere. The number density of gas particles is of the order of $10^{25}\,m^{-3}$ near the earth's surface decreasing to $10^{19}\,m^{-3}$ at the height of 100 km and $10^{13}\,m^{-3}$ at 400 km. The validity of the fluid dynamic approach in a rarefied gas depends on the spatial and time scales of the phenomena under consideration. If collisions occur frequently on spatial and time scales much smaller than those characterizing the phenomena of interest, so that we can define various averages corresponding to macroscopic quantities such as density, pressure, temperature, conductivities etc., the fluid dynamic approach is applicable. The atmospheric waves that are the subject of this book belong to this category; these waves have periods of at least several tens of minutes and wavelengths of kilometers. If we think of air particles as solid spheres of radius 10^{-10} m randomly moving with a velocity of 5×10^2 m sec^{-1}, the collision frequency (per particle) with particles of number density of 10^{13} m^{-3} is about 10^{-4} sec^{-1} and, in a volume of $(10^2\text{ m})^3$ and for a time of 10^2 sec, the total collision frequency is 10^{15}. On many occasions we can adopt fluid dynamic approach even for the very rarefied gases which exist in the outer space or in solar wind.

We postulate the following fluid dynamic equations for describing the atmosphere:

$$\frac{\mathrm{D}\boldsymbol{W}}{\mathrm{D}t}+2(\boldsymbol{\Omega}\times\boldsymbol{W}) = -\frac{1}{\rho}\nabla p+\boldsymbol{g}+\boldsymbol{F} \tag{1.1}$$

where $\mathrm{D}/\mathrm{D}t=\partial/\partial t+(\boldsymbol{W}\cdot\nabla)$, showing that the velocity changes with time $(\partial/\partial t)$ but also through advection $(\boldsymbol{W}\cdot\nabla)$, the air mass moving with velocity $\boldsymbol{W}$ suffering a spatial change. $\mathrm{D}/\mathrm{D}t$, D-operator, denotes a differentiation following the motion of the fluid. Mathematically $\mathrm{D}/\mathrm{D}t$ means the total derivative. Then, (1. 1) is a direct extension of Newton's equation of motion for a particle with the velocity $\boldsymbol{W}$ under the influence of forces which are given in the right-hand side. Since we are concerned with atmospheric motion relative to the earth or planets, it would be convenient to adopt a coordinate system rotating with the planets. The Coriolis force (some may say the Coriolis acceleration, if it is on the left-hand side in the equation of motion), $2(\boldsymbol{\Omega}\times\boldsymbol{W})$, is necessary to describe motion by coordinates rotating with angular velocity $\boldsymbol{\Omega}$. This apparent force complicates the treatment of atmospheric motion in that the motion is not parallel to the applied force. It is easily found by comparing the Coriolis force with the inertia force $\partial\boldsymbol{W}/\partial t$ that the Coriolis force becomes important for slowly varying motion i.e. motion with lower frequency than 2Ω. Thus, in the discussion of tides with period of one or half a day, the Coriolis force is important.

The density ρ varies with time and in space as follows:

$$\frac{\partial\rho}{\partial t}+\mathrm{div}(\rho\boldsymbol{W}) = 0. \tag{1.2}$$

(1. 2) shows that the density at one point varies with the divergence or convergence of the mass flow $(\rho\boldsymbol{W})$ as is expected. Thus, this equation shows the mass conservation in fluid dynamics. If any chemical reactions destroy gas particles, this equation must include such effects. But such a case is not considered in this book, except in the ion and electron continuity equations which will be given later.

(1. 2) can also be written as

$$\frac{\mathrm{D}\rho}{\mathrm{D}t}+\rho\,\mathrm{div}\boldsymbol{W} = 0.$$

The energy conservation law in thermodynamics says that the heat input Q per unit mass per unit time is consumed in elevating the temperature T and in the work done by the pressure p. Since the air mass advects, this must be given by the D-operator

$$Q = C_v \frac{\mathrm{D}T}{\mathrm{D}t} + p\,\frac{\mathrm{D}}{\mathrm{D}t}\left(\frac{1}{\rho}\right) \tag{1.3}$$

where C_v is the specific heat capacity per unit mass. There is a relation among T, ρ, and p with R as the gas constant i.e.

$$p = \rho R T \tag{1.4}$$

which is the equation of state for a perfect gas. (1. 4) is approximately true for the real atmosphere. Deviation from (1. 4) does occur when the phase of a gas changes. However, for dry air, which is our principal concern, this cannot happen. In (1. 3) Q can be either positive or negative heat sources. Radiative heating or cooling, joule heating or loss, thermal conduction, chemical heating etc. can be these sources. Equations (1. 1)–(1. 4) constitute a set of equations necessary to describe the behavior of the atmosphere in motion. Another expression is often used for (1. 3). This is obtained after eliminating T by (1. 4) in (1. 3)

$$\frac{\mathrm{D}p}{\mathrm{D}t} - C^2\frac{\mathrm{D}\rho}{\mathrm{D}t} = (\kappa-1)Q\rho \tag{1.3'}$$

where C is the sound speed as

$$C = \sqrt{\kappa R T} = \sqrt{\kappa\,\frac{p}{\rho}}. \tag{1.5}$$

κ is the ratio of heat capacities at constant pressure and constant volume, $\kappa=1.4$ for diatomic molecules.

Among the various forces included in $\boldsymbol{F}$, the viscous and hydromagnetic forces are important in the upper atmosphere, because the kinematic viscosity (proportional to molecular diffusion) becomes increasingly large with height and the atmosphere is ionized above 70 km.

Viscosity tends to diffuse the macroscopic momentum of the atomosphere among neutral particles composing the atmosphere, whilst ions tend to take the momentum from the atmosphere through collisions. In this respect, viscous and hydromagnetic forces are similar, contributing to attenuate the upper atmospheric motion. However, it will be found that there are other complex effects by these forces. The hydromagnetic force in $\boldsymbol{F}$ in (1. 1) is given as $(\boldsymbol{J}\times\boldsymbol{B}_0)/\rho$ where $\boldsymbol{J}$ is the electric current density, and $\boldsymbol{B}_0$ the geomagnetic field flux density. As to the viscous forces the general expression is fairly complicated (e.g. PITTEWAY and HINES, 1963), because it depends on the scale of motion as well as height-dependent coefficients. Among various simplifications usually done, a linear form proportional to $\boldsymbol{W}$ is called Rayleigh friction. This is easy to handle but misleading on some occasions. The loss of momentum by viscous and hydromagnetic forces accompanies a loss of energy i.e. joule heating and thermal conduction, respectively. These losses of energy must be simultaneously considered in Q in the energy equation in a self-consistent approach. Such an approach has recently been attempted, but the present situation is still far from being satisfactory. Radiative cooling is an important loss mechanism in the middle atmosphere. Its simple expression is given by a linear form proportional to temperature in (1. 3). This is called Newtonian cooling.

For discussing the dynamic energy flux the following relation may be helpful:

$$-\frac{\operatorname{div}(p\boldsymbol{W})}{\rho}=\frac{\mathrm{D}}{\mathrm{D}t}\left(\frac{W^2}{2}\right)-\boldsymbol{g}\cdot\boldsymbol{W}-\boldsymbol{F}\cdot\boldsymbol{W}+\frac{p}{\rho^2}\left(\frac{\mathrm{D}\rho}{\mathrm{D}t}\right) \tag{1.6}$$

by (1. 1) and (1. 2); the right-hand side implies the increase of kinetic energy ($W^2/2$) per unit mass per unit time, the work done by gravity ($\boldsymbol{g}\cdot\boldsymbol{W}$) and other forces $\boldsymbol{F}\cdot\boldsymbol{W}$ per unit mass per unit time, and the work done by pressure $(-p/\rho^2)(\mathrm{D}\rho/\mathrm{D}t)$ per unit mass per unit time. (1. 6) shows that this variation is supported by the energy flux ($p\boldsymbol{W}$) whose convergence rate into a unit volume through its surface is $[-\operatorname{div}(p\boldsymbol{W})]$, so $[-\operatorname{div}(p\boldsymbol{W})]/\rho$ is for a volume of unit mass. (1. 6) is modified by using the thermal energy relation (1. 3) as

$$Q-\frac{\operatorname{div}(p\boldsymbol{W})}{\rho}=\frac{\mathrm{D}}{\mathrm{D}t}\left(\frac{W^2}{2}+\frac{1}{\kappa-1}\frac{p}{\rho}\right)-\boldsymbol{g}\cdot\boldsymbol{W}-\boldsymbol{F}\cdot\boldsymbol{W}. \tag{1.7}$$

The fluid energy density per unit mass is defined by (1. 7) as (LAMB, 1945)

$$\frac{W^2}{2}+\frac{1}{\kappa-1}\frac{p}{\rho}. \tag{1.8}$$

The heat and dynamical energy flow on the left-hand side in (1. 7) are consumed in increasing the energy density (1. 8) and in doing work.

1.2 Hydrostatic Equilibrium

The earth's atmosphere as other planetary atmospheres cannot be truly static because of various disturbances constantly in existence. However, we can still consider the average state which is approximately static. Real states can be expressed as variations superposed on this static state. Such a static state can be given by neglecting $\partial/\partial t$, $\boldsymbol{W}$ and $\boldsymbol{F}$ in (1. 1) which reduces to

$$\frac{\mathrm{d}p_0}{\mathrm{d}z}=-\rho_0 g \tag{1.9}$$

where subscript 0 denotes static part. Because of no balancing force in the horizontal direction, p_0 is constant with latitudes and longitudes. The equation of state (for ideal gas) is $p_0=\rho_0 R T_0$ where R is the gas constant given by the Boltzmann constant σ and the mean molecular weight $\bar{m}$ as $R=(\sigma/\bar{m})$. Then, (1. 9) can be integrated as

$$p_0 = p_{00} \exp\left(-\int \frac{dz}{H}\right) \tag{1.10}$$
$$\rho_0 = \rho_{00} (T_{00}/T_0) \exp\left(-\int \frac{dz}{H}\right)$$

where H is the scale height as

$$H = \frac{RT_0}{g} \tag{1.11}$$

ρ_{00}, p_{00} and T_{00} are ρ_0, p_0 and T_0 at a reference level $z=0$.

In the earth's lower atmosphere turbulent motions cause the atmosphere to be thoroughly mixed i.e. the time scale for mixing is much less than the time scale for diffusive separation of the atmospheric species. Thus, in the lower atmosphere, the atmospheric species are distributed by (1. 10) with a common scale height H, representative of the mean molecular mass $\bar{m}$. However, diffusion becomes more effective with increasing altitude, and above some height turbulent mixing cannot be as fast as diffusion separation. In the earth's atmosphere it is expected that this change of regimes occurs around 100–110 km. Luminous clouds produced by chemical release from rockets in the ionosphere clearly show the cessation of turbulence around 100–110km. The lower atmosphere is sometimes called the homosphere, and the region where diffusion becomes of comparable or greater importance the homopause. The turbopause is also used to refer to the height where turbulence is observed to cease. Above the homopause or turbopause each atmospheric species is distributed in approximate hydrostatic equilibrium with the scale height determined by its mass. Since the scale height H is larger for lighter gases, (1. 10) shows that lighter gases O, He and H tend to dominate with increasing height.

If the atmosphere is isothermal and g is constant with height, H is uniform. Then, (1. 10) is simply integrated as

$$\frac{p_0}{p_{00}} = \frac{\rho_0}{\rho_{00}} = \exp\left(-\frac{z}{H}\right) \tag{1.12}$$

which shows

$$p_0, \rho_0 \to 0 \qquad \text{with } z \to \infty. \tag{1.13}$$

Thus, the isothermal atmosphere can be static.

As non-isothermal atmosphere, consider the case where $T_0 = T_{00}(1-\varepsilon_t z)$, the temperature decreasing linearly with height as $dT_0/dz = -\varepsilon_t T_{00}$ where $(-dT_0/dz)$ is called the temperature lapse rate. By (1. 10), then,

$$p_0 = p_{00} (1 - \varepsilon_t z)^{(1/\varepsilon_t H_0)} \tag{1.14}$$

where H_0 is H at a reference level with $T = T_{00} (z=0)$. Also

$$\rho_0 = \rho_{00} (1-\varepsilon_t z)^{[-1+(1/\varepsilon_t H_0)]}. \tag{1.15}$$

(1. 14) and (1. 15) show that this atmosphere is limited for $z<1/\varepsilon_t$. In the earth's troposphere, $\varepsilon_t T_{00} \simeq 6°$ km^{-1} which gives this limit z_0 as 50 km; the troposphere ends below this limit and the stratosphere starts with increasing temperature with height. If $\varepsilon_t H_0=1$, $\rho_0=\rho_{00}$ for any z. ρ_0 can even increase for $\varepsilon_t H_0>1$. Such an atmosphere is unstable and cannot exist except for very localized parts of the atmosphere.

When we consider the decrease of gravity with distance r as

$$g = g_0\left(\frac{a}{r}\right)^2 \tag{1.16}$$

where g_0 is g at the earth's surface, $r=a$, (1. 10) can be written as a function of r instead of z, giving for the isothermal atmosphere

$$p_0 = p_{00} \exp\left[-\left(\frac{a}{H_0}\right)\left\{1-\left(\frac{a}{r}\right)\right\}\right] \tag{1.17}$$

which shows

$$p_0 \to p_{00} \exp\left(-\frac{a}{H_0}\right) \neq 0 \quad \text{with } r \to \infty \tag{1.18}$$

implying that the atmosphere cannot be static.

In the earth's thermosphere, it has been found that an empirical formula approximates the mean temperature variation with height (JACCHIA, 1964), as

$$T(z) = T_\infty - [T_\infty - T_0(z_0)] \exp[-\mathscr{S}(z-z_0)] \tag{1.19}$$

where T_∞ is the exospheric temperature (Fig. 1. 1); z_0 is taken as 120 km, and $\mathscr{S}$ is an analytic function of T_∞. Later, (1. 19) has been modified by WALKER (1965) to enable analytical integration of (1. 9):

$$T(z) = T_\infty - [T_\infty - T(z_0)] \exp(-\eta \mathscr{Z}) \tag{1.20}$$

where $\mathscr{Z}$ is the geopotential height and is defined as

$$\mathscr{Z} = \frac{(z-z_0)(a+z_0)}{a+z} \tag{1.21}$$

with z and a being measured in kilometers; η is a constant chosen to give a good fit to (1. 19). It is found that (1. 19) and (1. 20) differ only slightly (less than 1% for $T_\infty=200$ K), if $z_0=120$ km and

$$\eta = \mathscr{S} + \frac{1}{a+120} = \mathscr{S} + 0.00015. \tag{1.22}$$

Now transforming z to $\mathscr{Z}$ by (1. 21), we can integrate (1. 10) to obtain

$$\rho_0 = \rho_0(z_0)\left[\frac{T(z_0)}{T_\infty \exp(\eta\mathscr{Z}) - T_\infty + T(z_0)}\right]^{(1+\zeta)} \exp(-\eta\mathscr{Z}) \tag{1.23}$$

where

$$\zeta = \frac{1}{\eta \mathfrak{H}} \tag{1.24}$$

$$\mathfrak{H} = \frac{RT_\infty}{g(z_0)}.$$

$\mathfrak{H}$ is a quantity similar to H but with $T=T_\infty$ and g at $z=z_0$. Note that, as (1. 18),

$$\rho_0 \to \rho_0(z_0) \left\{ \frac{T(z_0)}{T_\infty \exp\left[\eta(a+z_0)\right]} \right\}^{(1+\zeta)} e^{\eta(a+z_0)} \neq 0 \tag{1.25}$$

with $z\to\infty$ i.e. $\mathcal{Z}\to(a+z_0)$. These peculiar situations cannot be removed, even if the temperature decreases with altitude. An example belonging to this category is that where the atmosphere is constantly heated from below, and the heat is carried upwards by thermal conduction to produce a negative temperature gradient with height (Roberts and Soward, private communication, 1972). In the stationary state

$$4\pi\, r^2 \kappa_t \left(\frac{\mathrm{d}T_0}{\mathrm{d}r} \right) = -E_\infty = \text{const} \tag{1.26}$$

where κ_t is the thermal conductivity and E_∞ is the energy flux; $E_\infty > 0$. By (1. 4) and (1. 9),

$$\frac{\mathrm{d}p_0}{p_0} = \left(4\pi \frac{g_0 a^2}{RE_\infty} \kappa_t \right) \left(\frac{\mathrm{d}T_0}{T_0} \right) \tag{1.27}$$

which gives

$$p_0 = p_{00} \exp\left[\frac{4\pi a^2 g_0}{RE_\infty} \int_{T_{00}}^{T_0} \left(\frac{\kappa_t}{T_0} \right) \mathrm{d}T_0 \right] \tag{1.28}$$

for neutral gas $\kappa_t = \kappa_0 \sqrt{T_0}$ and

$$p_0 = p_{00} \exp\left[-\frac{8\pi g_0 a^2}{RE_\infty} \kappa_0 \left(\sqrt{T_{00}} - \sqrt{T_0} \right) \right] \tag{1.29}$$

and p_0 cannot vanish with $r\to\infty$ (corresponding to $T_0 \to T_\infty < T_{00}$). This causes the stellar wind to blow outwards. Thus, planetary atmospheres cannot be static in a rigorous sense, although, generally, p_∞ is very small for planets. Note that in the above the centrifugal force is not considered. It only enhances instability. However, the lower atmosphere can be considered static in the vertical direction because of strong gravitational attraction.

1.3 Atmospheric Temperature Distribution and Brünt-Väisälä Frequency

The earth's atmosphere has the temperature profile as shown in Fig. 1. 1. The decreasing temperature with height in the troposphere is produced mainly by adiabatic expansion of the convecting atmosphere. In the stratosphere the atmospheric temperature increases with height due to radiative heating by ozone O_3. The heat balance is established between ultraviolet absorption of O_3 and infrared radiation from water vapor (H_2O), carbon dioxide (CO_2) and O_3. The temperature attains its maximum of 270°K around 50 km and again decreases up to the mesopause around 80 km. In the mesosphere the temperature is determined by the balance between radiative heating due to molecular oxygen O_2 and infrared radiative cooling due to CO_2. Above 80 km, the atmospheric temperature increases monotonically with height due to radiative heating by dissociation of molecular oxygen in the lower thermosphere, and by ionization of O etc. Thus, radiation balance is the main controller of temperature above the troposphere. In the polar thermosphere and mesosphere, however, we have various other energy inputs which are almost constant from above, in the form of precipitating high energy particles and hydromagnetic disturbances. These affect, to a considerable degree, the thermal structure of the polar thermosphere and mesosphere. Atmospheric waves, the subject of this book, may supply a certain amount of energy input to the mesosphere and thermosphere.

Under such a temperature distribution the earth's atmosphere is in hydrostatic equilibrium. The stability of the equilibrium can be examined by considering a small displacement from equilibrium. Consider an air mass which is displaced upwards from the hydrostatic equilibrium position at z_0 in a quasi-static way by a small distance (Fig. 1. 2). For upward ξ the mass

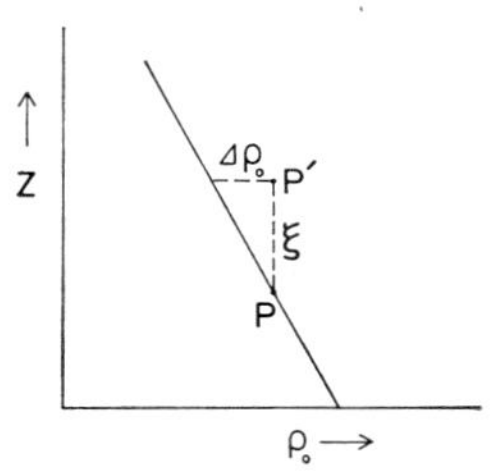

FIG. 1. 2. Displaced air mass from the equilibrium position as $P \rightarrow P'$.

expands adiabatically to decrease the density ρ_0 by $(-\Delta\rho_0)$ as $\Delta\rho_0 = (\Delta p_0 / C^2)$, Δp_0 being the pressure increment over ξ. Then, by (1. 9), $\Delta p_0 = -\rho_0 g \xi$, and the *buoyancy* force upon the air mass, which has moved by ξ, is proportional to the difference between the ambient air mass density $[\rho_0 + (\mathrm{d}\rho_0/\mathrm{d}z)\xi]$ at $z = z_0 + \xi$, and the displaced mass density from $z = z_0$. Since this buoyancy force equals the acceleration of the displaced mass, we get

$$\frac{\mathrm{d}^2\xi}{\mathrm{d}t^2} = -\omega_g{}^2 \xi \tag{1.30}$$

where

$$\omega_g{}^2 = -\left(\frac{1}{\rho_0}\frac{\mathrm{d}\rho_0}{\mathrm{d}z} + \frac{g}{C^2}\right)g = \left(\frac{\kappa-1}{\kappa}\frac{g}{H_0}\right) + \left(\frac{g}{T_0}\frac{\mathrm{d}T_0}{\mathrm{d}z}\right). \tag{1.31}$$

ω_g is named the Brünt-Väisälä frequency. Thus, for most temperatures and temperature gradients, the air mass oscillates vertically around the hydrostatic equilibrium point with a local angular frequency $\omega_g{}^2 > 0$.

However, if $\omega_g{}^2 < 0$ a displaced air parcel experiences no restoring force and will continue moving away from the equilibrium position. Then, the atmosphere is unstable. In terms of the temperature gradient, $\omega_g{}^2 \lesseqgtr 0$ corresponds to

$$-\frac{1}{T_0}\frac{\mathrm{d}T_0}{\mathrm{d}z} \gtreqless \left(\frac{\kappa-1}{\kappa}\right)\left(\frac{1}{H_0}\right) \tag{1.32}$$

respectively. If T_0 varies linearly, (1.32) is replaced by

$$\varepsilon_t \gtreqless \varepsilon_0 \tag{1.33}$$

where $\varepsilon_0 = [(\kappa-1)/\kappa]/H_0$; $\varepsilon_0 T_0 = [(\kappa-1)/\kappa](g/R)$ is the adiabatic lapse rate. The earth's atmosphere is stable in this respect except for the troposphere which has a lapse rate of 6.5°K km^{-1} in mid-latitude in equinox. Since the adiabatic lapse rate for dry air is 9.8°K km^{-1}, the troposphere should be stable for dry air. But this is not the case for wet air with water vapor which is lighter than molecular oxygen and nitrogen, the main consituents of dry air. Thus, convection occurs for wet air which also absorbs radiation resulting into a sub-adiabatic state (ESKINAZI, 1975).

1.4 Steady Horizontal Motion

The solar radiation heats the atmosphere, but its intensity varies with time, latitude, and longitude. Such a differential heating with latitude drives a mean meridional circulation which in the stratosphere and mesosphere consists of rising motion near the warm summer pole, a meridional drift at high levels to the cold winter hemisphere, and sinking in the coldest winter pole (e.g. HOLTON, 1975). Then, the Coriolis force tends to deflect the meridional drift generating a steady mean zonal wind. As a result we have westward winds (easteries) in summer hemisphere and eastward winds (westeries) in winter hemisphere. The overall situation of the mean (steady) zonal wind can be well approximated by the following simple approach (MURGATROYD, 1965). The motion is described by a balance between the latitudinal pressure gradient and the Coriolis force as

$$\frac{1}{\rho_0 a}\frac{\partial p_0}{\partial \theta} = 2\Omega W_0 \cos\theta \tag{1.34}$$

where W_0 is the steady eastward zonal velocity and θ is the co-latitude. The wind due to the balance between the pressure gradient and the Coriolis force

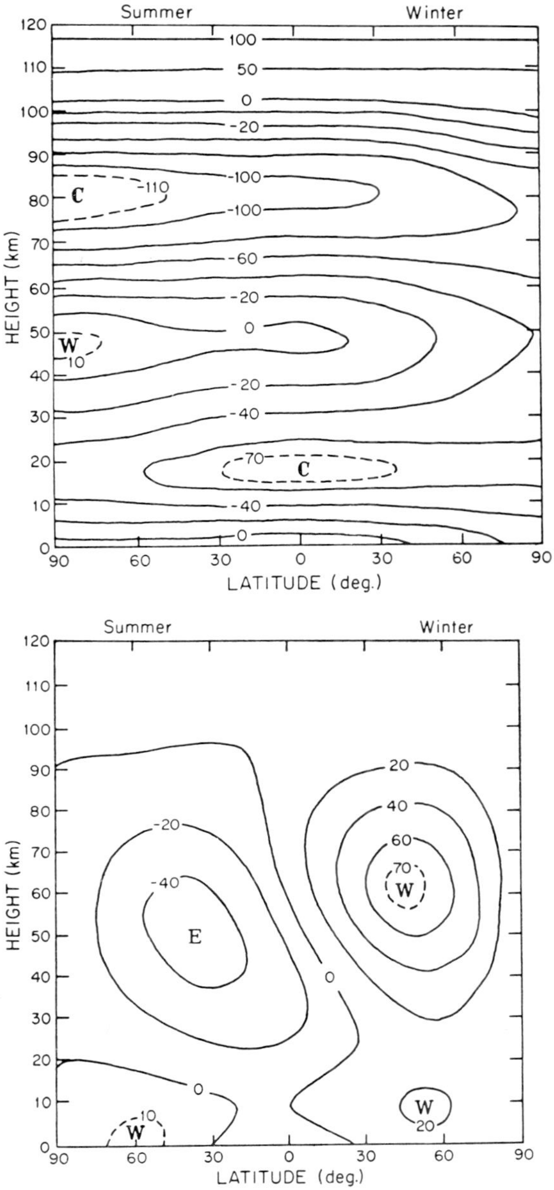

(a)

(b)

FIG. 1. 3. (a) Meridional cross section of T_0 (°C) in solstice. (b) Meridional cross section of steady zonal wind (m sec^{-1}) in solstice for T_0 in (a); W an E denote westery and eastery (LINDZEN and HONG, 1974).

is called the geostrophic wind. By comparing the non-linear term and the Coriolis force term in (1. 1) in the absence of $\boldsymbol{F}$ and $\partial \boldsymbol{W}/\partial t$ for steady motion, we understand that (1. 34) can be valid if

$$R_b = |W|/L\Omega \ll 1$$

where R_b is called the Rossby number; L is the scale of motion. For $|W| <$ 50 m sec^{-1}, $L \simeq 10^7$ m as in our concern, $R_b = 0.1$. Besides such a steady motion, atmospheric waves with much longer periods than the earth's rotation period, planetary waves, are approximately in the geostrophic balance. It is readily found that the geostrophic winds blow along isobars, clockwise around high pressure and counterclockwise around low pressure in the northern hemisphere, the direction being reversed in the other hemisphere.

We can assume a local hydrostatic equilibrium in the vertical direction. Then, differentiation of (1. 34) with respect to z gives the following thermal wind equation:

$$\frac{\partial W_0}{\partial z} - \left(\frac{1}{T_0}\frac{\partial T_0}{\partial z}\right) W_0 = \frac{g}{2\Omega \cos\theta}\left(\frac{1}{T_0}\frac{\partial T_0}{a\partial\theta}\right). \qquad (1.35)$$

For a given $T_0(\theta, z)$ we can calculate $W_0(\theta, z)$. The result is shown in Fig. 1. 3 (Lindzen and Hong, 1974). Such a steady motion should exist also in the thermosphere, but very little experimental evidence is available (Dickinson *et al.*, 1975).

These zonal winds and the corresponding p_0, T_0 and ρ_0 show that the earth's atmosphere is highly non-uniform both horizontally and vertically. The behavior of atmospheric waves which is of our main concern in this book should accordingly complex and in general only numerical approach or simulation can give solutions for realistic situations. But before going into such a complication one must well understand the basic behaviors of the atmospheric waves under simple situations. In certain cases these basic behaviors have not yet been fully clarified theoretically. In such cases complex realistic approaches should be misleading. And a tacit assumption would invalidate the whole result of complex treatments. It is also true that the observed facts in general can be interpreted in terms of the basic state with certain modifications. In the following chapters we shall pay much attention to simple cases relative to acoustic gravity waves and tides with a further discussion on more complex realistic situations. Mathematical details will be shown in many places to help readers understand precisely each theory and its base.

CHAPTER 2

ACOUSTIC GRAVITY WAVES (AGW's)

In this chapter, the acoustic gravity waves (AGW's) will be discussed. The waves were studied by MARTYN as early as 1950. But it is HINES (1960) who has pointed out the importance of these waves for understanding disturbances in the upper atmosphere.

Now, the perturbations on the basic state defined by p_0, T_0, ρ_0, and $\boldsymbol{W}_0$ will be considered in (1. 1)–(1. 3) as $p=p_0+p_1$, $T=T_0+T_1$, $\rho=\rho_0+\rho_1$, and $\boldsymbol{W}=\boldsymbol{W}_0+\boldsymbol{W}_1$. The basic condition for the treatment is that the perturbations are small and only the first order perturbation is taken into account. The condition is fairly well satisfied in the lower atmosphere up to 60 km above which certain errors may take place. In Section 2. 1, neither viscosity nor hydromagnetic force is considered and $\boldsymbol{F}=0$. Both effects become increasingly important with height. Nevertheless, this simple approach will be found basic and allowed in many practical applications with a little modification for viscosity and hydromagnetic effects. Note that the viscosity and hydromagnetic force prevent the waves from growing above certain limit. This would justify, to some extent, for the linear perturbation to be used fairly widely in the upper atmosphere. Comparing the term $(\partial \boldsymbol{W}/\partial t)$ with the Coriolis term $2(\boldsymbol{\Omega}\times\boldsymbol{W})$ in (1. 1), the latter term is found negligible for those waves which have frequencies much higher than 2Ω. Atmospheric waves which are called AGW's are defined to such frequency waves. Another simplification is introduced by considering those waves which have horizontal wavelengths much shorter than the earth's radius. Then, one can approximate the earth as to be flat. Since the gravity is vertical, any horizontal direction is not special. So the x-axis is arbitrarily chosen in the horizontal direction. The z-axis is taken positive upwards.

2.1 AGW's in an Isothermal Atmosphere

The simplification mentioned above gives the following equations to describe AGW's for $\boldsymbol{W}_0=0$ from (1. 1).

$$\rho_0 \frac{\partial W_x}{\partial t} = -\frac{\partial p_1}{\partial x} \tag{2.1}$$

$$\rho_0 \frac{\partial W_z}{\partial t} = -\frac{\partial p_1}{\partial z} - \rho_1 g. \tag{2.2}$$

The basic state, p_0 and ρ_0, are related by the hydrostatic equation (1. 10), which becomes (1. 12) for the present isothermal atmosphere. In (1. 3′) no heat input is considered here ($Q=0$, adiabatic state). When we discuss excitation later, certain heat inputs will be introduced. Linearization of (1.2) and (1. 3′) is straightforward except for a little care on

$$W_z \frac{\mathrm{d}p_0}{\mathrm{d}z} \text{ and } W_z \frac{\mathrm{d}\rho_0}{\mathrm{d}z}$$

which are non-zero. Thus, we have

$$\frac{\partial \rho_1}{\partial t} = -\rho_0 \frac{\partial W_x}{\partial x} - \rho_0 \frac{\partial W_z}{\partial z} - W_z \frac{\mathrm{d}\rho_0}{\mathrm{d}z} \tag{2.3}$$

$$\frac{\partial p_1}{\partial t} = C^2 \left(\frac{\partial \rho_1}{\partial t} + W_z \frac{\mathrm{d}\rho_0}{\mathrm{d}z} \right) - W_z \frac{\mathrm{d}p_0}{\mathrm{d}z}. \tag{2.4}$$

Equations (2. 1)–(2. 4) consist of four homogeneous equations in four unknowns $W_{x,z}$, ρ_1, and p_1. Thus, there should exist a relation, which is necessary for the four equations to be simultaneously valid, i.e. the dispersion relation. First, the Fourier transform about t gives

$$f = \frac{1}{2\pi} \int_{-\infty}^{\infty} f_\omega \exp(i\omega t) \mathrm{d}\omega \tag{2.5}$$

where $f = W_{x,z}$, ρ_1, and p_1. (2. 2) and (2. 3) eliminate ρ_1, giving $W_{z,\omega}$ as

$$W_{z,\omega} = \frac{-\rho_0 g \chi_\omega + i\omega(\partial p_{1,\omega}/\partial z)}{\omega^2 \rho_0 + g(\mathrm{d}\rho_0/\mathrm{d}z)} \tag{2.6}$$

where $\chi = \operatorname{div} \boldsymbol{W}$. (2. 4) (in the right-hand side) gives $(\rho_0 \chi)$ from the first two terms in parenthesis. Eliminating $W_{z,\omega}$ by (2. 4) and (2. 6),

$$\chi_\omega = i\omega \frac{(\omega^2 - g/H)p_{1,\omega} - g(\partial p_{1,\omega}/\partial z)}{\rho_0 C^2(\omega_g^2 - \omega^2)} \tag{2.7}$$

which is, in turn, substituted into (2. 6). Then,

$$W_{z,\omega} = -\left(\frac{i\omega}{\rho_0}\right) \frac{C^2 (\partial p_{1,\omega}/\partial z) + g p_{1,\omega}}{C^2(\omega_g^2 - \omega^2)}. \tag{2.7′}$$

(2. 1) gives $W_{x,\omega}$ in terms of $p_{1,\omega}$, and (2. 7′) gives $W_{z,\omega}$ also in terms of $p_{1,\omega}$. Thus, we can give χ_ω in terms of $p_{1,\omega}$ and its derivative, whilst χ_ω is given also by (2. 7). After equating these two χ_ω's and introducing Φ_ω as

$$\Phi_\omega = \mathrm{e}^{z/2H} p_1 \tag{2.8}$$

the following dispersion equation is obtained:

$$\frac{\partial^2\Phi_\omega}{\partial x^2}+A\frac{\partial^2\Phi_\omega}{\partial z^2}+B\Phi_\omega=0 \tag{2.9}$$

where

$$\begin{aligned} A &= \frac{\omega^2}{\omega^2-\omega_g{}^2} \\ B &= \frac{\omega^2-\omega_a{}^2}{\omega^2-\omega_g{}^2}\left(\frac{\omega^2}{C^2}\right) \\ \omega_a &= \kappa g/(2C). \end{aligned} \tag{2.10}$$

Some limiting case might interest readers. Increase of ω gives that $A\to 1$, $B\to\omega^2/C^2$, and the limiting situation of (2. 9) is

$$\frac{\partial^2\Phi_\omega}{\partial x^2}+\frac{\partial^2\Phi_\omega}{\partial z^2}+\frac{\omega^2}{C^2}\Phi_\omega=0 \tag{2.11}$$

which is a usual wave equation for acoustic waves, the phase traveling with velocity C. Generally, for $\omega>\omega_a>\omega_g$, the same type of wave equation is obtained as is known by transforming $z\to(z/\sqrt{A})$; both A and B are positive. However, for $\omega_a>\omega>\omega_g$

$$A=a^2>0,\ B=-b^2<0 \text{ and } z\to(z/a), \text{ and}$$

$$\frac{\partial^2\Phi_\omega}{\partial x^2}+\frac{\partial^2\Phi_\omega}{\partial z^2}-b^2\Phi_\omega=0 \tag{2.12}$$

where $a>0$ and $b>0$. (2. 12) is a diffusion equation and no wave solution can be found. Further, for $\omega<\omega_g$, we have $A=-a^2<0$, and $B=b^2>0$, and (2. 9) becomes

$$\frac{\partial^2\Phi_\omega}{\partial x^2}-a^2\frac{\partial^2\Phi}{\partial z^2}+b^2\Phi_\omega=0 \tag{2.13}$$

which is of hyperbolic type, containing a discontinuous characteristic. (2. 13) gives, thus, a peculiar dispersion for this region. The dispersion is best illustrated by dispersion diagrams as in Fig. 2. 1 (a) (e.g. HINES, 1960) where α and γ are the wave number by the space Fourier transformation

$$\Phi_\omega=\frac{1}{(2\pi)^2}\int_{-\infty}^{\infty}\int_{-\infty}^{\infty}\Phi_{\omega,\alpha,\gamma}\exp\left[-i(\alpha x+\gamma z)\right]\mathrm{d}\alpha\,\mathrm{d}\gamma. \tag{2.14}$$

Then, (2. 9) becomes an algebraic equation as

$$\frac{\alpha^2}{B}+\frac{\gamma^2}{(B/A)}=1. \tag{2.15}$$

Figure 2. 1 shows that for $\omega>\omega_a$, case (1) (acoustic wave region, AW region hereafter), the dispersion consists of ellipses, the size decreasing with decreasing ω; for $\omega_a>\omega>\omega_g$, case (2) (Evanescent wave region, EW region

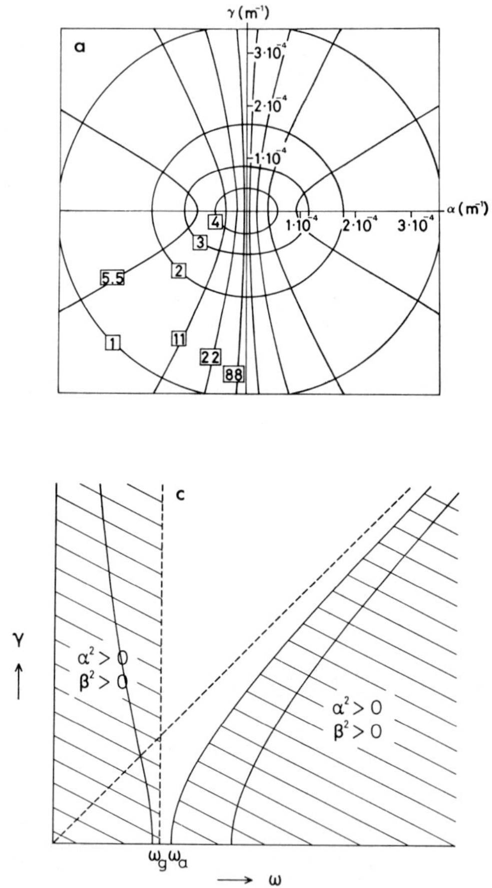

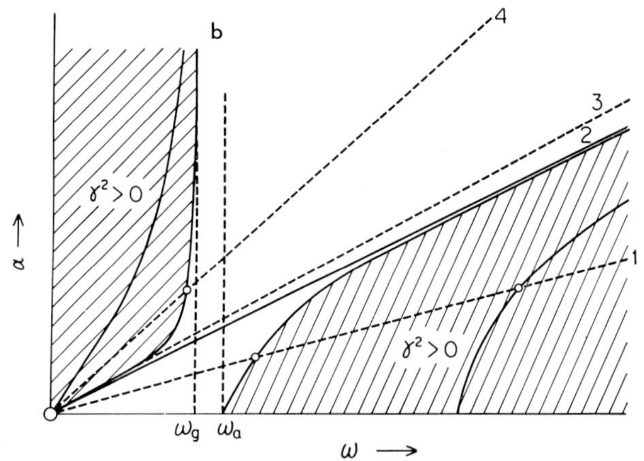

FIG. 2. 1. Dispersion diagram. (a) $\boldsymbol{k}$ $(\alpha, 0, \gamma)$ with $(2\pi/\omega)$ as parameter in minutes given by the numbers in the small boxes (HINES, 1960). (b) α versus ω with γ as parameter. The internal waves as $\gamma^2>0$ exist only in the shaded region. The slant lines 1, 2, 3 and 4 are given by $\omega=m\alpha C$ where $m>1$, $m=1$, $1>m>m_0$, and $m_0>m$, respectively. Slant line 2 is for the Lamb wave. $m_0=(\omega_g/\omega_a)$. The origin $(\omega=0)$ always gives an intersection between the slant lines and the dispersion curve. (c) γ versus ω with α as parameter. The vertical broken line is for $\beta\rightarrow\infty$, the slant broken line for $\omega=\gamma C$.

hereafter) no wave exists; for $\omega<\omega_g$, case (3) (Gravity wave region, GW region hereafter), the dispersion consists of hyperbolas which, with decreasing ω, approach the origin.

Generally, by (2. 15) the phase velocity is given by

$$V_p=\frac{\omega}{\sqrt{\alpha^2+\gamma^2}}=C\sqrt{\frac{\omega^2-\omega_g^2\sin^2\theta_k}{\omega^2-\omega_a^2}} \tag{2. 16}$$

where θ_k is the angle between the z-axis and the wave vector $\boldsymbol{k}(\alpha,\gamma)$. V_p is real only for $\sin\theta_k>(\omega/\omega_g)$ for $\omega<\omega_g$.

Another way of expressing the dispersion is to draw the diagram of dispersion as α versus ω, or γ versus ω as in Fig. 2.1 (b), (c). Then, (2. 15) is given in another form as

$$\gamma^2=\left(\frac{\omega_g^2}{\omega^2}-1\right)\alpha^2-\frac{\omega_a^2-\omega^2}{C^2} \tag{2. 17}$$

which shows that, with decreasing $\omega, \gamma^2\rightarrow(\omega_g^2/\omega^2)\alpha^2$ for $\gamma^2\gg\omega_a^2/C^2=1/(2H)^2$, $\boldsymbol{k}$ tending to be *vertical* for waves with the vertical wavelength much shorter than the scale height.

It is found in Fig. 2. 1 (c) that *the vertical phase velocity* (ω/γ) *is opposite to the vertical group velocity* $\partial\omega/\partial\gamma$ for GW's. This is common to all vertically traveling atmospheric waves including tides. This shows that the vertical phase velocity is opposite to the vertical energy flow, because the group velocity is equal to the energy flow velocity. Direct proof of this property of GW's is given by considering (1. 6) or (1. 7) where the energy flow is given by $p\boldsymbol{W}$. The perturbed parts of $(p\boldsymbol{W})$ consist of $p_0\boldsymbol{W}_1$ and $p_1\boldsymbol{W}_1$; $\boldsymbol{W}_0=0$. However, the energy flow which does not vanish after averaging over one period should be $(p_1\boldsymbol{W}_1)$. To each Fourier component we must adopt the real or imaginary part of both $p_{1,\omega}$ and $\boldsymbol{W}_{1,\omega}$ which are complex just for mathematical representation. It is easily shown that the averaging of these quantities is equivalent to forming $(1/2)\mathscr{R}(p_1W^c)$ where $\mathscr{R}$ denotes to take the real part and superscript c the conjugate complex. Then, the energy flow $\mathscr{F}_\omega$ is given by (2. 1) and (2. 7′), where $p_1\propto\exp[-(z/2H)-i\gamma z]$, as

$$\mathscr{F}_{x,\omega}=\frac{1}{2}\mathscr{R}(p_1W_x^c)=\frac{1}{2}\frac{\alpha}{\rho_0\omega}|p_{1,\omega}|^2$$

$$\mathscr{F}_{z,\omega}=\frac{1}{2}\mathscr{R}(p_1W_z^c)=-\left(\frac{\omega}{2\rho_0}\right)\frac{\gamma|p_{1,\omega}|^2}{(\omega_g^2-\omega^2)} \tag{2. 18}$$

which shows that $\mathscr{F}_{z,\omega}$ changes sign at $\omega=\omega_g$. For $\gamma<0$, which means that the vertical phase velocity is negative (all quantities $\propto e^{i(\omega t-\gamma z)}$), $\mathscr{F}_{z,\omega}<0$ for $\omega>\omega_g$, and $\mathscr{F}_{z,\omega}>0$ for $\omega<\omega_g$. This is consistent with the result which is obtained from the group velocity approach (Fig. 2. 1 (c)).

Almost all of GW's observed in the upper atmosphere have downward phase velocities. Thus, these waves should be internal and excited somewhere in the lower atmosphere and travel upwards. *Forward tilting* of the

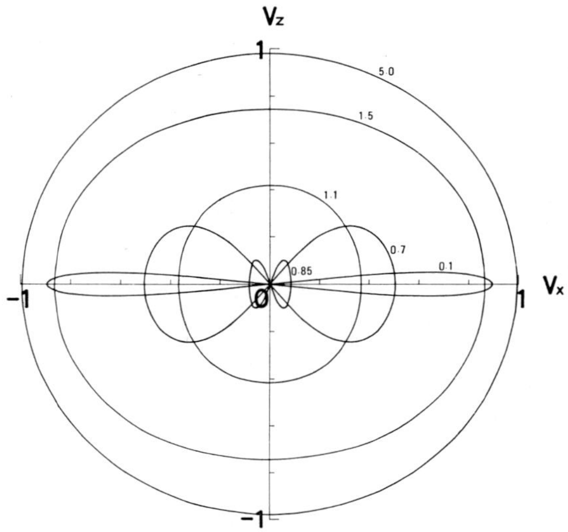

FIG. 2. 2. Group velocity $\boldsymbol{V}_g = \partial\omega/\partial\boldsymbol{k} = (\partial\omega/\partial\alpha,\ 0,\ \partial\omega/\partial\gamma)$; the frequency is given as parameters by (ω/ω_g) to each curve. Vectors from the origin to each point on the curves give $\boldsymbol{V}_g$; $\boldsymbol{V}_g$ is normalized as $\boldsymbol{V}_g = C\boldsymbol{V}$.

GW wave front is also due to this characteristic, resulting from a combined downward and horizontally forward phase velocity. Observed ionospheric disturbances (TID) usually have such a characteristic and can be identified with *internal* GW's, though some other possibility may exist (see Section 2. 3).

The group velocity $\boldsymbol{V}_g = (\partial\omega/\partial\alpha, \partial\omega/\partial\gamma)$ is given in terms of ω

$$V_g = \frac{(\omega^2-\omega_g{}^2)\sqrt{(\omega^2-\omega_g{}^2)(\omega^2-\omega_a{}^2)(\omega^2-\omega_g{}^2\cos^2\theta_g)}}{\omega[(\omega^2-\omega_g{}^2)^2+\omega_g{}^2(\omega_a{}^2-\omega_g{}^2)\sin^2\theta_g]}\,C \tag{2.19}$$

where θ_g is the angle between the z-axis and $\boldsymbol{V}_g$. The hodograph of $\boldsymbol{V}_g$ is shown in Fig. 2. 2. Here also the contrast between AW and GW regions is remarkable, the former containing ellipses and the latter a pair of two loops with the axis along the x-axis. *The energy flows increasingly horizontally with decreasing* ω *for GW's.* Thus, the internal GW's of long period can transport wave energy over long distances with little attenuation without any waveguide of the atmosphere structure. Since $\boldsymbol{k}$ tends to be vertical with decreasing frequency as mentioned above, the energy flow tends to *be orthogonal to* $\boldsymbol{k}$.

By (2. 1), (2. 4), (2. 7′), (2. 8), and (2. 17) the following polarization relations are obtained (HINES, 1960). The exponential factor as in (2. 8) is omitted for $W_{x,z} = e^{z/2H}\hat{W}_{x,z}$, $p_1 = e^{-z/2H}\hat{p}_1$, and $\rho_1 = e^{-z/2H}\hat{\rho}_1$:

$$\frac{\hat{W}_{x,\omega}}{X} = \frac{\hat{W}_{z,\omega}}{Z} = \frac{\hat{p}_{1,\omega}}{P p_0} = \frac{\hat{\rho}_{1,\omega}}{R\,\rho_0} \tag{2.20}$$

where

$$\begin{aligned} X &= \omega\alpha C^2\left[\gamma+i(g/C^2)\left(\frac{\kappa}{2}-1\right)\right] \\ Z &= \omega(\omega^2-\alpha^2C^2) \\ R &= \omega^2\gamma+i(\kappa-1)g\alpha^2-i(\kappa g\omega^2/2C^2) \\ P &= \kappa\omega^2\left[\gamma-i\left(1-\frac{\kappa}{2}\right)(g/C^2)\right]. \end{aligned} \tag{2.21}$$

By (2. 22) and (2. 21) with $\omega \to 0$ and $|\gamma| \gg (1/H)$

$$\frac{\hat{W}_x}{\hat{W}_z} \to -\left(\frac{\gamma}{\alpha}\right) \quad \text{or} \quad (\alpha\hat{W}_x+\gamma\hat{W}_z) \to 0 \tag{2.22}$$

which implies that GW's tend to be *transverse* to $\boldsymbol{k}$. Since $\boldsymbol{k}$ tends to be vertical, $\boldsymbol{W}$ tends to be *horizontal* i.e. *parallel to* $\mathscr{F}_\omega$, the energy flow.

(2. 22) also means div $\hat{\boldsymbol{W}}=0$, no compressibility for GW's with low frequency. This is proved directly by using the incompressibility condition as $\omega_g \to \omega_{g'} = \sqrt{\kappa}(g/C)$ and $(\omega_a^2-\omega^2)/C^2 \to (1/2H)^2$, because with $\kappa \to \infty$, $C \propto \sqrt{\kappa} \to \infty$. Then, in the dispersion diagram the GW domain remains with a little modification as $\omega_g \to \omega_g'$, whilst the AW domain disappears (Fig. 2. 1). Thus, the gravity is essential for GW's as distinguished from AW. This limiting situation gives by (2. 17)

$$\gamma^2 = \left(\frac{\omega_g'^2}{\omega^2}-1\right)\alpha^2-\frac{1}{4H^2} \tag{2.23}$$

which can approximate also the limiting case with very low frequency and will be discussed in Chapter 3. AGW's are GW's modified by compressibility as they really are.

In $\omega_a > \omega > \omega_g$ there are no vertically traveling waves (internal waves), but there exist those waves which travel as surface waves or external waves. One example is the *Lamb wave* which travels horizontally with $(\omega/\alpha)=C$ as given by slant line 2 in Fig. 2. 1 (b). Then, by (2. 17) $\gamma^2=-\gamma_0^2<0$ where $\gamma_0=-\sqrt{(\omega_a^2-\omega_g^2)}/C=(2-\kappa)/2\kappa H$. This shows further that

$$W_{z,\omega} = 0 \tag{2.24}$$

by (2. 7′) where $p_1=\Phi\, e^{-(z/2H)} \propto e^{-(z/2H)-\gamma_0 z}$ by (2. 8) and (2. 14). The Lamb wave can travel along the earth's surface over very long distance (more than the earth's circumference) with little attenuation because of no vertical energy flow by $e^{-\gamma_0 z}$.

In non-isothermal atmosphere, ω_a and ω_g vary with z. Internal AGW's at one height with ω_a and ω_g by (2. 16) (for real V_p) can be modified, or cannot even exist as internal waves at other heights with real ω_a and ω_g. Then, the original internal waves can be refracted or reflected between these heights. By (2. 16) and (2. 17), waves initially with $\omega<\omega_g$ and $\gamma>0$ reflect ($\gamma=0$) after traveling in regions with decreasing ω_g i.e. *increasing temperature*. However, except for small θ_k's internal GW's with low frequency have no reflection as will be found later in Section 2. 3. Small θ_k's imply generally small α's i.e. large horizontal phase velocity; GW's of large horizontal phase velocity can be reflected.

Besides temperature gradients, the background wind can be a cause of reflection and refraction. A simple explanation will be given. Consider a uniform wind W_0 in the x direction. The equation for perturbation for (2. 1)–(2. 4) can be readily obtained from (1. 1)–(1. 3) without neglecting W_0. However, a simpler method is adopted here. Since W_0 is constant, we can introduce a moving coordinate system (x', z', t') with W_0 (along the x-axis). On this system the equations for perturbation should be identical to (2. 1)–(2. 4) with $W_0=0$. Then, we transform these equations in the (x', z', t')-system to those in the (x, z, t)-system by

$$x = x' + W_0 t' \tag{2.25}$$

which gives

$$\frac{\partial}{\partial t} = \frac{\partial}{\partial t'} - W_0 \frac{\partial}{\partial x'}. \tag{2.26}$$

Thus, a Fourier transform about x and t by (2. 5) and (2. 14) gives

$$\omega \rightarrow (\omega - \alpha W_0) \tag{2.27}$$

which gives a Doppler shift in ω as is expected. (2. 25) is substituted into (2. 17)

$$\gamma^2 = \left[\frac{\omega_g{}^2}{(\omega-\alpha W_0)^2} - 1\right]\alpha^2 - \frac{\omega_a{}^2 - (\omega - \alpha W_0)^2}{C^2}. \tag{2.28}$$

Then, the frequency, at which reflection occurs for particular α, is shifted by (2. 27). This implies that even in isothermal atmosphere a reflection can occur, if the Doppler-shifted frequency produces $\gamma=0$ upon meeting the wind in the propagation. It is found that the wind can contribute to reflection and refraction. A singular situation can further arise, if $\omega \rightarrow \alpha W_0$. Then, $\gamma \rightarrow \infty$ which implies the vertical group velocity $\partial\omega/\partial\gamma$ tends to vanish as is readily proved by (2. 28). The level with this singularity is called a critical level and has been discussed by BOOKER and BRETHERTON (1967) and others. This is an important mechanism leading to strong interactions between winds and atmospheric waves, especially planetary waves which are waves with periods of several days and travel fairly slowly, thereby the condition $\omega = \alpha V_0$ being frequently satisfied.

Generally, if W_0 varies with height, it is found that $(\mathrm{d}^2 W_0/\mathrm{d}z^2)$ is only effective on the propagation of GW's. HINES and REDDY (1967) have discussed on the effects of wind on the transmission of GW's in details. CHARNEY and DRAZIN's work on this problem for planetary waves (1951) is well-known. DICKINSON (1969) has contributed to a further development of the theory of wave flow interaction. But these results will not be discussed any further in this book.

As to dissipative forces in ***F*** in (1. 1) we have molecular viscosity and hydromagnetic effects. The latter, however, is not very effective for waves with period shorter than a few hours (Section 3. 2) because of low collision frequencies with ions per neutral particle as 10^{-5} sec^{-1} or less. Molecular viscosity dissipates AGW's in upward propagation to the thermosphere. Thermal conduction also dissipates AGW's approximately by an equal amount as molecular viscosity. A simple estimate of the attenuation has been done by HINES (1960) who equated the attenuated energy over one period to the

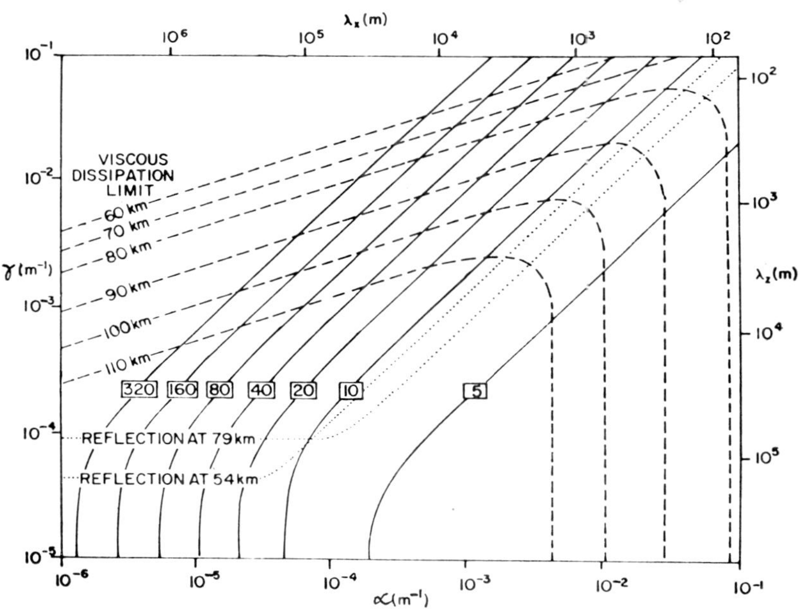

FIG. 2. 3. Propagation modes at meteor heights. The periods, measured in minutes, are shown in boxes on the constant period contours (solid lines). The limits of the permitted spectrum, as determined by viscous damping, are shown for heights of 60, 70, 80, 90, 100, and 110 km (broken curves); modes lying above and to the right for these curves are excluded. The modes subject to reflection at heights of 54 and 79 km (due to temperature variation) are also shown (dotted lines); modes lying below these curves cannot proceed from the lower atmosphere to the upper (HINES, 1960; the diagram is redrawn).

wave energy (1. 8) and gave

$$2\pi \mu_V{}^2 \alpha^2 (\omega_g{}^2/\omega^3) = 1 \tag{2.29}$$

to decide propagating modes at meteor heights (μ_V=kinematic viscosity). Generally, with increasing heights only large scale waves can survive the attenuation (Fig. 2. 3).

Besides attenuation, viscosity and thermal conduction are effective in reflection (e.g. YANOWITCH, 1967) and dispersion (e.g. LAMB, 1945).

2.2 Simple Examples of AGW Excitation in the Upper Atmosphere*

In the previous section, the basic behaviors, especially, the peculiar dispersions, of AGW in isothermal atmosphere are presented. It is the idea of the present author to give some more materials to help readers better understand them. For instance, how the AGW fronts look like, how they are different from those water wave-fronts which are often seen on the water surface in ponds, when one throws a pebble into a pond, these questions would be hopefully answered by the following examples. The last example pertains to bow waves of AGW's, very different bow waves from the usually observed on boats. The first two models of source are harmonic oscillators in time, the first being a horizontally infinite line source along the y-axis, the second a point source. The other two models are a point and finite line source in horizontally uniform motion. The atmosphere is not bounded and filled only with directly excited waves.

The relevant equations are similar to (2. 1)–(2. 4) except that the y-direction, the forcing, and the heating are now considered as follows (e.g. KATO *et al.*, 1977):

$$\begin{aligned} \rho_0 \frac{\partial W_x}{\partial t} &= -\frac{\partial p_1}{\partial x} + F_x \\ \rho_0 \frac{\partial W_y}{\partial t} &= -\frac{\partial p_1}{\partial y} + F_y \\ \rho_0 \frac{\partial W_z}{\partial t} &= -\rho_1 g - \frac{\partial p_1}{\partial z} + F_z \end{aligned} \tag{2.30}$$

$$\frac{\partial \rho_1}{\partial t} + W_z \frac{\mathrm{d}\rho_0}{\mathrm{d}z} + \rho_0 \left(\frac{\partial W_x}{\partial x} + \frac{\partial W_y}{\partial y} + \frac{\partial W_z}{\partial z} \right) = 0 \tag{2.31}$$

$$\frac{\partial p_1}{\partial t} + W_z \frac{\mathrm{d}p_0}{\mathrm{d}z} - C^2 \left(\frac{\partial \rho_1}{\partial t} + W_z \frac{\mathrm{d}\rho_0}{\mathrm{d}z} \right) = (\kappa - 1) Q \rho_0. \tag{2.32}$$

* Readers can skip this section without losing the main point of the author's explanation on AGW. This section is mainly for theoretical interest.

Following a very similar line to that in Section 2. 1, we reach the following basic equation as (2. 9)

$$\frac{\partial^2\Phi_\omega}{\partial x^2}+\frac{\partial^2\Phi_\omega}{\partial y^2}+A\frac{\partial^2\Phi_\omega}{\partial z^2}+B\Phi_\omega = P_\omega \tag{2.33}$$

where

$$P_\omega = \frac{e^{-z/2H}}{(\omega^2-\omega_g{}^2)}\Bigg[\omega^2\left(\frac{\partial F_{x,\omega}}{\partial x}+\frac{\partial F_{y,\omega}}{\partial y}+\frac{\partial F_{z,\omega}}{\partial z}\right)$$

$$-\frac{\omega^2}{\kappa H}F_{z,\omega}-\omega_g{}^2\left(\frac{\partial F_{y,\omega}}{\partial y}+\frac{\partial F_{x,\omega}}{\partial x}\right)\Bigg]$$

$$-\frac{i\omega(\kappa-1)e^{-(z/2H)}}{C^2(\omega^2-\omega_g{}^2)}\left[\left(\omega^2-\frac{\kappa g^2}{C^2}\right)Q_\omega+g\frac{\partial Q_\omega}{\partial z}\right]. \tag{2.34}$$

Generally, in (2. 34) P_ω depends on the precise time and space structure of $\boldsymbol{F}$ and Q. Here, the space dependence of all source terms is assumed to be of a delta function. The result is also basic for realistic shape of the source through space integration, because the present solution is a Green function of the general solution of (2. 33).

a) *Infinite-line harmonic-oscillator*

Consider P_ω as

$$P_\omega = P_0\delta(x)\delta(z). \tag{2.35}$$

The Fourier transformation (2. 14) for P_ω gives Φ_ω; only one frequency ω is considered now. By (2. 35) and (2. 33)

$$\Phi_{\omega,\alpha,\gamma} = -\frac{P_0}{D(\omega,\alpha,\gamma)} \tag{2.36}$$

where

$$D(\omega,\alpha,\gamma) = \alpha^2+A\gamma^2-B. \tag{2.37}$$

$D\equiv 0$ gives the dispersion relation (2. 17). Thus, the Fourier inverse relation gives the solution, which consists of three cases, depending on (1) $\omega>\omega_a$, (2) $\omega_a>\omega>\omega_g$, and (3) $\omega_g>\omega$:

1) $\omega>\omega_a$ (AW region)

We use the path integral method on the complex ω-plane for integration. Then, by (A. 2. 1)–(A. 2. 4) the γ-integration leads to

$$\Phi_\omega = iP_0\frac{\pi^2}{a}H_0{}^{(2)}\left(b\sqrt{x^2+\frac{z^2}{a^2}}\right) \tag{2.38}$$

where $H_0{}^{(2)}$ is the Hankel function of order zero. Multiplying (2. 38) by $\exp(i\omega t)$, and in the limit $x\to\infty$ and $z\to\infty$,

$$\Phi e^{i\omega t}\to\exp\left[i\left(\omega t-b\sqrt{x^2+\frac{z^2}{a^2}}\right)\right] \tag{2.39}$$

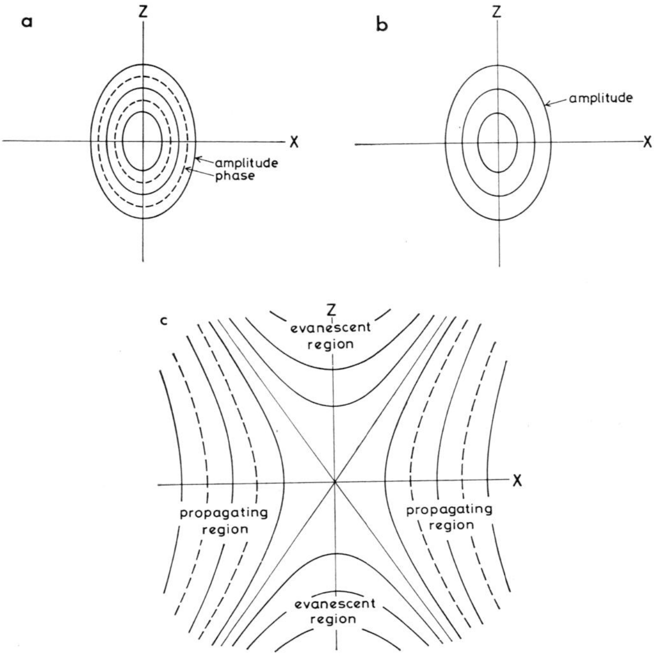

FIG. 2. 4. (a) Φ_ω in $\omega > \omega_a$, for an infinite line source along y. (b) Φ_ω in $\omega_a > \omega > \omega_g$ for an infinite line source along y. (c) Φ_ω in $\omega_g > \omega$, for an infinite line source along y. Amplitudes and phases are denoted by solid and broken lines, respectively.

which shows an outgoing wave whose wave front is given by

$$x^2 + \frac{z^2}{a^2} = \text{const} > 0. \qquad (2.40)$$

(2. 40) gives a family of ellipses which tend to be a circle with $\omega \to \infty$ i.e. $a \to 1$ (Fig. 2. 4 (a)). This would belong to usual concept of waves and their wave front.

2) $\omega_a > \omega > \omega_g$ (EW region)

In this range, a similar mathematical deduction as in 1) gives

$$\Phi_\omega = \frac{2\pi}{a} K_0 \left(b \sqrt{x^2 + \frac{z^2}{a^2}} \right) \qquad (2.41)$$

where K_0 is the imaginary Bessel function and (2. 41) shows an evanescent wave with iso-intensity surface

$$x^2 + \frac{z^2}{a^2} = \text{const}. \qquad (2.42)$$

(2. 42) is a family of ellipses as (2. 40). But the intensity now decreases monotonously with the distance from the source (Fig. 2. 4 (b)).

3) $\omega_g > \omega$ (GW region)

We have again propagating waves as

$$\Phi_\omega = -\frac{\pi^2}{a} H_0^{(2)}\left(b\sqrt{x^2 - \frac{z^2}{a^2}}\right) \tag{2.43}$$

which is valid for

$$x^2 - \frac{z^2}{a^2} > 0 \tag{2.44}$$

and also evanescent waves as

$$\Phi_\omega = -i\frac{2\pi}{a} K_0\left(b\sqrt{\frac{z^2}{a^2} - x^2}\right) \tag{2.45}$$

which is valid for

$$\frac{z^2}{a^2} - x^2 > 0. \tag{2.46}$$

It is found that the propagating wave can exist only in a restricted region given by (2. 44). The wave front is $x^2-(z^2/a^2)=\text{const}>0$, a family of hyperbolas with the major axis along the x-axis (Fig. 2. 4(c)). The asymptotic behavior is

$$\Phi e^{i\omega t} \sim \exp\left[i\left(\omega t - b\sqrt{x^2 - \frac{z^2}{a^2}}\right)\right], \tag{2.47}$$

showing an outgoing wave from the source. On the boundary $x^2-(z^2/a^2)=0$, $\Phi\to\infty$. Figure 2. 4(c) illustrates that phase progresses *downwards* for $z>0$ and *upwards* for $z<0$. Since the energy flows out from the source at the origin, the flow should be *upward* for $z>0$ and *downward* $z<0$. Thus, the figures help us intuitively understand the peculiarity of GW's relative to the vertical energy flow and phase progression which are opposite to each other, as has been discussed in the preceding section. Another interesting point appears in comparison between Figs. 2.1 (a) and 2.4 (a) and (c), each ellipse and pair of hyperbola in the wave number space (α, γ) in Fig. 2. 4 (a) corresponding to a family of those with the same ω or $(2\pi/\omega)$ in the real space (x, z) in Figs. 2. 4 (a) and (c). In AW region $(\omega>\omega_a)$ where the ellipses appear in both spaces, the ratio between the two axes of the ellipse in one space is inverse to that of the other space. In GW region $(\omega<\omega_g)$, where the hyperbolas appear, the gradient of the asymptotes with the same ω or $(2\pi/\omega)$ is also inverse to each other implying that the angle of the asymptotes to the abscissa is compliment to each other.

b) Point harmonic oscillator

Now consider

$$P = P_0\,\delta(x)\delta(y)\delta(z) \tag{2.48}$$

which gives by a Fourier transform

$$\Phi_{\omega,\beta,\gamma} = -\frac{P_0}{D(\omega, \alpha, \beta, \gamma)} \tag{2.49}$$

where β is the wave number component along the y-axis and

$$D(\omega, \alpha, \beta, \gamma) = \alpha^2 + \beta^2 + A\gamma^2 - B. \tag{2.50}$$

The Fourier inverse transformation gives solution ((A.2.5)–(A.2.14)) as follows:

1) $\omega > \omega_a$ (AW region)

$$\Phi_\omega = -\frac{\exp\left[-ib\sqrt{x^2+y^2+(z^2/a^2)}\right]}{4\pi a\sqrt{x^2+y^2+(z^2/a^2)}}. \tag{2.51}$$

This represents those propagating waves which have the wave front of ellipsoid of rotation around the z-axis as

$$x^2+y^2+\frac{z^2}{a^2} = \text{const.} \tag{2.52}$$

When ω increases, the ellipsoid tends to be a sphere because $a^2 \to 1$. This is again in our usual concept of waves.

2) $\omega_a > \omega > \omega_g$ (EW region)

$$\Phi_\omega = -\frac{\exp\left[-b\sqrt{x^2+y^2+(z^2/a^2)}\right]}{4\pi a\sqrt{x^2+y^2+(z^2/a^2)}}. \tag{2.53}$$

This represents those evanescent waves which decay with distance and which have iso-intensity surface as (2. 52).

3) $\omega_g > \omega$ (GW region)

$$\Phi_\omega = \frac{\exp\left[-ib\sqrt{x^2+y^2-(z^2/a^2)}\right]}{4\pi a\sqrt{x^2+y^2-(z^2/a^2)}} \quad \text{for} \quad \left(x^2+y^2-\frac{z^2}{a^2}\right) > 0. \tag{2.54}$$

This represents again those propagating waves which have the wave front of hyperboloid of one sheet as obtained by rotating a hyperbola, $x^2-(z^2/a^2)=$ constant, around the z-axis. For $x^2+y^2-(z^2/a^2)<0$, no propagation occurs, and

$$\Phi_\omega = \frac{\exp\left[-b\sqrt{-(x^2+y^2)+(z^2/a^2)}\right]}{4a\sqrt{-(x^2+y^2)+(z^2/a^2)}} \tag{2.55}$$

which becomes infinitely large on the boundary as

$$x^2+y^2-\frac{z^2}{a^2} = 0 \tag{2.56}$$

which gives two cones with their axes along the z-axis (Fig. 2.5). With decreasing ω, the propagating waves by (2. 54) are increasingly narrower about the horizontal direction. The result here is a three-dimensional version of the preceding two-dimensional example. Both examples are highly simplified but may find an application in the upper atmosphere as with excitation by electric current in the auroral zone (line source) and by nuclear explosion (point source) (Row, 1967).

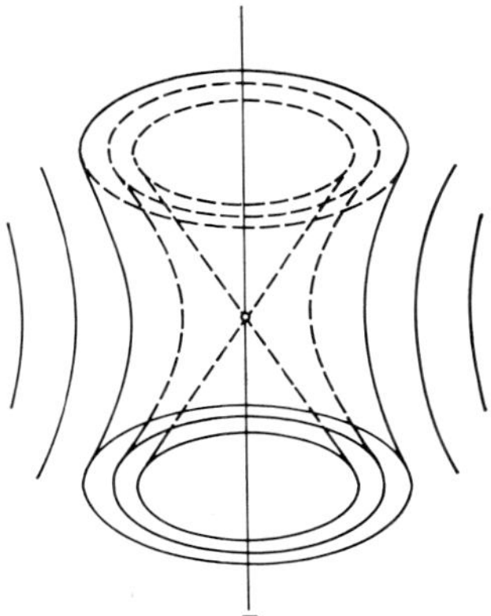

FIG. 2. 5

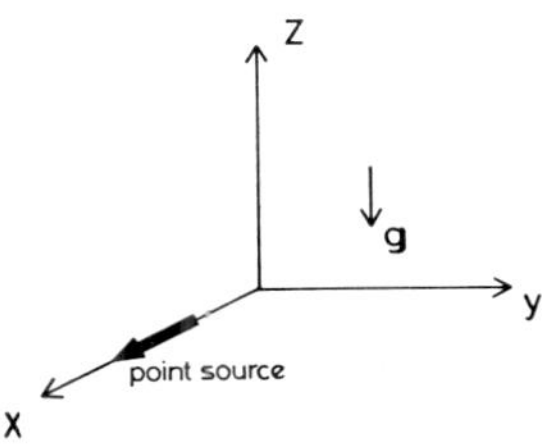

FIG. 2. 6

FIG. 2. 5. Φ_ω in $\omega_g > \omega$ for a point source at the origin ; the front makes a hyperboloid of one sheet.
FIG. 2. 6. Moving point source.

c) *Uniformly moving point source* (KATO *et al.*, 1977)

This example treats emission from a source which is moving along the x-direction with a constant velocity V_0 (Fig. 2. 6). No time variation occurs except for the uniform displacement of the source which is a function of $(x-V_0t)$ as well as y and z as

$$f = f(x-V_0t, y, z) \tag{2.57}$$

where $f=Q$ and $\boldsymbol{F}$. This type of emission of light is well-known as the Cherenkov emission in atomic physics. The waves produced by this mechanism also are called the bow wave in fluid dynamics, because the ship's bow can produce these waves. The spectrum spreads now continuously over all frequencies instead of one frequency in the preceding examples,

$$f_{\omega,\alpha,\beta,\gamma} = \iint_{-\infty}^{\infty}\iint f\, e^{-i(\omega t-\alpha x-\beta y-\gamma z)}\, dt dx dy dz \tag{2.58}$$

which gives

$$f_{\omega,\alpha,\beta,\gamma} = 2\pi\delta(\omega-\alpha V_0) f_{\alpha,\beta,\gamma} \tag{2.59}$$

where $f_{\alpha,\beta,\gamma}$ is the Fourier transformation of $f(x, y, z)$. Consider

$$f(x, y, z) = (2\pi)^3\delta(x)\delta(y)\delta(z) f_0$$

where f_0 is a constant giving the force density for excitation. Thus

$$f_{\alpha,\beta,\gamma} = (2\pi)^3 f_0. \tag{2.60}$$

Of various terms which correspond to each in P_ω in (2. 34), we shall calculate only the following case:

$$P_{\omega,\alpha,\beta,\gamma} = -\frac{(2\pi)^4\omega}{(\omega^2-\omega_g{}^2)}\delta(\omega-\alpha V_0)P_0 \tag{2.61}$$

where P_0 is a constant giving the source intensity per unit volume in association to f_0 in (2. 60). Note that dimension of P_0 is velocity cubed. Thus, by (2. 61)

$$\Phi_{\omega,\alpha,\beta,\gamma} = \frac{(2\pi)^4\omega\delta(\omega-\alpha W_0)P_0}{(\omega^2-\omega_g{}^2)D(\omega,\alpha,\beta,\gamma)} \tag{2.62}$$

where D is given by (2. 50). The inverse Fourier transformation gives the solution. Note that other terms in (2. 34) are different in the power of ω, which implies a differentiation or integration with respect to t by (2.5), or the differentiation with respect x, y, z. These terms can be taken into account through differentiation and integration of the resultant form of the present case. Note that by (2. 57) the operation with respect to t is identical to that to x except for a constant factor.

There are three distinct cases depending on $m>1$, $1>m>m_0$, and $m<m_0$, where m is the Mach number and $m_0=\omega_g/\omega_a$. First, one can get the analytic expression for these three cases in a spacewise inverse Fourier transformation about (α, β, γ) as

$$\Phi_\omega = \left(\frac{2\pi}{V_0}\right)e^{-i(\omega x/V_0)}G_\omega(\omega, y, z)P_0 \tag{2.63}$$

$$G_\omega = \int_{-\infty}^{\infty}\int_{-\infty}^{\infty}\omega\frac{e^{-i(y\beta+z\gamma)}}{D'(\omega,\beta,\gamma)}\,d\beta d\gamma \tag{2.64}$$

$$D' = -(\omega^2-\omega_g{}^2)D\left(\omega, \frac{\omega}{V_0}, \beta, \gamma\right) \tag{2.64'}$$

by (2. 33). The explicit expression of G_ω and the corresponding Φ_ω are given as follows; readers should refer to List of Symbols wherever necessary to know about symbols:

Case 1. $m>1$, supersonic motion

One can use the path integral method on the complex plane for the β integration in (2. 64). Then, the γ-integration leads to (OBERHETTINGER, 1957)

$$G_\omega = -\frac{2\pi i}{\sqrt{\omega_g{}^2-\omega^2}}K_0(\mathscr{W}) \quad \text{for } 0<|\omega|<\Omega_c$$

$$= -\frac{\pi^2}{\sqrt{\omega_g{}^2-\omega^2}}\times\begin{cases}H_0{}^{(2)}(\mathscr{W}) & \text{for } \Omega_c<\omega<\omega_g\\(-H_0{}^{(1)}(\mathscr{W})) & \text{for } -\Omega_c>\omega>-\omega_g\end{cases}$$

$$= -\frac{2\pi s_g(\omega)}{\sqrt{\omega^2-\omega_g{}^2}} K_0(\mathscr{W}) \quad \text{for } \omega_g < |\omega| < \Omega_0$$

$$= \frac{i\pi^2}{\sqrt{\omega^2-\omega_g{}^2}} \times \begin{cases} H_0{}^{(2)}(\mathscr{W}) & \text{for } \Omega_0 < \omega \\ H_0{}^{(1)}(\mathscr{W}) & \text{for } -\Omega_0 > \omega \end{cases} \tag{2.65}$$

where

$$\mathscr{W} = \frac{\sqrt{|m^2-1|}}{V_0}\,\hat{r}\sqrt{\frac{|(\omega^2-\Omega_0{}^2)(\omega^2-\Omega_c{}^2)|}{|\omega^2-\omega_g{}^2|}} \tag{2.65'}$$

$$\hat{r} = \sqrt{y^2+z^2}.$$

Note that $\Omega_0 > \omega_a$, $\Omega_c < \omega_g$. The final result Φ is in Fig. 2.7 where normalization is done as

$$\begin{aligned} U &= (\pi V_0 \Phi / S_0) \\ \tau' &= \omega_g t \\ R &= \frac{\sqrt{|m^2-m_0{}^2|}}{V_0}\omega_a \hat{r} \\ \theta_R &= \text{arc tan}\,(z/y). \end{aligned} \tag{2.66}$$

t is reckoned from the instant at which the source passes the point of the closest approach to observer. In the calculation, the following numerical value is adopted; $m=1.1$, $m_0=0.2$, $R=1.0$, $\theta_R=0$. The present numerical value of m_0, which should be about 0.9 in realistic situation in the earth's atmosphere, allows us to identify the frequency of the excited wave. Note that under the realistic situation ω_a and ω_g are too colse to identify the characteristic frequencies in Fig. 2.7 (bottom). It seems fairly easy to know that the lower characteristic frequency in the wave is very close to ω_g. The higher frequency superposed is found to be very close to Ω_0.

Case 2. $1 > m > m_0$, subsonic motion 1

Similarly to *Case* 1

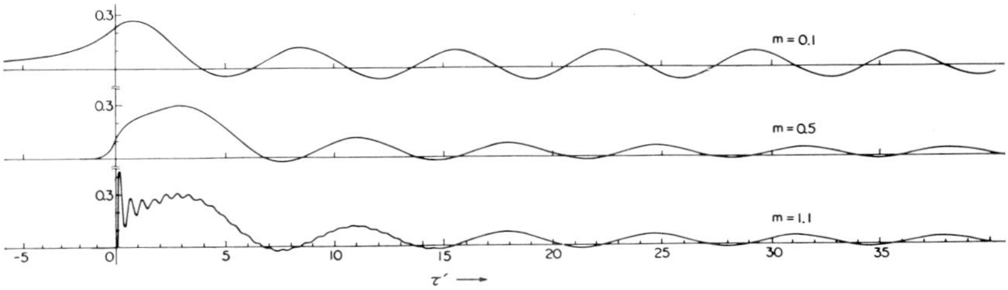

FIG. 2. 7. Gravity waves radiated from a moving source. Bottom, $m>1$, $R=1.0$, $\theta_R=0$ (supersonic); middle, $1>m>m_0$, $R=1.0$, $\theta_R=0$ (subsonic 1); top, $m_0>m$, $R=1.0$, $\theta_R=0$ (subsonic 2) (KATO *et al.*, 1977).

$$G_\omega = -\frac{2\pi i}{\sqrt{\omega_g{}^2-\omega^2}} K_0(\mathscr{W}') \quad \text{for } 0<\omega<\Omega_c$$

$$= -\frac{\pi^2}{\sqrt{\omega_g{}^2-\omega^2}} H_0{}^{(2)}(\mathscr{W}') \quad \text{for } \Omega_c<\omega<\omega_g$$

$$= -\frac{2\pi s_g}{\sqrt{\omega^2-\omega_g{}^2}} K_0(\mathscr{W}') \quad \text{for } \omega_g<\omega \tag{2.67}$$

where G_ω for $\omega<0$ is obtained as in *Case* 1 by replacing $H_0{}^{(2)}$ by $(-H_0{}^{(1)})$ and K_0 by K_0 for each $|\omega|$ and

$$\mathscr{W}' = \frac{\sqrt{1-m^2}}{V_0}\hat{r}\sqrt{\frac{|(\omega^2+\Omega_0{}^2)(\omega^2-\Omega_c{}^2)|}{|\omega^2-\omega_g{}^2|}}. \tag{2.68}$$

Numerical integration about ω gives Φ as in Fig. 2. 7 where the same values of m_0, R and θ_R (but $m=0.5$) to those in *Case* 1 are used. Note that Φ now is almost monochromatic with a frequency very close to ω_g and the wave arrives earlier than the source.

Case 3. $m<m_0$, subsonic motion 2

$$G_\omega = \frac{\pi^2}{\sqrt{\omega_g{}^2-\omega^2}} H_0{}^{(1)}(\mathscr{W}) \quad \text{for } 0<\omega<\Omega_{c,0}$$

$$= \frac{2\pi i}{\sqrt{\omega_g{}^2-\omega^2}} K_0(\mathscr{W}) \quad \text{for } \Omega_{c,0}<\omega<\Omega_{0,c}$$

$$= -\frac{\pi^2}{\sqrt{\omega_g{}^2-\omega^2}} H_0{}^{(2)}(\mathscr{W}) \quad \text{for } \Omega_{0,c}<\omega<\omega_g$$

$$= -\frac{2\pi s_g(\omega)}{\sqrt{\omega^2-\omega_g{}^2}} K_0(\mathscr{W}) \quad \text{for } \omega_g<\omega \tag{2.69}$$

where $\Omega_{c,0}$ equals the smaller of Ω_c and Ω_0, and $\Omega_{0,c}$ the larger. $\mathscr{W}$ is given by (2. 65′), and G_ω for $\omega<0$ is obtained as in *Case* 1 and 2 by $H_0{}^{(1)} \rightleftarrows [-H_0{}^{(2)}]$, $K_0 \to K_0$ for $\omega \to |\omega|$. Φ with $m=0.1$ is Fig. 2. 7. Note that the wave is just monochromatic with a frequency very close to ω_g and arrives, as for $1>m>m_0$, earlier than the source.

d) Uniformly moving finite lines source (KATO *et al.*, 1977)

The source here is a line with finite length moving horizontally and orthogonally to the line with a constant velocity V_0. The source is just an ensemble over the line length of the point source in the preceding example. The solution is therefore obtained by integrating Φ_p which is Φ in (c) the previous example (Fig. 2. 7) as

$$\Phi_L(t, x, y, z) = \int_{-L}^{L} \Phi_p(t, x, y-y', z)\mathrm{d}y' \tag{2.70}$$

where the source length is $2L$. Figure 2. 8 illustrates some results in which the observer stands at the origin and $L=10\,l$ and $3\,l$ where

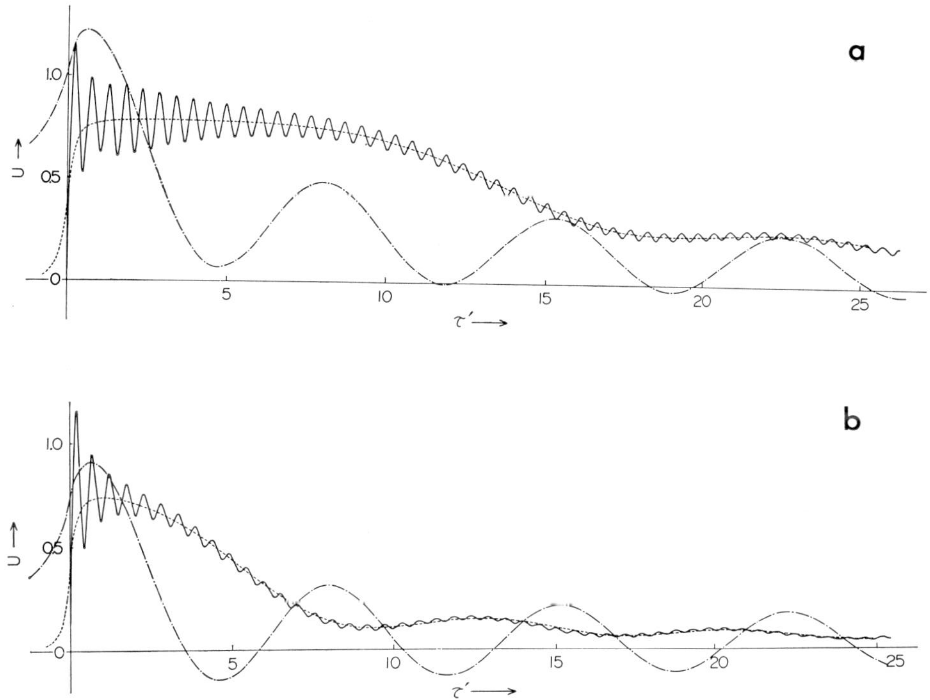

FIG. 2. 8. Gravity waves radiated from a moving finite line source: supersonic motion corresponding to Fig. 2. 7 bottom (solid curve); subsonic motion 1 corresponding to Fig. 2. 7 middle (dotted curve), subsonic motion 2 corresponding to Fig. 2. 7 top (chain curve), (a) $L=10l$, (b) $L=3l$ (KATO *et al.*, 1977).

$$l = \frac{V_0}{\Omega_0 \sqrt{m^2-1}} = \frac{mC}{\omega_a \sqrt{m^2-m_0{}^2}}$$

$$\Phi = \frac{8\pi P_0 U}{\omega_a \sqrt{m^2-m_0{}^2}}.$$

It is noticed that, for $m>1$, i.e. a supersonic motion, a high frequency oscillation is remarkable, the frequency being Ω_0. For $1>m>m_0$ i.e. subsonic motion 1 (the broken curve), no such frequency is found. For $m_0>m$, subsonic motion 2 (the chain curve), the oscillation with a frequency Ω_0 again appears, but in this case Ω_0 is fairly small because of small m. Note that in all cases, a careful examination detects a superposed oscillation with a frequency very close to ω_g after certain initial period, as is discussed below.

ABREU NETO *et al.* (1972) have discussed the generation of atmospheric waves from a moving point source. They have considered only far fields using the saddle point integration method. However, their result can asymptotically well explain the present one which is obtained by elaborate numerical integration by the Gaussian method in Fourier inverse transformation on (2. 63). For $m>1$, the frequency should tend to consist of Ω_c, ω_g, and Ω_0, which are consistent with the result in Fig. 2.7 where $\Omega_c=0$ ($\theta_R=90°$). As to $1>m>m_0$ and $m_0>m$, the diagram by ABREU NETO *et al.*, can explain the result in Fig. 2. 7. Note that $\omega_g/\Omega_0 \simeq 1.1$ in Fig. 2. 7 (the top) and the frequency separation of the two waves expected seems hardly detectable. Other cases with $\theta_R \neq 0$, which are not shown here, are also consistent with

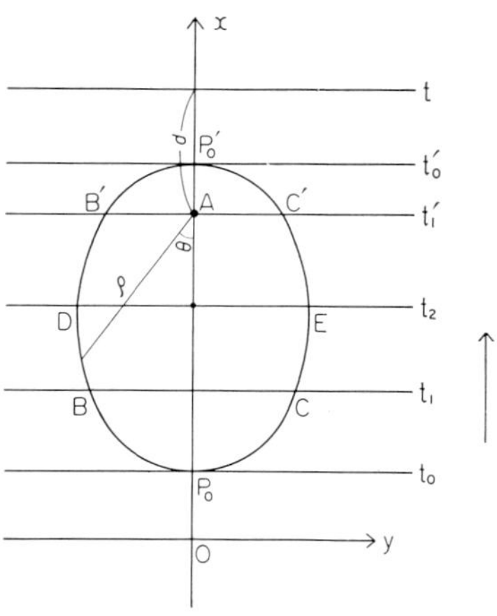

FIG. 2. 9. Effective area for GW emission from a line source along the y-axis, the motion being along the x-axis, the observation being at A. The contribution comes from only those elementary waves which emit at the cross points of the line and the ellipse at each retarded time; $\hat{\rho}$ is denoted by ρ in the diagram (KATO *et al.*, 1977).

their approximation.

As to the line source, the situation is a little more complicated than that of point source. Waves radiated from all points on the line may contribute to the overall radiation field as shown in Fig. 2. 8. Now consider those waves which arrive at A at time t (Fig. 2. 9), the waves having been emitted at earlier times from various points along the line source as from B and C at t_1 ($t_1<t$), from D and E at t_2 ($t_1<t_2<t$), and from B′ and C′ at t_1' ($t_1<t_2<t_1'<t$). The source has passed the observer A to reach the location as denoted in Fig. 2. 9 by t with a distance d from A. A is taken on the x-axis in the plane including the finite line source. Location of A outside the plane does not change the conclusion. Introducing a polar coordinate $(\hat{\rho}, \hat{\theta})$ where the origin is at A and $(\pi-\hat{\theta})$ is the angle between $\hat{\rho}$ and the x-axis, the following simple relation is valid, because the source arrives at the location, distant by d from A, during the time in which the wave from point $(\hat{\rho}, \hat{\theta})$ arrives at A:

$$\frac{\hat{\rho}}{C_s} = \frac{\hat{\rho}\cos\hat{\theta}+d}{V_0} \tag{2.71}$$

where C_s is the velocity of wave, which, for the present discussion, is assumed constant (without dispersion); for the present purpose, which is to find the effective length of the source, C_s should be a group velocity. Equation (2. 71) gives

$$\hat{\rho} = \frac{\hat{d}}{1-\delta_\gamma \cos\theta} \tag{2.72}$$

where

$$\hat{d} = d\left(\frac{C_s}{V_0}\right), \quad \delta_r = \frac{C_s}{V_0}.$$

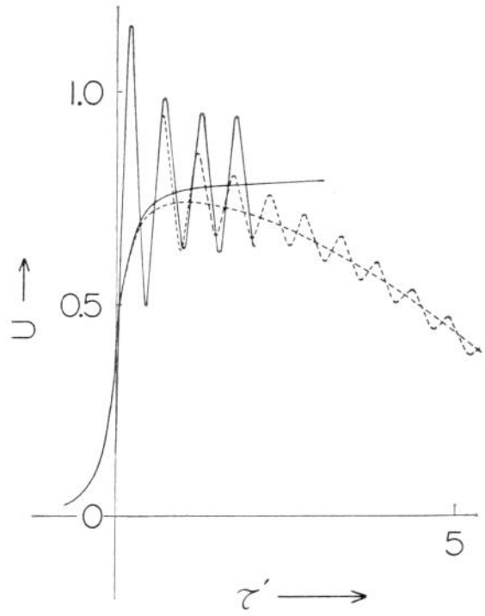

FIG. 2. 10. GW radiated from a moving infinite line source (solid curve) and finite line source (broken curve). The wave-like curves are for supersonic motion and the non-wave-like curves for subsonic motion 1. $L=3\,l$ (KATO *et al.*, 1977).

For $C_s < V_0$, $\delta_r < 1$, and (2. 72) gives an ellipse which has one focus at A as illustrated in Fig. 2. 9. This is the case of supersonic motion. The intersections between the ellipse and the line source, P_0, B, C, D, and E, give retarded instants for the emission of those waves which arrive at A at t. The parts of the line source outside the ellipse are ineffective for the disturbance at A at t. Therefore, if the source is longer than the width of the ellipse $\hat{d}/\sqrt{1-\delta_r^2}$, one can replace, at will, the finite line source by an infinite line source leaving the radiation field unchanged. Then, the problem is two-dimensional by removing the y-dependence and putting $\beta=0$ in (2. 50) and (2. 62). Calculations with $m=1.1$ for the two-dimensional case have been made, and the result is illustrated in Fig. 2. 10 together with the initial part for the case of a finite line source in Fig. 2. 9 for comparison. Note that the two-dimensional disturbance consists of an enhanced constant level and a characteristic oscillation with frequency very close to Ω_0. In Fig. 2. 9, the ellipse enlarges with t, and eventually engulfs the finite source after t_L at which $\hat{d}/\sqrt{1-\delta_r^2} = L$, and then, for $L=3\ l$, in Fig. 2. 8,

$$\tau' = \omega_g t_L = \frac{3m_0\sqrt{1-\delta_r^2}}{m\sqrt{m^2-m_0^2}} \simeq 0.3.$$

For $t<t_L$, the infinite and finite line sources produce the identical disturbance at A as in Fig. 2. 10.

The end effect appears after t_L for a finite source, the field being not simple, but much later the source goes far away from the observer and tends to behave as a point source, which radiates a wave with a characteristic frequency as in Fig. 2. 7.

For subsonic motion as $C_s > V_0$, (2. 72) gives a hyperbola, because $\delta_r > 1$ in (2. 72). The hyperbola has its major axis along the x-axis and is open

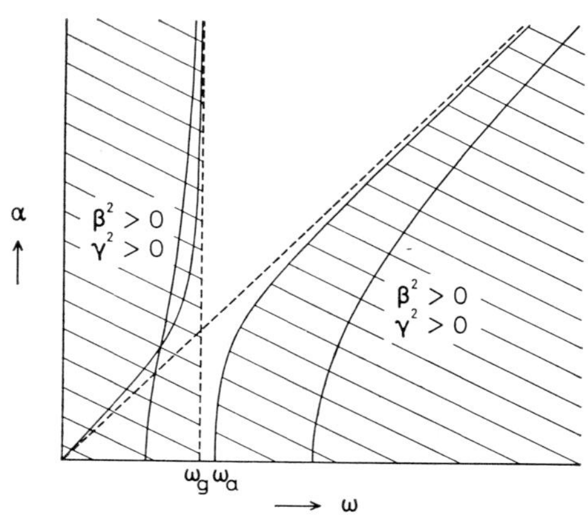

FIG. 2. 11. Dispersion of GW; α versus ω with both β and γ as parameters.

for $x \to \pm\infty$ implying a possible contribution to the radiation field observed at A and t over the entire length of the source. It turns out, therefore, that the end effect of the line source should now be important. However, it is well-known that *such* subsonic motion cannot induce bow waves. Note that this is not the case for gravity waves, which have the peculiar dispersion shown in Figs. 2. 1 and 2. 11. It is readily found that, no matter how small V_0 is, there are those wave components which have horizontal phase velocity (ω/α) equal to V_0, the condition, which produces constructive interference. In Fig. 2. 1(b), broken lines 3 and 4 correspond to subsonic situation. Generally, a similar situation as in Fig. 2. 9 holds valid for any V_0. Numerical calculation favors the argument as in Fig. 2. 10 where, both in supersonic motion and subsonic motion 1, the radiation received at A for an initial period as $\tau' < 0.4$ is similar to that by infinite line source. Note that in subsonic motion 1 the infinite line source produces a disturbance which is a sudden increase of Φ upon the arrival of the source to remain the same in later times. The situation is readily understood by the fact that the two-dimensional wave component now concerned has phase velocity equal to V_0 only with $\omega \to 0$ for $1 > m > m_0$ (broken line 3 in Fig. 2. 1(b)). On the other hand, at later times, the finite line produces an end effect complicating the situation. However, still later the source goes so far away that the finite line tends to behave as a point. Then, the field for the finite line source tends to be that for a point source, as shown by comparing the middle curve in Fig. 2. 7 and the dotted curve in Fig. 2. 8. For subsonic motion 2 $(m_0 > m)$ the radiation field Φ is not available for an infinite line source since mathematical difficulties occur concerning the convergence of the inverse Fourier transformation.

It is thus concluded that *GW can easily be radiated in association with almost any motion of the source* whilst acoustic waves can be radiated *only by supersonic motion* as discussed by WILSON (e.g. 1969) and SWIFT (1973) for the auroral infrasonic waves. Detail ionospheric observation on gravity waves is very necessary. BROCHE *et al.* (1976) and BERTIN *et al.* (1977) measured electron density fluctuations during an eclipse for the purpose of detecting gravity waves generated by the moon shadow motion.

2.3 Traveling Ionospheric Disturbances (TID) and Other Observed AGW's

Since MARTYN (1950) discussed traveling ionospheric disturbances (TID) in terms of cellular waves, many works on TID's have been done both experimentally and theoretically. Among others, HINES (e.g. 1960) has contributed much to establish beyond doubt that TID's are the manifestation of GW's in the neutral atmosphere which travel through the ionosphere over very long distance.

There are various methods of observation of TID's. Figure 2. 12 il-

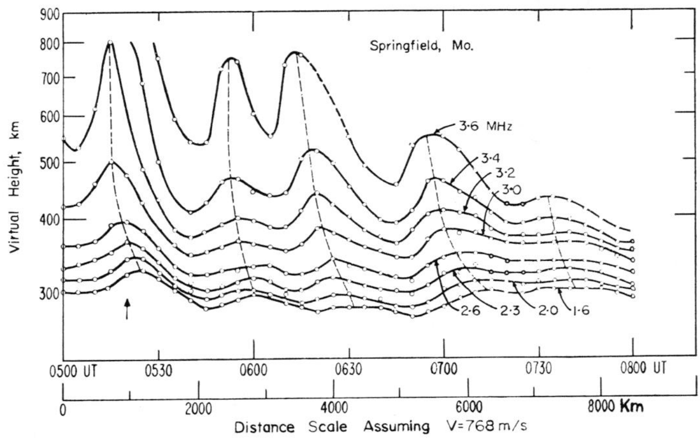

FIG. 2. 12. LS TID: isoionic contours in the virtual height versus time plane obtained at Springfield, *Mo* during a large TID for various frequencies of radio signals (GEORGES, 1968).

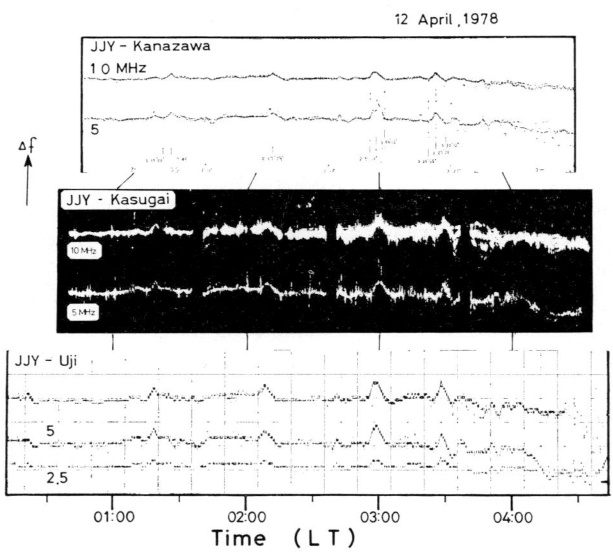

FIG. 2. 13. TID by HF Doppler (Ogawa and Tsutsui, private communication, 1978). The standard radio signal (JJY) is used for observation at 3 locations, Kanazawa (36°33′ N, 136°40′ E), Kasugai (35°16′ N, 137°01′ E) and UJI (34°54′ N, 135°48′ E).

lustrates an observed TID which is now known as large scale TID's (LSD) by using pulsed radio waves with six frequencies, each of these waves being reflected back at a particular height at which the electron density has its corresponding plasma frequency $f_p = f$ where

$$f_p(\text{Hz}) = 9.0\sqrt{n_e(m^{-3})}. \tag{2.73}$$

The electron density perturbation is mainly produced by wind shear of GW's (Section 4. 2) and height-dependent loss factor of electron density. Monitoring the time necessary for the round trip of each pulse, one can obtain the (vertical) height variation with time of the equal electron density corresponding to each frequency as is shown in Fig. 2. 12. The figure clearly shows that the phase of the wave-like disturbances propagates downwards, a characteristic, which is peculiar to GW's excited somewhere below (see Figs. 2. 1 (c) and 2. 4(c)). A similar method detected TID's as early as the 1950's (e.g. MUNRO, 1958). Using radio waves of highly stabilized frequency, the HF Doppler method can detect a change of reflection height of the ionosphere. The Doppler shift equals twice the doppler shift due to vertical motion of the reflection height. Observation of these Doppler shifts tells about TID's, as can be seen in Fig. 2. 13. Networks of the HF Doppler observation can give information about the horizontal transmission of GW's, and especially about the dispersion and the arrival direction (HERRON, 1974).

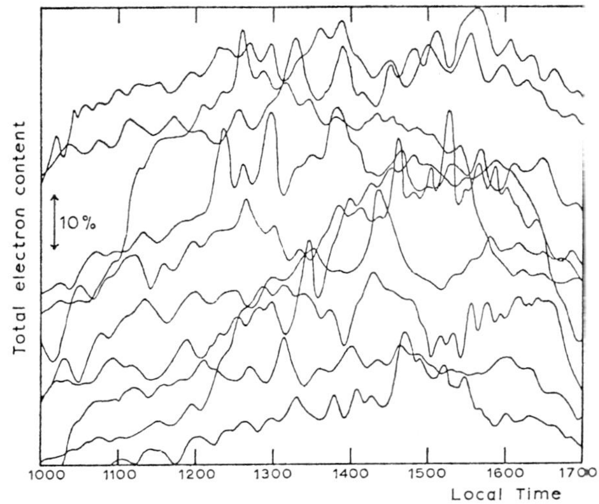

FIG. 2. 14. MS TID: wave-train type, observed in the total electron content from the Faraday rotation of radio signal from satellites (TITHERIDGE, 1968).

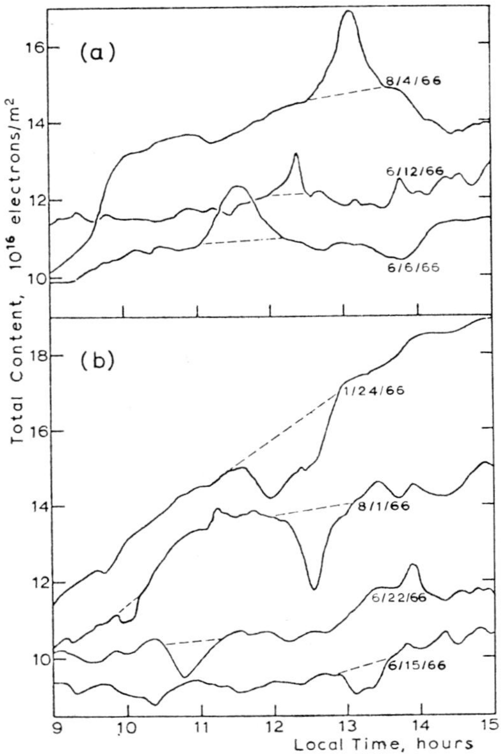

FIG. 2.15. MS TID: isolated pseudo-wave type observed in the total electron content; (a) positive disturbance, (b) negative one (TITHERIDGE, 1971).

When one measures the polarization of radio beacons with linear polarization from man-made satellites, synchronous with the earth's rotation over the equator, the polarization direction varies (with time) with the total electron density between the satellite and the receiver on the ground. This Faraday rotation method has been found effective to measure TID's (Figs. 2.14 and 2.15).

Incoherent scatter radars (Section 3.4) are very powerful in observing TID s in that the radars can measure simultaneously various parameters as n_e, $T_{e,i}$ and $W_{\parallel}$.

These observations have made it clear that there are two types of TID's, one being large scale (LS) TID's and the other being medium scale (MS). LS TID's have long periods as 2–4 hr, lasting for a few cycles, and travel horizontally as fast as 700 m sec^{-1}. MS TID's are more often observed than

LS's. The periods are shorter and the horizontal propagation is slower (e.g. HEARH and YEH, 1977). MS TID's have further two types, one consisting of many cycles of wave trains (Fig. 2. 14) and the other being an isolated wave (Fig. 2. 15). These two types occur nearly equally in frequency.

It is felt that these different types of TID's are excited differently. After many years of observation and debates, LS TID's have been identified, beyond any doubt, with the ionospheric response to gravity waves in the neutral atmosphere, which are produced by joule heat and by the Ampere mechanical force at the auroral zone during magnetic substorms (e.g. CHIMONAS and HINES, 1970; TESTUD, 1970). The occurrence depends on the intensity of the magnetic activity. TITHERIDGE (1971) has shown $K_p>5$ to be a necessary condition, and TESTUD (1970) has shown, from his obsevation of the F region neutral wind, that the occurrence frequency of large gravity waves reaches (25/27) for $\sum_{24} K_p>16$, but (1/16) for $\sum_{24} K<16$, where K_p is the 3-hourly magnetic index and $\sum_{24}$ its sum for one day. The energy flux associated with such joule-heating-induced intense gravity waves amounts to 10^{-4}–10^{-3} Jm^{-2} sec^{-1} and cannot be neglected in the heat budget of the F region, where the waves, excited in the E region, are absorbed through viscosity and ion drag (BLUMEN and HENDL, 1969; TESTUD, 1970). Thus, the thermosphere temperature and composition may be modified through the passage of the intense gravity waves to a considerable degree (PRASAD and FURMAN, 1975).

As to MS TID's, FRANCIS (1974) has discussed their excitation similar to that of LS TID's at high latitudes. But the energy input region is more restricted in size as will be shown below.

Other possibilities of excitating TID's have been discussed. CHIMONAS (1970) has pointed out that the equatorial electrojet can generate GW's giving rise to TID's. However, he noticed that the $(\boldsymbol{J}\times\boldsymbol{B}_0)$ force by the equatorial electrojet is vertical and is less effective than the almost horizontal force by the auroral electrojet. It is now well-known that various meteorological disturbances generate GW's in the lower atmosphere (JORDAN, 1972), and the generated GW's travel upwards to the ionosphere producing TID's. Among meteorological disturbances, tornadoes have been confirmed to be a source by HUNG and SMITH (1979). TSUTSUI and OGAWA (1973) have identified some TID's observed by HF Doppler with GW's generated in association with tyhoons. Also, by HF Doppler observation, some TID's have been associated to GW's caused by disturbance in the lower stratospheric westerly jet-flow (OKUZAWA *et al.*, 1978).

As to unusual sources of AGW's, there are the sun's shadow motion during eclipses or the auroral electrojet motion (e.g. WILSON, 1969) and nuclear detonations (e.g. ROW, 1967). These are discussed in the previous section. Even artificial excitation can be possible (PERKINS and FRANCIS, 1974).

Some sporadic E can be produced by GW's through their wind shear (see Section 4. 2), although its horizontal propagation remains unknown.

Meteor radars, which track meteor trails in motion with the ambient air between 80–100 km height, show that there are always components corresponding to GW's, their dissipation heating the thermosphere (HINES, 1965).

Air glow observations have detected GW's (NOXON, 1978). The rotational temperature variation in the air glow of O_2 around 95 km clearly progressed to that of OH around 85 km. The period was 1–2 hr.

The noctilucent clouds (NLC) are beautiful clouds which form around

FIG. 2. 16. NLC 15 July, 1965 from Grande Prairie, Alberta (HAURWITZ and FOGLE, 1969).

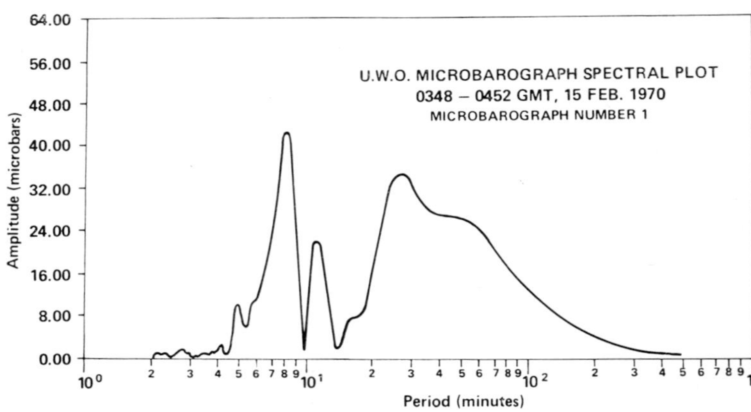

FIG. 2. 17. Pressure spectrum observed by microbarograph (BEER, p. 22, 1974).

the mesopause, where the temperature is low enough to produce cloud formation (Fig. 2. 16). NLC are observed only in latitude between 45°–75°, in summer, and for the solar zenith angle between (−6°) and (−16°) in twilights. Their formation seems to depend fairly critically on the local temperature. Study of NLC has revealed that the observed wavy structures in NLC are produced by GW's both internal and external (HAURWITZ and FOGLE, 1969). Recently, powerful VHF radars have been found effective to detect velocity fluctuations due to GW's in the mesosphere (RASTOGI and WOODMAN, 1974; FUKAO *et al.*, 1979) and the stratosphere (RUSTER *et al.*, 1978; FUKAO *et al.*, 1978). These radars can measure small vertical velocities of a few cm sec^{-1} in every 10 sec. On the ground, GW's are constantly recorded by microbarograph (Fig. 2.17). Since GW's can have the amplitude of the order 10 m sec^{-1} in the mesosphere, they should play important role in the general circulation of the mesosphere. Observation of GW's over various heights in a systematic manner should be an important future problem.

The behavior of AGW's in realistic atmosphere (e.g. FRIEDMAN, 1966) is fairly complex and the general treatment should be difficult. What follows is an introduction of some studies done mainly for TID's which are generated in the upper atmosphere by joule heating and the Ampere force of the auroral electrojets. However, in many respects, the results are generally helpful to understand the propagation of AGW's in realistic atmosphere. First, Subsection 2. 3.1 considers freely propagating waves which are reflected on the earth's surface, but receive little refraction and no reflection in the upper atmosphere. Waves with horizontal velocity (ω/α) much slower than local sound velocity belong to this category.

Secondly, in Subsection 2. 3. 2 the guided waves traveling faster will be discussed mainly in relation to LS TID's. The discussion will include, however, other waves found in nuclear explosion and Krakatora eruption in 1823 etc.

Thirdly, in Subsection 2. 3. 3 a simulation about an observed TID on a particular day will be introduced and compared to the real observation of the same TID.

2. 3. 1 *Freely propagating Waves and MS TID's*

These waves follow similar dispersion to that, as in the preceding section, in isothermal unbounded atmosphere. The earth's surface just introduces reflection waves which satisfy, together with the incident waves, the boundary condition that $W_{z,\omega}=0$ at $z=0$, the surface. This connects each elementary wave given by (ω, α, γ) to that by $(\omega, \alpha, -\gamma)$, since by (2. 7′) $W_{z,\omega}=0$ at $z=0$ only for combined waves of $p_{1,\omega}\propto[\pm e^{i(\omega t-\alpha x\pm\gamma z)}]$ (Snell's law). Thus, the $(\omega-\alpha)$-diagram is identical to that shown by Fig. 2. 1(b). Further, existence of the thermosphere can be approximated by a region with a higher temperature atmosphere above a lower temperature region as given by Fig. 2. 18(a);

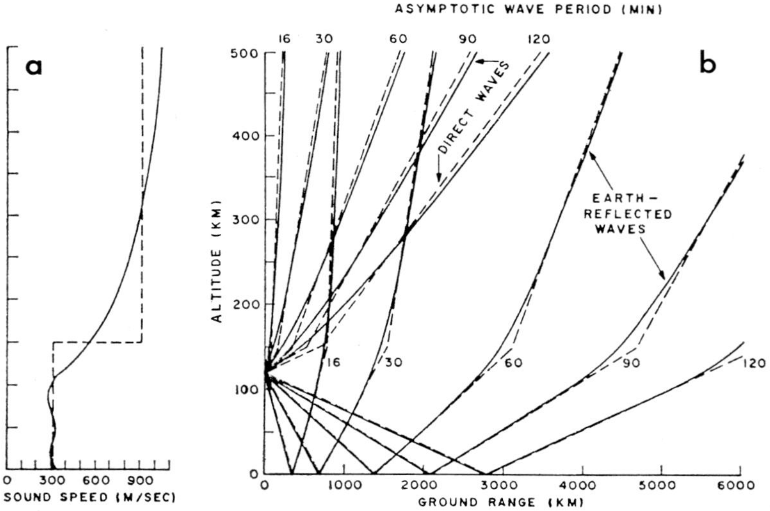

FIG. 2. 18. (a) Sound speed profile in a realistic atmosphere (solid curve) and a simplified model (broken curve) (FRANCIS, 1974). (b) Ray-path obtained by $dx/dz=[(\omega_g^2/\omega^2)-1]^{1/2}$ for D-and R-waves for which $(\omega/\alpha)^2 \ll C^2$ for each case in (a) (FRANCIS, 1974).

each region is isothermal. Along the interface between these two regions the horizontal velocity, (ω/α), is continuous (Snell's law). This connects, as at the earth's surface, each elementary wave given by $(\omega,\ \alpha,\ \gamma_{I})$ to that by $(\omega, \alpha, \gamma_{II})$ where $\gamma_{I,II}$ are the vertical wave number in the lower and upper regions, respectively. Thus, the $(\omega-\alpha)$-diagram is also identical to Fig. 2. 1(b). Each curve in the diagram corresponds to γ_{I} in the lower region and γ_{II} in the upper region respectively, where $\gamma_{I} \neq \gamma_{II}$.

The approximation of the upper atmosphere by these two isothermal regions can be justified by calculating the ray paths of various waves in this model and also in a more realistic one with a continuous temperature distribution (Fig. 2. 18(b)).

FRANCIS (1974) has attempted to find shapes of these freely propagating waves produced by a particular source. He adopted a simple source proportional to $\delta(x)\delta(z-z_0)H(t)$ where $\delta(x)$ and $\delta(z-z_0)$ are delta functions and $H(t)$ is step function; the x-axis is positive towards the south, and z_0 is 120 km. This is a mdoel for explaining MS TID's, which satisfy the condition that γ is real by $(\omega/\alpha)^2 \ll C^2$ by (2.17). The mathematical approach was linear and similar to that in the previous section, Fourier transformation being done, with respect to t, x, and z. But the solution of the homogeneous equation without source should also be taken into account, because there are two boundaries, one at $z=0$ the earth's surface, and the other at $z=150$ km.

The electron density was calculated from the following equation of continuity:

$$\frac{\partial n_e}{\partial t}+\mathrm{div}\,(n_e\boldsymbol{V}_i) = Q_p - L_s \tag{2.74}$$

where Q_p is production due to photo-ionization, and L_s is a loss term due to attachment ($N_2+O^+ \rightarrow N+NO^+$) to be followed by dissociative recombination ($NO^+ + e \rightarrow N+O$) i.e.

$$L_s = \frac{\alpha_r\beta_a n_e^2}{(\alpha_r n_e+\beta_a)} \tag{2.75}$$

in which β_a is the attachment coefficient and α_r the recombination coefficient. HOOKE (e.g. 1968) has discussed on peculiar effects of GW's on Q_p and L_s. But these effects were neglected in Francis' treatment of (2. 74), arguing that the effect could not be large, because the density perturbations in neutral particles, which affect Q_p and L_s, are generally smaller than that in charged particles, and, moreover, the variation tends to cancel between Q_p and β_a, by $\beta_a \ll \alpha_r n_e$ (2. 75) giving $L_s \simeq \beta_a n_e$.

As to $\boldsymbol{V}_i$, $\boldsymbol{V}_i = \boldsymbol{W}_{/\!/}$ was assumed i.e. ions move only along $\boldsymbol{B}_0$ with winds. (4. 1) shows that this is true, if (1) $\Omega_i \gg \nu_i$, (2) the electric field and the general flow $\boldsymbol{W}_0$, are neglected, and (3) the ambipolar diffusion (see (4. 21)) is neglected. Of these three conditions (1) and (3) are generally contradictory, because the ambipolar diffusion can be neglected for $\Omega_i \ll \nu_i$ against (1); (1) and (3) can be consistent only for 150 km$<z<$230 km. Condition (1) implies that the convergence of ionization due to the wind shear of GW's tends to take place simply at the nulls of northward winds overlying southward winds, in the northern hemisphere (see Section 4. 2). Generally, there are two factors to produce the electron density disturbances by GW's in TID's. One is the wind shear and the other the decreasing loss with height of ionization which is transported to higher altitudes by GW's.

The result of Francis' calculation is shown in Fig. 2. 19 where both velocity W_x and the perturbed density n_e' are shown for various time and x's at $z=200$ and 300 km. In principle, these results can explain well the existence of the two type MS TID's. GW's are exited at $z=120$ km by the $(\boldsymbol{J}_0\times\boldsymbol{B}_0)$-force where the given $\boldsymbol{J}_0$ is westward, so $(\boldsymbol{J}_0\times\boldsymbol{B}_0)$ is southward i.e. equatorward; the total current, concentrated as a delta function around $x=0$ and $z=120$ km, amounts to 10^5 A.

It was found that waves of two kinds could be detected at the points distant enough from the source to validate the stationary phase method which was adopted. Waves of one kind (D-wave) are those which have reached observers (above 200 km), with no chance of reflection at the ground, but with a little refraction at the boundary between the lower and upper region at $z=$ 150 km. Waves of the other kind (R-wave) are those which have experienced

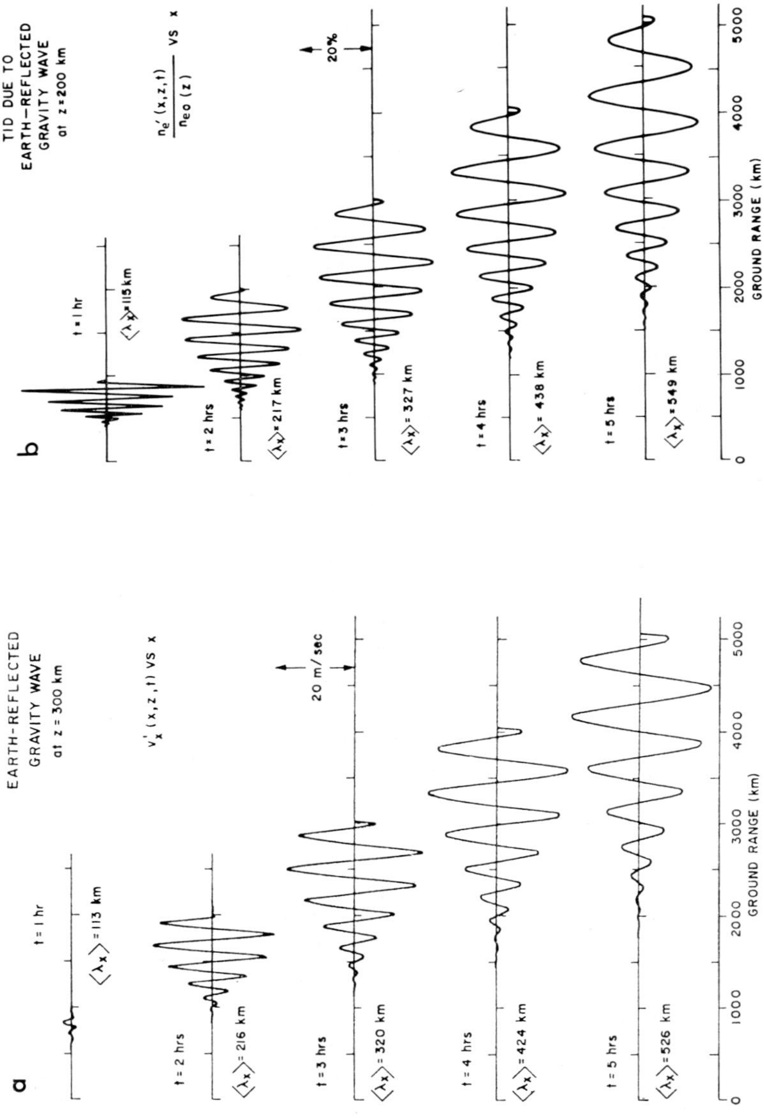
a
EARTH-REFLECTED GRAVITY WAVE at z = 300 km
$v_x'(x,z,t)$ VS x
20 m/sec
t = 1 hr
$\langle\lambda_x\rangle$ = 113 km
t = 2 hrs
$\langle\lambda_x\rangle$ = 216 km
t = 3 hrs
$\langle\lambda_x\rangle$ = 320 km
t = 4 hrs
$\langle\lambda_x\rangle$ = 424 km
t = 5 hrs
$\langle\lambda_x\rangle$ = 526 km
0
1000
2000
3000
4000
5000
GROUND RANGE (km)
b
TID DUE TO EARTH-REFLECTED GRAVITY WAVE at z = 200 km
$\frac{n_e'(x,z,t)}{n_{eo}(z)}$ VS x
20%
t = 1 hr
$\langle\lambda_x\rangle$ = 115 km
t = 2 hrs
$\langle\lambda_x\rangle$ = 217 km
t = 3 hrs
$\langle\lambda_x\rangle$ = 327 km
t = 4 hrs
$\langle\lambda_x\rangle$ = 438 km
t = 5 hrs
$\langle\lambda_x\rangle$ = 549 km
0
1000
2000
3000
4000
5000
GROUND RANGE (km)

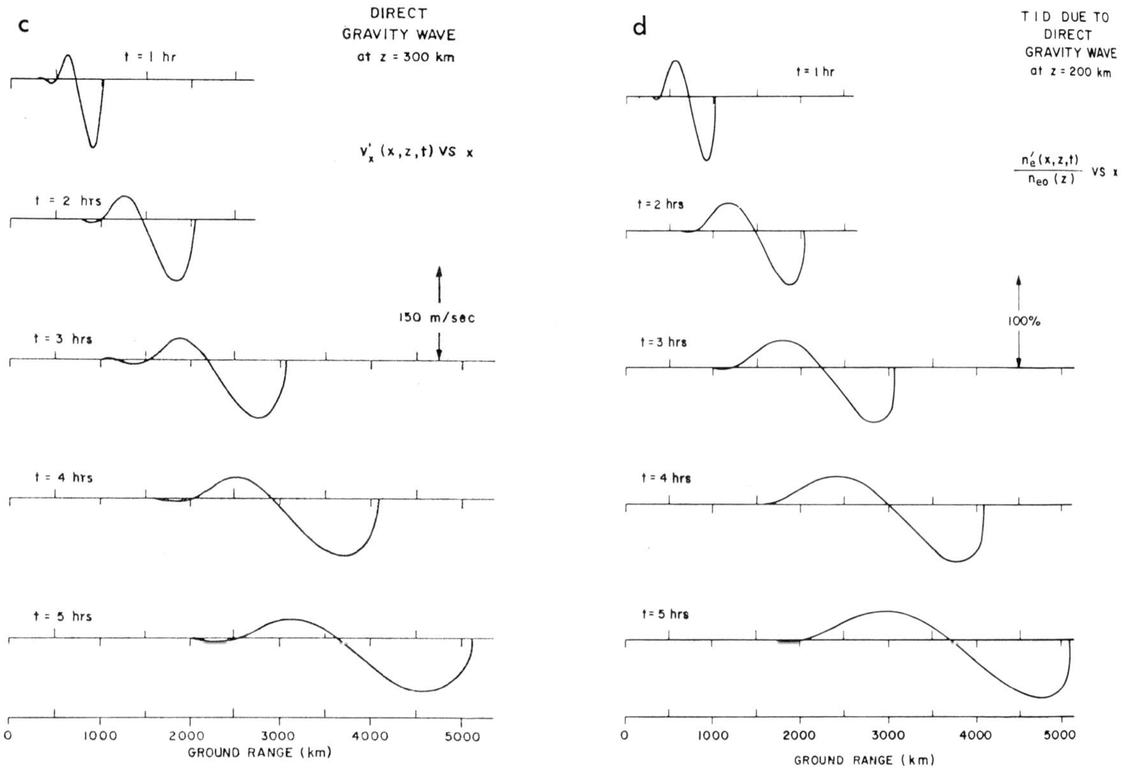

FIG. 2. 19. (a) Horizontal velocity fluctuation W_x (V_x' in the diagram) for R waves at $z=300$ km versus x ($t=1$, 2, 3, 4, 5 hr after switch-on of the source). (b) Electron density fluctuation similar to (a) except at $z=200$ km. (c) Horizontal velocity fluctuation similar to (a) except for D-waves. (d) Electron density fluctuation similar to (b) except for D-waves (FRANCIS, 1974).

a reflection at the ground and also a refraction at the boundary (Fig. 2.18(b)). No reflection takes place at the boundary at 150 km for MS TID's because of slow propagation of short period GW's.

D-waves are little dispersed during the passage to observers (Fig. 2. 19), and can keep their original shape as nearly a pulse. On the other hand, R waves are much more dispersed during the travel and almost lose their original shape to become quasi-monochromatic waves. The lack of attenuation with distance from the source, common to both D- and R-waves, is due to the anisotropic group velocity of GW's as in (2. 19). The radiated power is equal to the group velocity times the energy density, and the energy density is independent of the direction of the energy flow. Thus, the energy flux will be proportional to the group velocity. For GW's as Fig. 2. 2 shows, that the energy flux increases with decreasing elevation angle of the radiation direction from the source. The longer distance from the source implies the smaller elevation angle, so that more energy goes to the more horizontally distant points. This can compensate for a decrease of the amplitude with distance from the source.

The anisotropy of the energy flux is enhanced with (increasing) periods. And the more energy goes to the lower frequency components of GW's with decreasing elevation.

Thus, with increasing distance (decreasing elevation) the longer period of GW's becomes predominant as in Fig. 2. 19. The results by Francis were instructive to know the basic physics of the problem, though his model cannot stand detailed comparison with observation. Note that if the source were placed in the troposphere, the D- and R-waves would be almost identical, both being a single abbreviated wave train. Thus, the observed isolated TID's due to GW's must be generated in the upper atmosphere.

2. 3. 2 *Guided waves and LS TID's*

These waves are the products of boundaries or sharp gradients of parameters relative to the AGW propagation. The waves are surface waves (external waves) or trapped waves (internal waves). Along the earth's surface (at $z=0$), we have a surface wave by (2. 20)

$$W_{z,\omega} = AZ\, e^{i(\omega t - \alpha x) - \gamma_i z} \qquad (2.76)$$

where $\gamma_i^2 = -\gamma^2$ and A is the amplitude and $\gamma_i > 0$, because the wave intensity should decrease with height $z > 0$. The boundary condition $W_{z,\omega} = 0$ is now satisfied only by $Z = \omega(\omega^2 - \alpha^2 C^2) = 0$ which gives

$$\frac{\omega}{\alpha} = C. \qquad (2.77)$$

This is the dispersion of the Lamb wave with a sound speed C and without any vertical velocity at all for $z > 0$. Note that the overall amplitude tends to increase due to the decreasing air density as $e^{z/2H}$ by (2. 7′) and (2. 8). We call this mode L_0' (Fig. 2. 20).

Introduction of the discontinuous boundary as in Fig. 2. 18(a) produces another surface wave. A simple approach by THOME (1968) was to place the interface at $z=0$ and no other boundary elsewhere. The condition at the interface at $z=0$ is a continuity of both $p_{1,\omega}$, $W_{z,\omega}$, and α. They are

$$\left(\frac{\mathrm{D}p}{\mathrm{D}t}\right)_{\mathrm{I}} = \left(\frac{\mathrm{D}p}{\mathrm{D}t}\right)_{\mathrm{II}} \tag{2.78}$$

$$\left(\frac{\mathrm{D}W_{z,\omega}}{\mathrm{D}t}\right)_{\mathrm{I}} = \left(\frac{\mathrm{D}W_{z,\omega}}{\mathrm{D}t}\right)_{\mathrm{II}} \tag{2.79}$$

$$\alpha_{\mathrm{I}} = \alpha_{\mathrm{II}} \tag{2.80}$$

where $\mathrm{D}/\mathrm{D}t$ is required, because the interface can move freely; subscripts I, II denote region I and II respectively. Since the treatment is linear, (2. 79) results just in $(W_{z,\omega})_{\mathrm{I}}=(W_{z,\omega})_{\mathrm{II}}$ which is used in substitution of (2. 7′) into (2. 78) to eliminate $W_{z,\omega}$ in terms of $p_{1,\omega}$. Note that at the interface p_0 is continuous but ρ_0 cannot be continuous, because $T_{\mathrm{I}} \neq T_{\mathrm{II}}$. Such a model is unrealistic but still can be allowed dynamically. The dispersion relation is given now by

$$\frac{\Gamma_{i\mathrm{I}}-(\alpha^2 g/\omega^2)}{\Gamma_{i\mathrm{II}}-(\alpha^2 g/\omega^2)} = \frac{\omega^2-\alpha^2 C_{\mathrm{I}}^2}{\omega^2-\alpha^2 C_{\mathrm{II}}^2} \tag{2.81}$$

where

$$\Gamma_{i\mathrm{I}} = \frac{1}{2H_{\mathrm{I}}} - i\gamma_{\mathrm{I}}$$

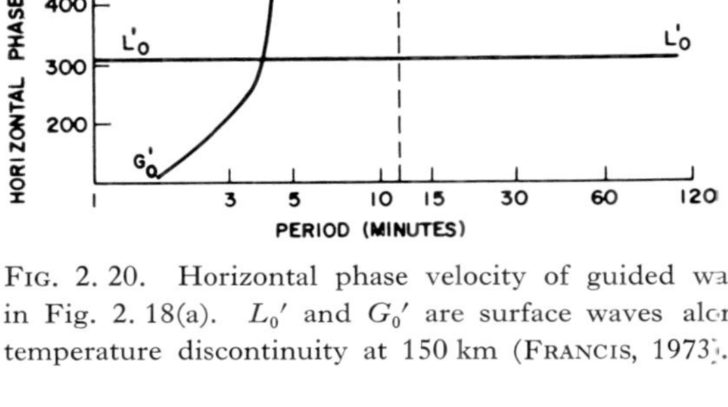

FIG. 2. 20. Horizontal phase velocity of guided waves for a simple model in Fig. 2. 18(a). L_0' and G_0' are surface waves along the ground and the temperature discontinuity at 150 km (FRANCIS, 1973).

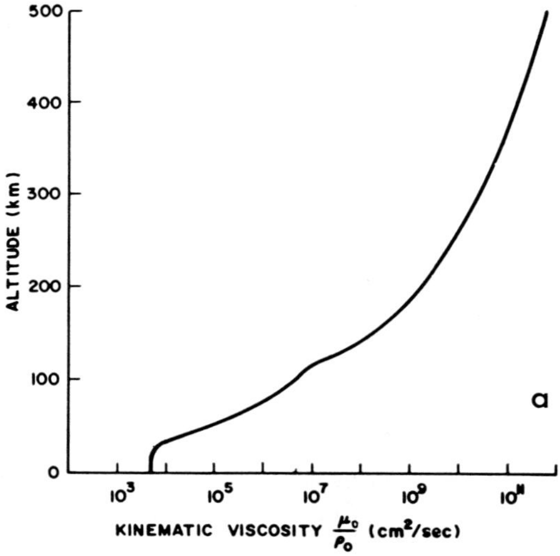

FRANCIS: ACOUSTIC-GRAVITY MODES

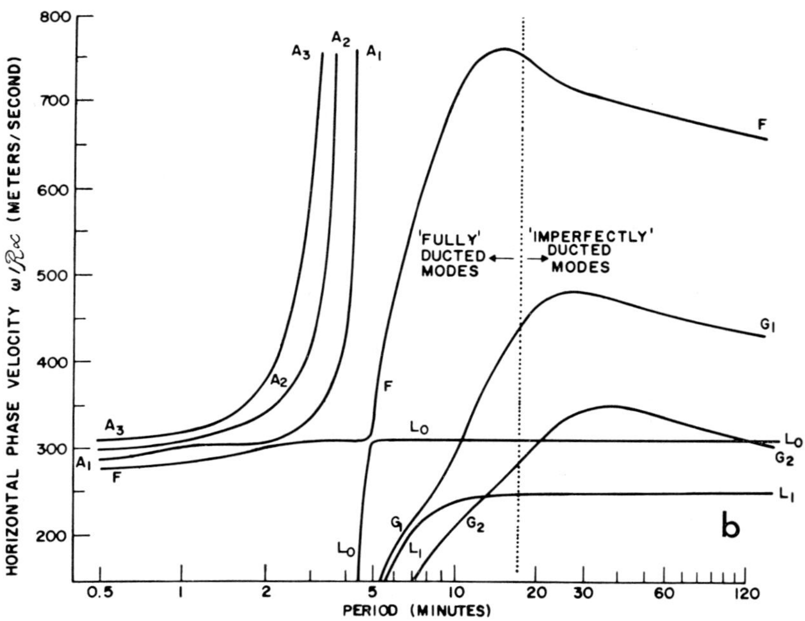

FIG. 2.21

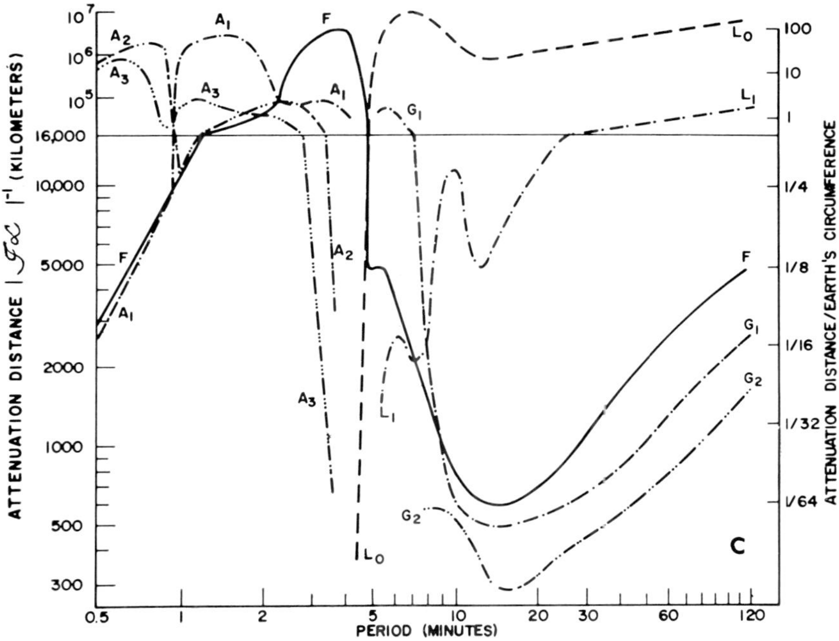

Fig. 2. 21. (a) Kinematic viscosity versus height. (b) Horizontal phase velocity of guided modes for a realistic model in Fig. 2. 18 (a) (Francis, 1973). $\mathscr{R}$ denotes real part. (c) Horizontal attenuation distance similar to (b). $\mathscr{I}$m denotes imaginary part.

$$\Gamma_{\text{iII}} = \frac{1}{2H_{\text{II}}} + i\gamma_{\text{II}}. \tag{2.82}$$

$\gamma_{\text{I,II}}$ are given by (2. 17) in two regions I and II. Then, (2. 81) and (2. 82) give the dispersion for this wave whose horizontal phase velocity (ω/α) is close to C_{II}, the sound speed in the higher temperature region and for long periods, but (ω/α) rapidly falls for short periods close to the Brünt-Väisälä period of the lower temperature region. We call this mode G_0' (Fig. 2. 20). If both the earth's surface and the temperature discontinuity exist, the situation is the superposition of these two separate cases with an additional appearance of higher order modes of G_0'. These modes resemble G_0' for low frequency limits, but their phase velocity falls around the Brünt-Väisälä frequency of the higher temperature region.

Francis (1973) has calculated the dispersion in a realistic dissipative atmosphere (Fig. 2. 21). He attempted to find solutions as $e^{i(\omega t - \alpha x)}$ by using the lower boundary conditions which are $W_{z,\omega}=0$ at the earth's surface and radiation conditions for large $z>0$. The radiation condition means

$$\mathscr{F}_{z,\omega} = \frac{1}{2}\mathscr{R}(p_{1,\omega} W_{z,\omega}{}^{c}) > 0$$

in (2. 18) (see Subsection 3. 1. 5), but Francis found that other conditions as the free surface conditions and rigid surface conditions made no difference because of strong dissipation of waves for large z. The problem is identical to find free oscillation in atmospheric tide (Subsection 3. 1. 5). For a given ω, these boundary conditions are satisfied only for a sets of discrete α's, eigenvalues which are complex numbers because of dissipative atmosphere. In Subsection 3. 1. 5 the eigenvalues h's are real numbers in non-dissipative atmosphere.

Similarities between the two results in Figs. 2. 20 and 2. 21(b) are found in basic features, though in realistic dissipative case the original L_0' for short period and the original G_0' mode for long period are connected around the

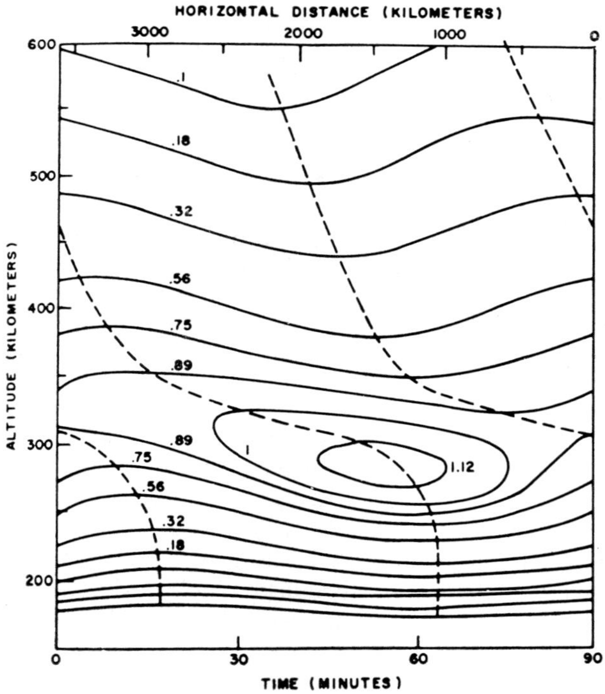

FIG. 2. 22. Electron density contour of TID caused by the fundamental mode with a period 90 min and a horizontal velocity 677 m sec^{-1}. The number to each curve denotes relative values of electron density. The broken lines are contours of constant phase corresponding to maxima and minima (FRANCIS, 1973).

Brünt-Väisälä frequency of the lower atmosphere to make the F mode (fundamental mode). The dispersion appears for the F and higher modes $G_{1,2}$, their phase velocity decreasing with period, an indication of imperfectly ducted modes. These F and $G_{1,2}$ modes in the thermosphere are rather internal GW's for long periods, and their attenuation distance are shorter than other modes, the distance being about 5,000 km for the F mode with a period of 2 hr (Fig. 2. 21 (c)). This is consistent with the fact that LS TID's across the equator have not been observed. The point, which seemed inconsistent with observation, was the observed forward tilting of the wave front, a characteristic of *pure* internal waves excited below the observed height (Fig. 2. 12). It was not to be found for guided modes. However, a solution has been found for considering the increasing horizontal phase velocity with height due to increasing temperature with height. Francis calculated electron density disturbance by these guided modes and found consistent with the observation (Fig. 2. 22). Note that there are many higher modes because of finite temperature gradient in the thermosphere supporting many modes. This corresponds to observed varieties of LS TID's.

However, a question has been raised by RICHMOND (1978) as to whether the F and $G_{1,2}$ modes were really the surface wave modes modified by viscosity and finite temperature gradient from the original simple form by THOME (1968). Richmond concluded that the realistic temperature profile is not suitable to produce surface waves discussed by Thome, and $G_{1,2}$ modes should be almost internal, and they travel in making total reflection at the earth's surface and partial reflection in the upper atmosphere due to viscosity and heat conduction effects. He obtained by a simple model the horizontal phase velocity and attenuation distance similar to those by Francis.

As to $A_{1,2,3}$ the acoustic modes, there are two wave guides, one around the mesopause and the other around the tropopause. In the short period limit, (ω/α) becomes the sound speed of the mesosphere. The attenuation distance (Fig. 2. 21(c)) is longer than the earth's circumference distance. These modes damp for the frequencies approaching ω_a of the lower atmosphere. This can explain observation of sound waves by nuclear explosion and others.

The L_0 mode is the Lamb wave for long periods and has the longest attenuation distance of about hundred times the earth's circumference, but in reality the surface friction and the eddy cause the distance to be much shorter. The same mode appears in the acoustic range as a part of the F mode. The L_1 mode has no dispersion for long periods, the horizontal phase velocity amounting to a speed of 254 m sec^{-1}. This is a trapped GW between the earth's surface and the mesopause, a problem which was discussed by HINES (1960) about the wave guide for long distance transmission of GW's.

How these guided modes are excited cannot be known in detail without introducing a heat input in the calculation. But it seems necessary for surface waves that the source is fairly closely located to the height or regions which are characteristic to each mode because of expected inefficient mode conversion from internal to external waves. It is likely, therefore, that if the LS TID's were due to surface waves, the auroral electrojet upon enhancement (as $\boldsymbol{J}\times\boldsymbol{B}_0$ or heat) expands up to 150 km around which the temperature gradient is rapid.

2. 3. 3 *Simulation of TID*

For direct comparison between observation and theory, simulation based on the theory is often very powerful. Works along this line have been attempted by RICHMOND and MATSUSHITA (1975) and ROBLE *et al.* (1978). They simulated GW's which are generated by the Ampere force and joule heating (RICHMOND and MATSUSHITA, 1975), and only by joule heating (ROBLE *et al.*, 1978), in the auroral zone. The work by Roble *et al.* is more advanced in that they

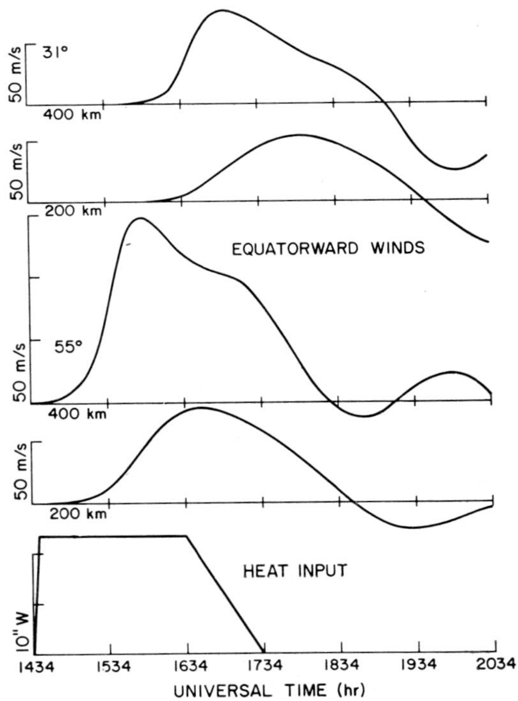

FIG. 2. 23. Time histories of calculated equatorward wind velocities at Arecibo (31 °) and at Millstone Hill (55 °) at altitudes of 400 and 200 km. At the bottom is the total heating rate imposed over the northern polar region. The effects of waves reflected from the equator (or, viewed differently, waves originating in the southern polar region) are apparent some time after 4 hr at 31 ° and after 5 hr at 55 ° (ROBLE *et al.*, 1978).

were fairly successful in reproducing a particular observation of TID's by incoherent scatter radars at Millstone Hill and Arecibo. What follows will be the main points of their work. The events they tried to simulate were those disturbances which originated in the sudden commencement (SC) of a magnetic storm at 1434 UT on September 18 in 1974. The previous day was very quiet. The model which gave the best fit in simulation was as follows:

1) The heat source (placed at the north pole) is joule heating of a model electric current which is distributed, with respect to z, as proportional to Pedersen conductivity, σ_1 (See Section 4. 1) and, with respect to θ, as proportional to $e^{-(\theta/\theta_0)^2}$ where $\theta_0 = 14°$. The total power amounts to 2.4×10^{11} W consistent with the estimate by COLE (e.g. 1962), more than 10^{-6} Wm^{-3}. Neglect of $(\boldsymbol{J} \times \boldsymbol{B}_0)$ could be incorrect (HUNSUCKER, 1977).

2) The heating occurs fairly suddenly, reaching a constant level in 5 min and lasting for 2 hr. Then, the heating decreases, in 1 hr, to the original zero level (Fig. 2. 23).

3) Only the (θ, z) dependency of variables is considered for the disturbances, although W_ϕ and J_ϕ are allowed to exist; W_ϕ is produced by $(\boldsymbol{J}_0 \times \boldsymbol{B}_0)$ i.e. $\tilde{\beta}\boldsymbol{W}$ (Subsection 3. 3. 2).

4) Non-perturbed horizontal wind in the absence of the GW's is obtained by a model based on satellite observations (ROBLE and DICKINSON, 1974) for Millstone Hill, and by a tidal wind model for Arecibo (HARPER, 1977b). The $(\boldsymbol{E} \times \boldsymbol{B}_0)$-drift velocity based on observation of the previous day i.e. September 17, 1974 is taken into account in addition to the tidal wind. The winds due to the disturbance were then superposed (Fig. 2. 24).

5) Electric field perturbation was not considered outside the source

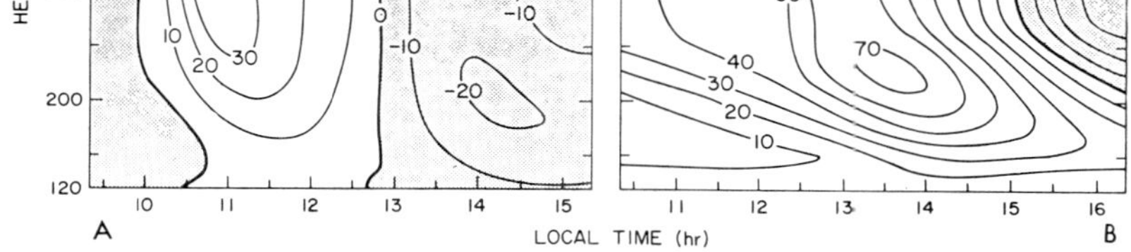

FIG. 2. 24. Total vertical upward drift. The spacing is 10 m sec^{-1}. (a) Millstone Hill, (b) Arecibo (ROBLE *et al.*, 1978).

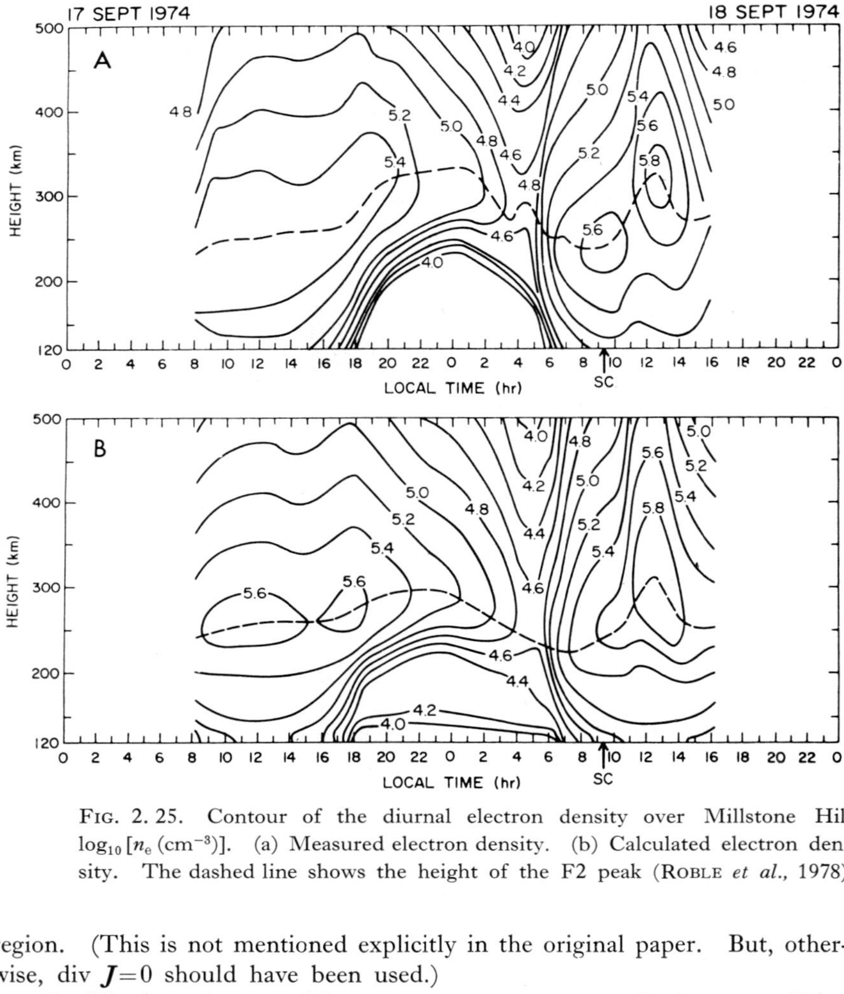

FIG. 2. 25. Contour of the diurnal electron density over Millstone Hill $\log_{10}[n_e\,(\mathrm{cm}^{-3})]$. (a) Measured electron density. (b) Calculated electron density. The dashed line shows the height of the F2 peak (ROBLE *et al.*, 1978).

region. (This is not mentioned explicitly in the original paper. But, otherwise, div $\boldsymbol{J}=0$ should have been used.)

6) The boundary condition was to assume no perturbation at $z=80$ km about $W_{\theta,\phi}$, p_1, and the gas energy density. Electron energy and ambi-polar diffusion were assumed to equal the measured values at $z=500$ km. The above six conditions decide the main features of the model used.

The solution was numerical, dividing the atmosphere above 80 km into 49 layers in the vertical and 121 elements in the horizontal (latitudes). Figures 2. 25 and 2. 26 show some of their results. The overall agreement between the simulation and the observation seems excellent. The sudden increase of the maximum electron density of the F2 region was reproduced

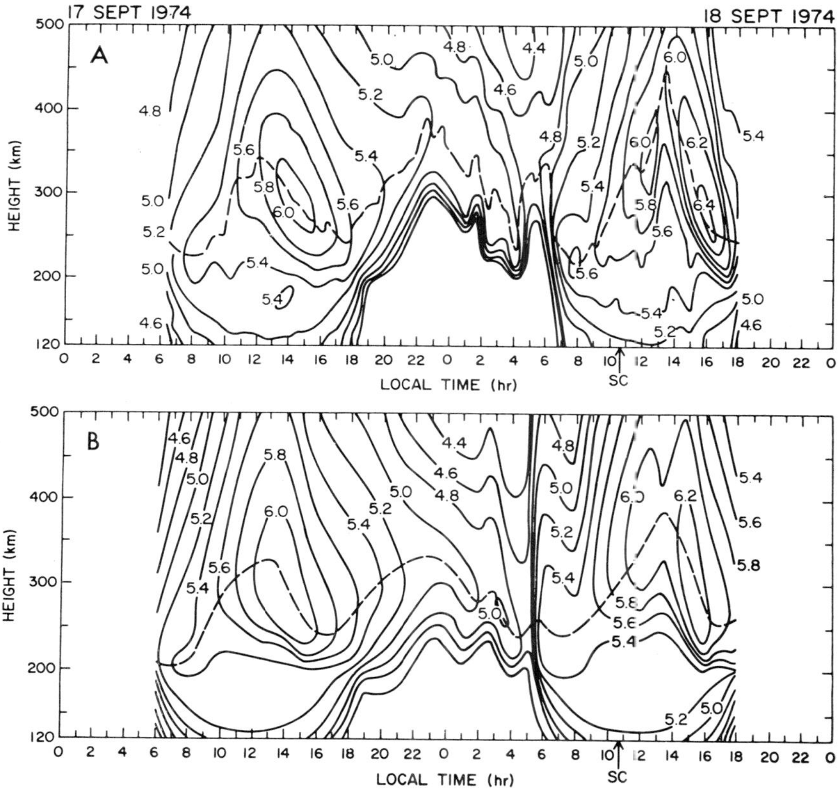

FIG. 2. 26. Same as Fig. 2. 25 except at Arecibo.

very well at Millstone Hill and fairly well at Arecibo. These increases were found to correspond to the sudden increase of southward velocity which would push ionization upwards, along field lines, to heights where L_s in (2. 75) is smaller than at the original heights without disturbances. The electron temperature decreased in association with the electron density increase, implying that the thermal energy available upon ionization, which is to be the same without disturbances, is now partitioned among the increased number of electrons, decreasing the partition per electron. In a related simulation, RICHMOND and MATSUSHITA (1975) found that the front of the disturbances moves at a nearly constant speed of 750 m sec^{-1}. It is not clear whether the waves excited are mainly internal or external as discussed in Subsection 2.3.2 partly because of the lower boundary condition. However, the velocity obtained by RICHMOND and MATSUSHITA (1975) was in phase over fairly wide height range (Fig. 2. 27). In this connection a simulation for MS TID's seems desirable.

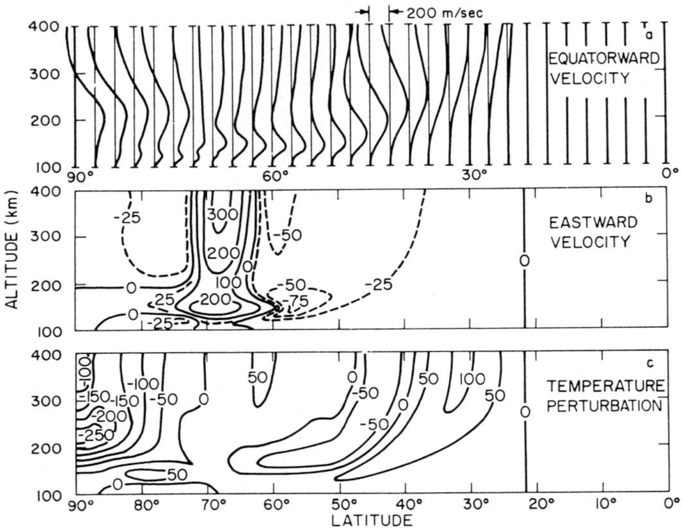

FIG. 2. 27. Snapshots of the disturbance at time 2 hr after the switch-on of the source. (a) Equatorward wind velocity profiles every 3° in latitudes. (b) Eastward neutral wind velocity. The spacing is 25 m sec^{-1}. (c) Neutral temperature perturbation. The spacing is 50° (RICHMOND and MATSUSHITA, 1975).

CHAPTER 3

ATMOSPHERIC TIDES

In the previous chapter we have considered waves in a stationary plane atmosphere. However, the earth is a rotating sphere. Both the sphericity and the rotation produce an additional dispersion to that which has been discussed in Chapter 2. These two factors become important for waves of frequency comparable to or smaller than 2Ω due to the Coriolis force, and of horizontal wavelength comparable to or longer than a due to the earth's sphericity. Atmospheric tides and planetary waves belong to this category, although the plane atmosphere approximation can be allowed for the planetary waves to some extent. The approach presented was developed in the 1960's mainly for tidal theory.

The tides are defined as the oscillation with periods of one day or its submultiples. If the day is solar, we have solar tides, and if the day is lunar, we have lunar tides. It is impossible to know whether the tidal oscillation, as just defined, is global or local from local observations. Only global observation can tell about those tidal *waves* which travel westwards with the sun for solar tidal waves and with the moon for lunar tidal waves. These tidal waves must have a westward phase velocity equal to the earth's rotational velocity relative to the sun for solar tide and the moon for lunar tide. For the diurnal tidal wave with period of one day, the wave number, which is defined as the ratio between the earth's circumference and the eastward wavelength, must be one. Note that the wave number can be defined at any latitude, because for a constant phase velocity with latitude of the tidal waves, the wavelength and the earth's circumference should be proportional. For semidiurnal and terdiurnal tidal waves of this kind, the wave number must be two and three, respectively. Thus, these waves oscillate only with local time at any latitude with reference to the sun or the moon. In this book, unless otherwise noted, the tides mean tidal waves of this kind. The tides which do not travel with the sun or the moon, are called local or non-migrating tides.

Atmospheric tides were discussed very early in the history of natural science. Pressure variations which were observed at ground level were largely *solar semidiurnal.* However, the tidal force due to gravity should be mainly *lunar semidiurnal* as is found in the oceanic tide. The observed lunar pressure tide is about (1/20) of the solar tide. Thus, it was thought that the atmospheric tide would be caused by heating due to solar radiation rather than gravity. Then, however, the question arose as to how the solar thermal tide could be mainly semidiurnal, when the solar heating was mainly diurnal. Newton commented on the atmospheric tide. KELVIN (1882) put forward the famous resonance theory to explain the larger semidiurnal pressure tide. However, the resonance theory, further developed by theoretical physicists such as TAYLOR (1936), PEKERIS (1937, 1939) met with a difficulty, when the observed atmospheric temperature profile did not agree with that necessary for resonance. Maximum stratospheric temperatures were observed to be lower than those necessary for the resonance. It was then realized that the problem was to understand why the diurnal tide is smaller and less regular than the semidiurnal tide.

The need for studying the diurnal tide occurred also in other field of geophysics. The solar daily geomagnetic variation on quiet days could be explained by an electric current system in the ionosphere. The driving force of the current is the atmospheric tidal winds blowing across the geomagnetic field in the ionosphere. This tidal dynamo was found to require predominantly diurnal winds over semidiurnal ones in the ionosphere.

Experimentally, there was an important work of precise analysis of the diurnal pressure tide data (HAURWITZ, 1965). The theory of diurnal tides was established mainly by KATO (1966a, b, d) and LINDZEN (1967). Our understanding about other components of the atmospheric tide and also global ocean tide progressed as well in the 1960's. Extensive calculation for obtaining the eigen-functions of Laplace's tidal equation has been done by FLATTERY as his Ph. D. thesis (1967), and LONGUET-HIGGINS (1968). Large computers which become widely available for the first time in the 1960's contributed much to the progress in these studies. The theory, which was then established, is called the classical theory. This still gives the basis in recent theories and observations.

Note that almost all of equations in Section 3. 1 are given in general forms for those atmospheric waves, which travel along latitudes eastwards or westwards, including planetary waves.

3.1 Classical Tidal Theory

3. 1. 1 *Basic equations*

The theory is constructed from the following equations (see (1. 1)–(1. 3) and (1. 9))

$$\frac{\partial W_\theta}{\partial t} - 2\Omega W_\phi \cos\theta = -\frac{1}{a}\frac{\partial}{\partial\theta}\left(\frac{p_1}{\rho_0}\right) \tag{3.1}$$

$$\frac{\partial W_\phi}{\partial t} + 2\Omega W_\theta \cos\theta = -\frac{1}{a\sin\theta}\frac{\partial}{\partial\phi}\left(\frac{p_1}{\rho_0}\right) \tag{3.2}$$

$$\frac{\partial p_1}{\partial z} = -g\rho_1 \tag{3.3}$$

$$\frac{\partial \rho_1}{\partial t} + W_z \frac{\mathrm{d}\rho_0}{\mathrm{d}z} + \rho_0\chi = 0 \tag{3.4}$$

$$\frac{\partial p_1}{\partial t} - g\rho_0 W_z + \rho_0 C^2 \chi = (\kappa - 1) Q\rho_0. \tag{3.5}$$

Of these equations (3. 1)–(3. 3) are the linearized equations of motion corresponding to (2. 1) and (2. 2), where neither the Coriolis force nor the earth's sphericity was considered. But (2. 2) is more accurate in considering a small acceleration of W_z which is neglected in (3. 3); (3. 4) is identical to (2. 3) except for the three dimensional χ in (3. 4) instead of the two dimensional χ in (2. 4); (3. 5) takes into account Q, the heating, whereas (2.4) is only for an adiabatic state. Note that there is no exact justification for (3.2) and (3.3) to neglect $\Omega \sin\theta$, the horizontal component of the earth's rotational velocity $\boldsymbol{\Omega} = \Omega(-\sin\theta, 0, \cos\theta)$; we can call the approximation historical (ECKART, 1960). This approximation enables us to reduce the basic equation to be self-adjoint. Another approximation is that gravity is assumed constant with height, whereas it actually decreases by about 3% from the ground to 100 km.

The fundamental equation describing the tide will be given in terms of $\hat{\chi}$ where

$$\hat{\chi} = \chi - \frac{Q}{C_p T_0}. \tag{3.6}$$

Note that in Chapter 2, GW's are described in terms of Φ which is proportional to p_1 (see (2.8)). There is, however, a close relation between χ and p_1 (by (3.5))

$$\hat{\chi} = -\frac{1}{\rho_0 C^2}\frac{Dp}{Dt}. \tag{3.7}$$

The Fourier transformation with respect to t as in (2.5) on $W_{\theta,\phi}$ and p_1 gives by $f = \omega/2\Omega$

$$W_{\theta,\omega} = \frac{\omega}{4a\Omega^2(f^2 - \cos^2\theta)}\left[i\frac{\partial}{\partial\theta} + \frac{\cot\theta}{f}\frac{\partial}{\partial\phi}\right]\left(\frac{p_{1,\omega}}{\rho_0}\right) \tag{3.8}$$

$$W_{\phi,\omega} = \frac{i\omega}{4a\Omega^2(f^2 - \cos^2\theta)}\left[i\frac{\cos\theta}{f}\frac{\partial}{\partial\theta} + \frac{1}{\sin\theta}\frac{\partial}{\partial\phi}\right]\left(\frac{p_{1,\omega}}{\rho_0}\right)$$

which further gives by differentiating $W_{\theta,\phi}$ to obtain χ_ω

$$\chi_\omega - \frac{\partial W_{z,\omega}}{\partial z} = \frac{i\omega}{4a^2\Omega^2} F\left(\frac{p_{1,\omega}}{\rho_0}\right) \tag{3.9}$$

where

$$F = \frac{1}{\sin\theta}\frac{\partial}{\partial\theta}\left[\frac{\sin\theta\left\{\frac{\partial}{\partial\theta} - i\left(\frac{\cot\theta}{f}\right)\left(\frac{\partial}{\partial\phi}\right)\right\}}{f^2-\cos^2\theta}\right] + \frac{i\left(\frac{\cot\theta}{f}\right)\left(\frac{\partial^2}{\partial\theta\partial\phi}\right) + \left(\frac{1}{\sin^2\theta}\right)\left(\frac{\partial^2}{\partial\phi^2}\right)}{(f^2-\cos^2\theta)}. \tag{3.10}$$

Differentiation of p_1 obtained from (3. 7), with respect to z yields

$$i\omega\frac{\partial p_{1,\omega}}{\partial z} = gW_z\frac{\mathrm{d}\rho_0}{\mathrm{d}z} + g\rho_0\frac{\partial W_{z,\omega}}{\partial z} - C^2\rho_0\frac{\partial\hat{\chi}_\omega}{\partial z} - \kappa\hat{\chi}\frac{\mathrm{d}p_0}{\mathrm{d}z}$$

whose left-hand side is rewritten by (3. 3) as $(-i\omega g\rho_1)$ and, in turn, by (3. 4) as $g\,[W_{z,\omega}(\mathrm{d}\rho_0/\mathrm{d}z)+\rho_0\chi]$. Thus, we have

$$\frac{\partial W_{z,\omega}}{\partial z} = -(\kappa-1)\hat{\chi}_\omega + \frac{C^2}{g}\frac{\partial\hat{\chi}_\omega}{\partial z} + \frac{Q_\omega}{C_pT_0}. \tag{3.11}$$

By (3.7) again we obtain

$$i\omega\left(\frac{p_{1,\omega}}{\rho_0}\right) = gW_{z,\omega} - C^2\hat{\chi}_\omega \tag{3.11'}$$

which is differentiated by z. Then, $(\partial W_{z,\omega}/\partial z)$ is eliminated by (3. 11) to obtain $(p_{1,\omega}/\rho_0)$ in terms of $\hat{\chi}_\omega$ and its derivatives. Finally, this result is substituted in the right-hand side of (3. 9) to eliminate $(p_{1\omega}/\rho_0)$. $(\partial W_{z,\omega}/\partial z)$ in (3. 9) is substituted from (3. 11). Thus, the basic equation in $\hat{\chi}_\omega$ is given as

$$C^2\frac{\partial^2\hat{\chi}_\omega}{\partial z^2} + \left(\frac{\mathrm{d}C^2}{\mathrm{d}z} - \kappa g\right)\frac{\partial\hat{\chi}_\omega}{\partial z} + \frac{g}{4a^2\Omega^2}F\left[\left\{g(1-\kappa) - \frac{\mathrm{d}C^2}{\mathrm{d}z}\right\}\hat{\chi}_\omega\right]$$

$$= -\frac{g^2}{4a^2\Omega^2}F\left[\frac{Q_\omega}{C_pT_0}\right]. \tag{3.12}$$

We are concerned with those waves which travel in the east-west in a non destructive manner. Thus, $\hat{\chi}_\omega$ should satisfy the periodic condition

$$\hat{\chi}_\omega \propto \exp(\mathrm{i}m\phi) \tag{3.13}$$

where $m=0, 1, 2, \ldots$

The first and second derivatives about ϕ in F are then replaced by (im) and $(-m^2)$, respectively; F is a second order differential operator in θ depending on ω and m, denoted by $F_{\omega,m}$.

Consider the following differential equation:

$$F_{\omega,m}\Theta + \frac{4a^2\Omega^2}{gh}\Theta = 0. \tag{3.14}$$

If $F_{\omega,m}$ is self-adjoint, and either $\Theta=0$ at $\theta=0$ and $\pi/2$, or $\Theta=0$ at $\theta=0$ and $d\Theta/d\theta=0$ at $\theta=\pi/2$, h can take discrete real eigenvalues as $h_{\omega,m,n}$, each of which gives an eigenfunction $\Theta_{\omega,m,n}(\theta)$ where n is integer. Note that the condition $\Theta=0$ at $\theta=0$ is necessary to have a finite velocity at $\theta=0$ as is found in (3. 8) and (3. 30). The condition at $\theta=\pi/2$ is to divide solutions into symmetric and anti-symmetric ones. Generally, $\Theta_{\omega,m,n}$ constitutes an orthogonal complete set, i.e. any regular function in $(0, \pi)$ can be expanded in $\Theta_{\omega,m,n}$, the convergence being guaranteed. Therefore,

$$\begin{aligned} \hat{\chi}_\omega &= \sum_{m,n} Y_{\omega,m,n}\Theta_{\omega,m,n} \exp(\mathrm{i}m\phi) \\ Q_\omega &= \sum_{m,n} q_{\omega,m,n}\Theta_{\omega,m,n} \exp(\mathrm{i}m\phi) \end{aligned} \tag{3. 15}$$

where $Y_{\omega,m,n}$ and $q_{\omega,m,n}$ depend on z. The orthogonality of each $\Theta_{\omega,m,n}$ splits (3. 12), upon substituting (3. 15), into two equations, one of which is (3. 14), and the other of which is

$$C^2\frac{d^2Y_{\omega,m,n}}{dz^2}+\left(\frac{dC^2}{dz}-\kappa g\right)\frac{dY_{\omega,m,n}}{dz}+\frac{Y_{\omega,m,n}}{h_{\omega,m,n}}\left[\frac{dC^2}{dz}+g(\kappa-1)\right]$$
$$=\frac{gq_{\omega,m,n}}{h_{\omega,m,n}C_pT_0}. \tag{3. 16}$$

Here (3. 14) has been known as the *Laplace tidal equation* for $\omega=m\Omega$ describing the tide over an imaginary global ocean of uniform depth h; $\Theta_{\omega,m,n}$ is named the Hough functions. (3. 16) is known as the Wilkes equation; (3. 16) is transformed by $ds=dz/H$ and $Y_{\omega,m,n}=\hat{Y}_{\omega,m,n}\,e^{(s/2)}$ (see (2. 8)) as

$$\frac{d^2\hat{Y}_{\omega,m,n}}{ds^2}+N^2\hat{Y}_{\omega,m,n}=\left(\frac{\kappa-1}{\kappa}\right)\frac{q_{\omega,m,n}\exp(-s/2)}{\kappa g h_{\omega,m,n}} \tag{3. 17}$$

where

$$N=\sqrt{-\frac{1}{4}+\frac{1}{h_{\omega,m,n}}\left(\frac{dH}{ds}+\frac{\kappa-1}{\kappa}H\right)}. \tag{3. 17'}$$

It still remains unknown as to whether $F_{\omega,m}$ behaves as self-adjoint for $\omega<2\Omega$, because $(f^2-\cos^2\theta)$ in the denominator of $F_{\omega,m}$ vanishes between $0<\theta<(\pi/2)$ for $\omega<2\Omega$ i.e. $f<1$. No proof for the diurnal tide has ever been given except for a numerical experiment by KATO (1966d) who showed the orthogonality among $\Theta_{\omega,m,n}$ for $\omega=m\Omega$, $m=1, 2$ and several n's. The theory of the atmospheric oscillation is thus based on this hypothesis.

In order to find possible modes of the atmospheric waves for specified (ω, m), we should first determine by (3. 14) $h_{\omega,m,n}$ and the corresponding $\Theta_{\omega,m,n}$ which satisfies the boundary conditions at $\theta=0$ and $\pi/2$. Then, for each $h_{\omega,m,n}$ we should determine by (3. 17) the vertical structure $\hat{Y}_{\omega,m,n}$ which satisfies the lower and upper boundary conditions. We can call the waves, obtained in this approach, the forced waves because of specified (ω,m). The tidal waves have mainly been studied by this approach.

3.1.2 *Horizontal structure of forced waves*

We shall now proceed to solve (3.14) by the following steps:

(*i*) *Eigenvalues*

Let $\Theta_{\omega,m,n}$ be finite and regular in $0 \leqq \theta \leqq \pi$ and expand by the associated Ledgendre function $P_{i,n}(\theta)$ (see (A.3.1)–(A.3.4)) as

$$\Theta_{\omega,m,n} = \sum_{j=m} \Gamma_{j,n}{}^{(\omega,m)} P_{j,m}(\theta) \tag{3.18}$$

where $\Gamma_{j,n}{}^{(\omega,m)}$ is constant for θ, ϕ and z, but depends on $h_{\omega,m,m}$. We follow HOUGH's approach (1898) and define two auxiliary functions $\Theta_{n,1}{}^{(\omega,m)}$ and $\Theta_{n,2}{}^{(\omega,m)}$ as

$$\begin{aligned} \Theta_{n,1}{}^{(\omega,m)} &= \sum_{j=m} \alpha_{j,n}{}^{(\omega,m)} P_{j,m}(\theta) \\ \Theta_{n,2}{}^{(\omega,m)} &= \sum_{j=m} \beta_{j,n}{}^{(\omega,m)} P_{j,m}(\theta) \end{aligned} \tag{3.19}$$

where $\alpha_{j,n}{}^{(\omega,m)}$, $\beta_{j,n}{}^{(\omega,m)}$ are constants as $\Gamma_{j,n}{}^{(\omega,m)}$ in (3.18). To avoid singularity due to $(f^2-\cos^2\theta)$ in the denominator in (3.14) we put

$$\Theta_{\omega,m,n} = \left(\mathscr{D} + \frac{m}{f}\mu \right) \Theta_{n,1}{}^{(\omega,m)} + \frac{m^2}{f^2}(f^2-\mu^2)\Theta_{n,2}{}^{(\omega,m)} \tag{3.20}$$

where $\mu = \cos\theta$ and $\mathscr{D} = (1-\mu^2)(\mathrm{d}/\mathrm{d}\mu)$. Further, we may connect $\Theta_{n,1}{}^{(\omega,m)}$ and $\Theta_{n,2}{}^{(\omega,m)}$ as

$$\left(\mathscr{L} + \frac{m}{f} \right) \Theta_{n,1}{}^{(\omega,m)} = 2\left(\frac{m}{f}\right)^2 \mu \Theta_{n,2}{}^{(\omega,m)} \tag{3.21}$$

where $\mathscr{L} = (\mathrm{d}\mathscr{D}/\mathrm{d}\mu) - m^2/(1-\mu^2) = (\mathscr{D}^2 - m^2)/(1-\mu^2)$. (3.19) and (3.21) give

$$\begin{aligned} \left\{ j(j+1) - \left(\frac{m}{f}\right) \right\} \alpha_{j,n}{}^{(\omega,m)} &= -2\frac{m^2}{f^2}\left\{ \frac{(j-m)}{2j-1} \beta_{j-1,n}{}^{(\omega,m)} \right. \\ &\quad \left. + \frac{(j+m+1)}{2j+3} \beta_{j+1,n}{}^{(\omega,m)} \right\}. \end{aligned} \tag{3.22}$$

For $m=1, f=1/2$ and $j=1$, we have $\beta_{2,n}{}^{(1,1)} = 0$. (3.22′)

Also, (3.18)–(3.21) give

$$\begin{aligned} \Gamma_{j,n}{}^{(\omega,m)} &= m^2 \beta_{j,n}{}^{(\omega,m)} + \frac{1}{2} \frac{(j-m)\{(j-1)(j-2)+(m/f)\}}{(2j-1)} \alpha_{j-1,n}{}^{(\omega,m)} \\ &\quad + \frac{1}{2} \frac{(j+m+1)\{(j+2)(j+3)+(m/f)\}}{(2j+3)} \alpha_{j+1,n}{}^{(\omega,m)}. \end{aligned} \tag{3.23}$$

In (3.14), substitution of (3.20) into $F_{\omega,m}\Theta$, and of (3.18) into $(4a^2\Omega^2/gh)\Theta$ result in

$$\frac{f^2}{m^2} \frac{4a^2\Omega^2}{gh} \Gamma_{j,n}{}^{(\omega,m)} = \left\{ j(j+1) + \left(\frac{m}{f}\right) \right\} \beta_{j,n}{}^{(\omega,m)}. \tag{3.24}$$

By (3.22′), $\Gamma_{2,n}{}^{(1,1)} \equiv 0$ for $h \neq \infty$. (3.24′)

(3. 22)–(3. 24) are first degree simultaneous equations with respect to $\alpha_{j,n}^{(\omega,m)}$ $\beta_{j,n}^{(\omega,m)}$ and $\Gamma_{j,n}^{(\omega,m)}$ where $j=m, m+1, \ldots$ for a specified set of (ω, m). We can get an equation in h by eliminating all $\alpha_{j,n}^{(\omega,m)}$, $\beta_{j,n}^{(\omega,m)}$ and $\Gamma_{j,n}^{(\omega,m)}$, the dispersion equation which decides modes of oscillations; ω, m and n distinguish infinite number of the modes which correspond to the roots $h_{\omega,m,n}$ thus obtained. (3. 22)–(3. 24) are infinite in dimensions, but in actual calculation j is to be terminated at some finite number $j=j_0>m$. The dispersion equation is also given in an elegant form of an infinite continued fraction by (A. 3. 5)–(A. 3. 14)

$$M_{m,m}^{(\omega)}-\frac{a_{m,m}}{N_{m+1,m}^{(\omega)}-}\,\frac{a_{m+1,m}}{M_{m+2,m}^{(\omega)}-}\,\frac{a_{m+2,m}}{N_{m+3,m}^{(\omega)}-}\cdots\cdots=0 \tag{3.25}$$

where

$$M_{j,m}^{(\omega)}=f^2\left\{j(j+1)-\frac{m}{f}\right\}-\frac{gh\,j^2(j+1)^2}{4a^2\Omega^2}$$

$$N_{j,m}^{(\omega)}=j(j+1)-\frac{m}{f}$$

$$a_{j,m}=\frac{j^2(j+2)^2(j-m+1)(j+m+1)}{(2j+1)(2j+3)} \tag{3.26}$$

$$j\geqq m.$$

We can terminate (3. 25) at a certain term as in the direct solution of the simultaneous equations (3. 22)–(3. 24). The solutions are shown in Fig. 3. 1 for wave number $m=1$ and 2 for various $f\gtrless 0$.

(*ii*) *Eigenfunction* (*Hough functions*)

By (3. 22)–(3. 24) we have

$$\frac{\Gamma_{j+2,n}^{(\omega,m)}}{\Gamma_{j,n}^{(\omega,m)}}=\frac{(2j+3)(2j+5)}{j^2(j+1)^2(j+m+1)(j+m+2)}\,\varepsilon_{j+2,m}^{(\omega,n)}F_{j+1,m}^{(\omega,n)} \tag{3.27}$$

where

$$\varepsilon_{j,m}^{(\omega,n)}=\frac{a_{j-1,m}}{M_{j,m}^{(\omega)}-}\,\frac{a_{j,m}}{N_{j+1,m}^{(\omega)}-}\,\frac{a_{j+1,m}}{M_{j+2,m}^{(\omega)}-}\cdots\cdots$$

$$F_{j,m}^{(\omega,n)}=\frac{a_{j-1,m}}{N_{j,n}^{(\omega)}-}\,\frac{a_{j,m}}{M_{j+1,m}^{(\omega)}-}\,\frac{a_{j+1,m}}{N_{j+2,m}^{(\omega)}-}\cdots\cdots \tag{3.28}$$

and

$$\varepsilon_{j,m}^{(\omega,n)}=\frac{a_{j-1,m}}{M_{j,m}^{(\omega)}-F_{j+1,m}^{(\omega)}},$$

$$F_{j,m}^{(\omega,n)}=\frac{a_{j-1,m}}{N_{j,m}^{(\omega)}-\varepsilon_{j+1,m}}. \tag{3.29}$$

Note that (3. 25) stipulates that the series (3. 27) converges with $j\to\infty$, and we can terminate in calculation of (3.28) at a certain term as $\varepsilon_{j,m}^{(\omega)}$ or $F_{j,m}^{(\omega)}=0$ for $j\geq j_0$. Figure 3. 2(A) illustrates $\Theta_{\omega,m,n}$ for the diurnal and semidiurnal

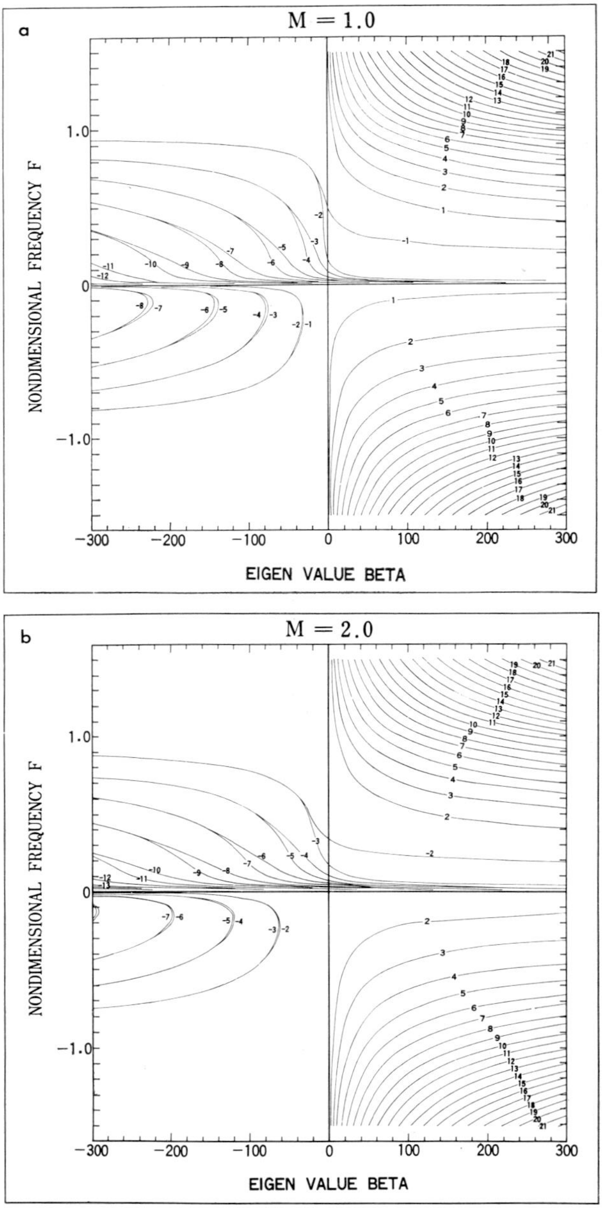

FIG. 3. 1. BETA (ω, m, n) versus F; $\text{BETA}=4a^2\Omega^2/gh_{\omega,m,n}$, $F=\omega/2\Omega$. (a) $m=1$, wave-number one, (b) $m=2$, wave-number two: The number to each curve shows n.

tide as $\omega=m\Omega$, $m=1, 2$. Table 3. 1 gives $h_{\omega,m,n}$ and $\Gamma_{j,n}{}^{(\omega,m)}$ of $f=1/2$, $m=1$, and $f=1$, $m=2$.

Before discussing in detail the dispersion, it seems convenient to know $(p_{1,\omega}/\rho_0)$, $\boldsymbol{W}_\omega$ in terms of $\hat{\chi}_\omega$ i.e. $\Theta_{\omega,m,n}$ and the corresponding $\hat{Y}_{\omega,m,n}$. They are as follows: By (3. 9), (3. 11) and (3. 14)

$$\frac{p_{1,\omega}}{\rho_0}=\sum_{m,n}\frac{p_{\omega,m,n}}{\rho_0}\Theta_{\omega,m,n}\exp(\mathrm{i}m\phi) \tag{3. 30}$$

where

$$\frac{p_{\omega,m,n}}{\rho_0}=-i\frac{\kappa g h_{\omega,m,n}}{\omega}\exp(s/2)\left(\frac{\mathrm{d}\hat{Y}_{\omega,m,n}}{\mathrm{d}s}-\frac{1}{2}\hat{Y}_{\omega,n,n}\right). \tag{3. 31}$$

It is now found that, since $(p_{1,\omega}/\rho_0)$ is proportional to $\Theta_{\omega,m,n}$ in (3. 30), $W_{\theta,\phi,\omega}$ is finite at $\theta=0$ in (3. 8) only if $\Theta_{\omega,m,n}=0$ at $\theta=0$, the boundary condition which gives h in (3. 14).

By (3.11′) and (3.31),

$$W_{z,\omega}=\sum_{m,n}W_{z,\omega,m,n}\Theta_{\omega,m,n}\exp(\mathrm{i}m\phi) \tag{3. 32}$$

where

$$W_{z,\omega,m,n}=\kappa h_{\omega,m,n}\exp(s/2)\left[\left(\frac{H}{h_{\omega,m,n}}-\frac{1}{2}\right)\hat{Y}_{\omega,m,n}+\frac{\mathrm{d}\hat{Y}_{\omega,m,n}}{\mathrm{d}s}\right] \tag{3. 33}$$

and $\Theta_{\omega,m,n}$ is given by (3.18) in terms of $\Gamma_{j,n}{}^{(\omega,m)}$. While $W_{z,\omega}$ consists of terms proportional to only $\Theta_{\omega,m,n}$, $W_{\theta,\omega}$, and $W_{\phi,\omega}$ are more complicated, containing derivatives of $\Theta_{n,1}{}^{(\omega,m)}$ and $\Theta_{n,2}{}^{(\omega,m)}$. By (3.8) and (3.20) we obtain

$$W_{\theta,\omega}=-\frac{i\omega}{4a\Omega^2\sin\theta}\sum_{m,n}\left(\frac{m}{f}\right)^2\left(\frac{p_{\omega,m,n}}{\rho_0}\right)\left[\Theta_{n,1}{}^{(\omega,m)}+\left(\mathscr{D}-\frac{m}{f}\mu\right)\right.$$
$$\left.\times\Theta_{n,2}{}^{(\omega,m)}\right]\exp(\mathrm{i}m\phi). \tag{3. 34}$$

By (3.19), (3.22), and (3.24), finally (3.34) becomes

$$W_{\theta,\omega}=-i(a\omega)\sum_{m,n}\sum_{j=m}\left(\frac{p_{\omega,m,n}}{\rho_0 g h_{\omega,m,n}}\right)[A_j{}^{(\omega,m)}\Gamma_{j-1\ n}{}^{(\omega,m)}$$
$$+B_j{}^{(\omega,m)}\Gamma_{j+1,n}{}^{(\omega,m)}]\left[\frac{P_{j,m}(\theta)}{\sin\theta}\right]\exp(\mathrm{i}m\phi) \tag{3. 35}$$

where

$$A_j{}^{(\omega,m)}=\frac{1}{j(j-1)+(m/f)}\left[\frac{2(m/f)^2}{j(j+1)-(m/f)}+j+\frac{m}{f}-1\right]\left(\frac{j-m}{2j-1}\right)$$
$$B_j{}^{(\omega,m)}=\frac{1}{(j+1)(j+2)+(m/f)}\left[\frac{2(m/f)^2}{j(j+1)-(m/f)}-j+\frac{m}{f}-2\right]\left(\frac{j+m+1}{2j+3}\right). \tag{3. 36}$$

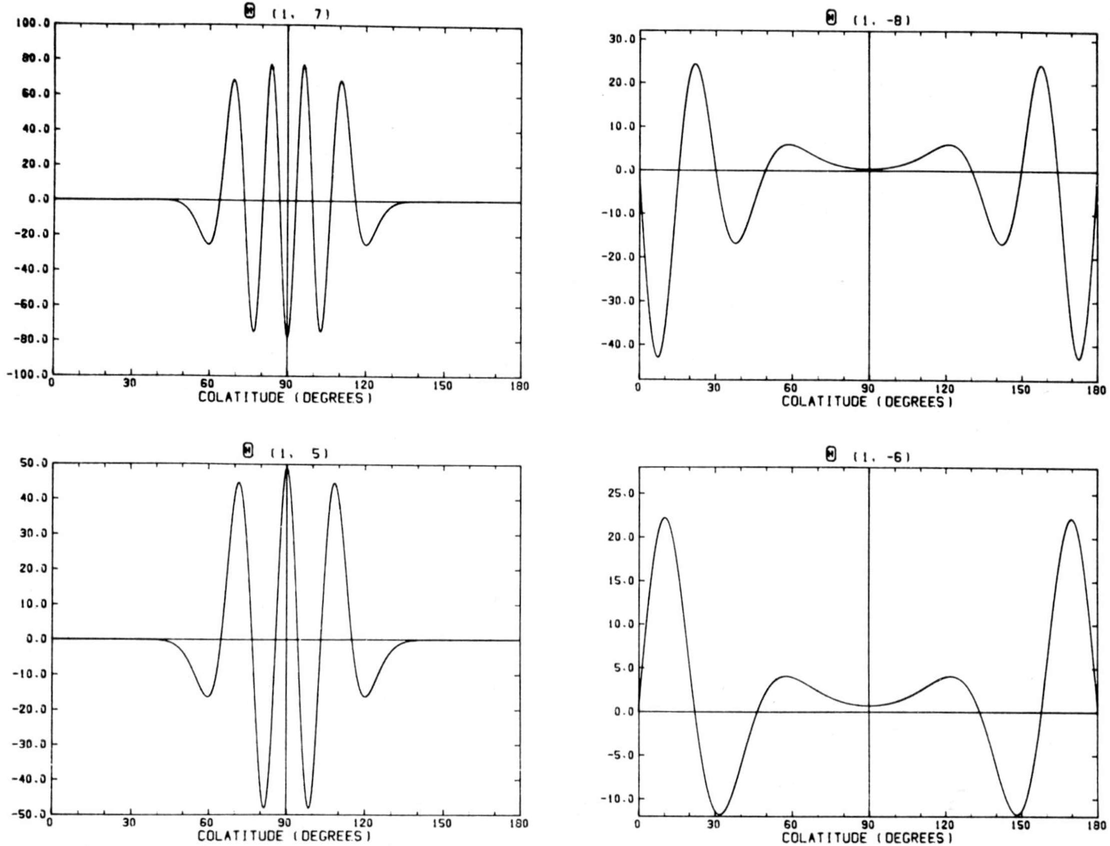
(1, 7)
(1, -8)
(1, 5)
(1, -6)
COLATITUDE (DEGREES)
0
30
60
90
120
150
180
100.0
80.0
60.0
40.0
20.0
0.0
-20.0
-40.0
-60.0
-80.0
-100.0
30.0
20.0
10.0
-10.0
-20.0
-30.0
-40.0
50.0
40.0
30.0
20.0
10.0
-10.0
-20.0
-30.0
-40.0
-50.0
25.0
20.0
15.0
10.0
5.0
-5.0
-10.0

(a)

FIG. 3.2 (A)

(a)

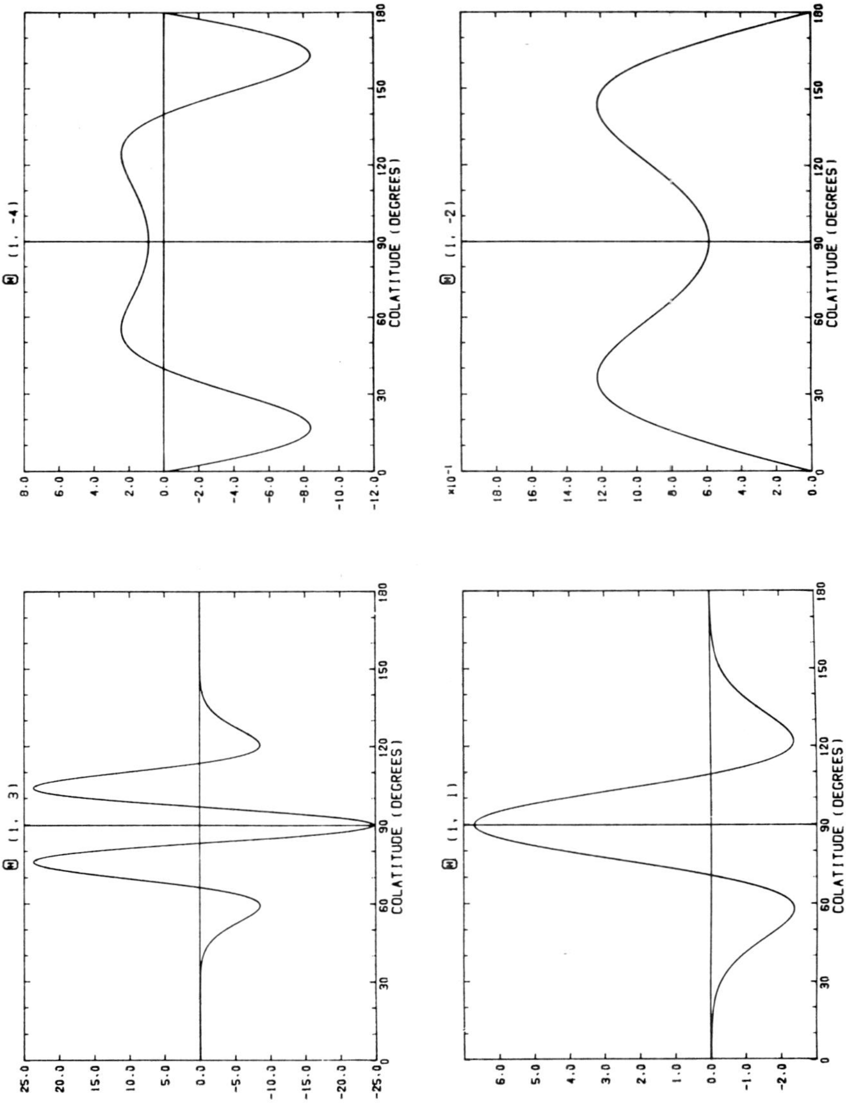
(1, -4)
(1, -2)
(1, 3)
(1, 1)
COLATITUDE (DEGREES)
COLATITUDE (DEGREES)
COLATITUDE (DEGREES)
COLATITUDE (DEGREES)
x10⁻¹

FIG. 3.2 (A)

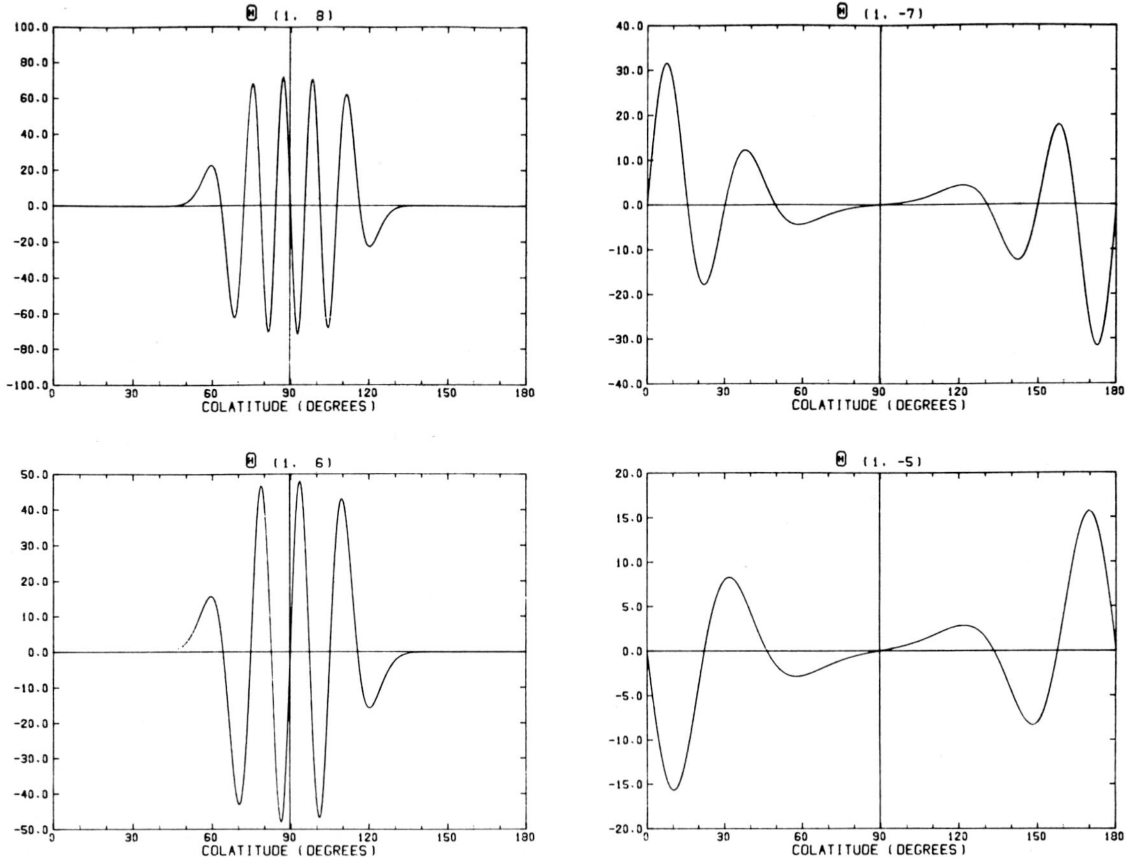
Θ (1. 8)
COLATITUDE (DEGREES)
Θ (1. -7)
COLATITUDE (DEGREES)
Θ (1. 6)
COLATITUDE (DEGREES)
Θ (1. -5)
COLATITUDE (DEGREES)

(b)

FIG. 3.2 (A)

(b)

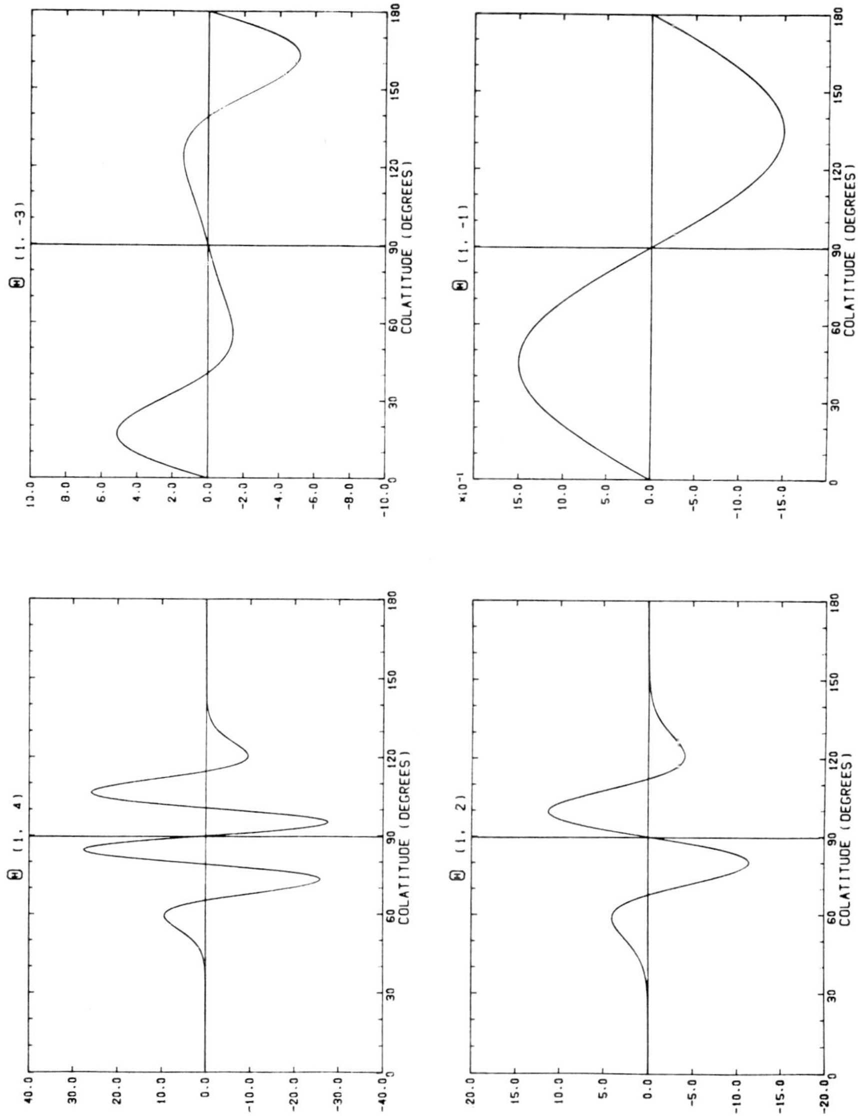

FIG. 3. 2 (A)

(c)

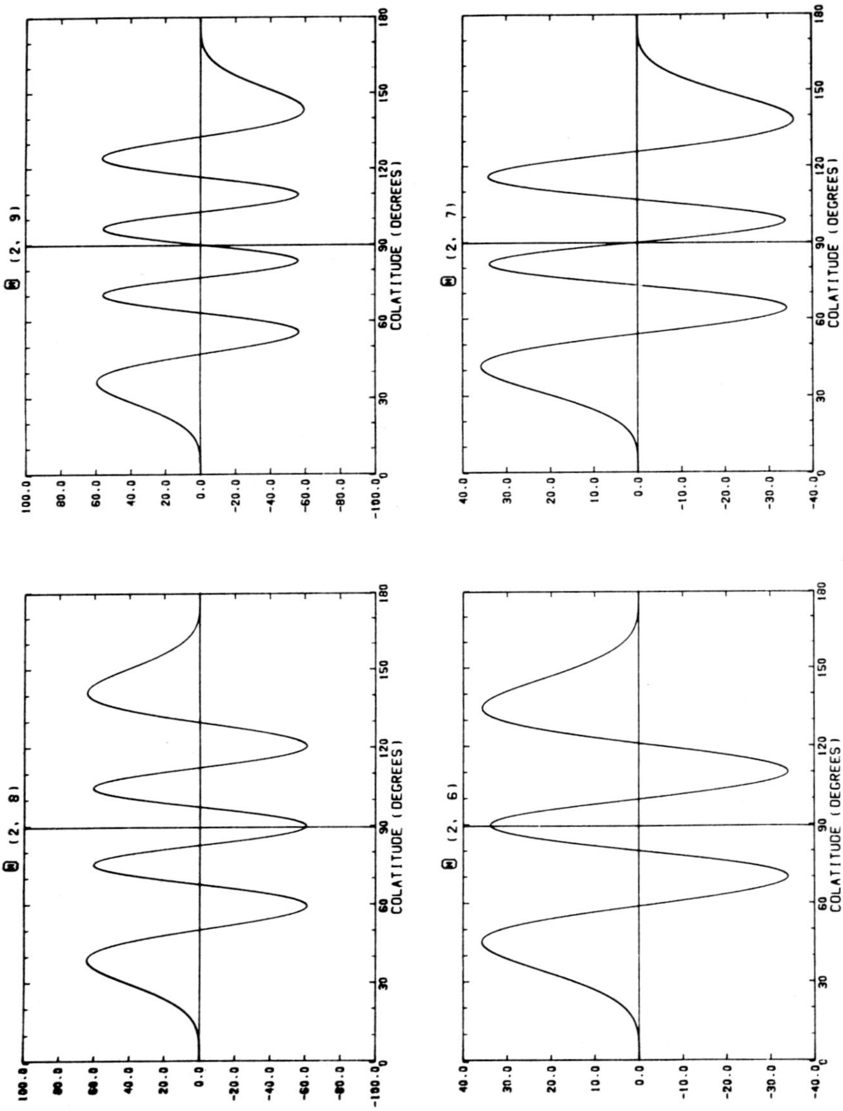

FIG. 3.2 (A)

(c)

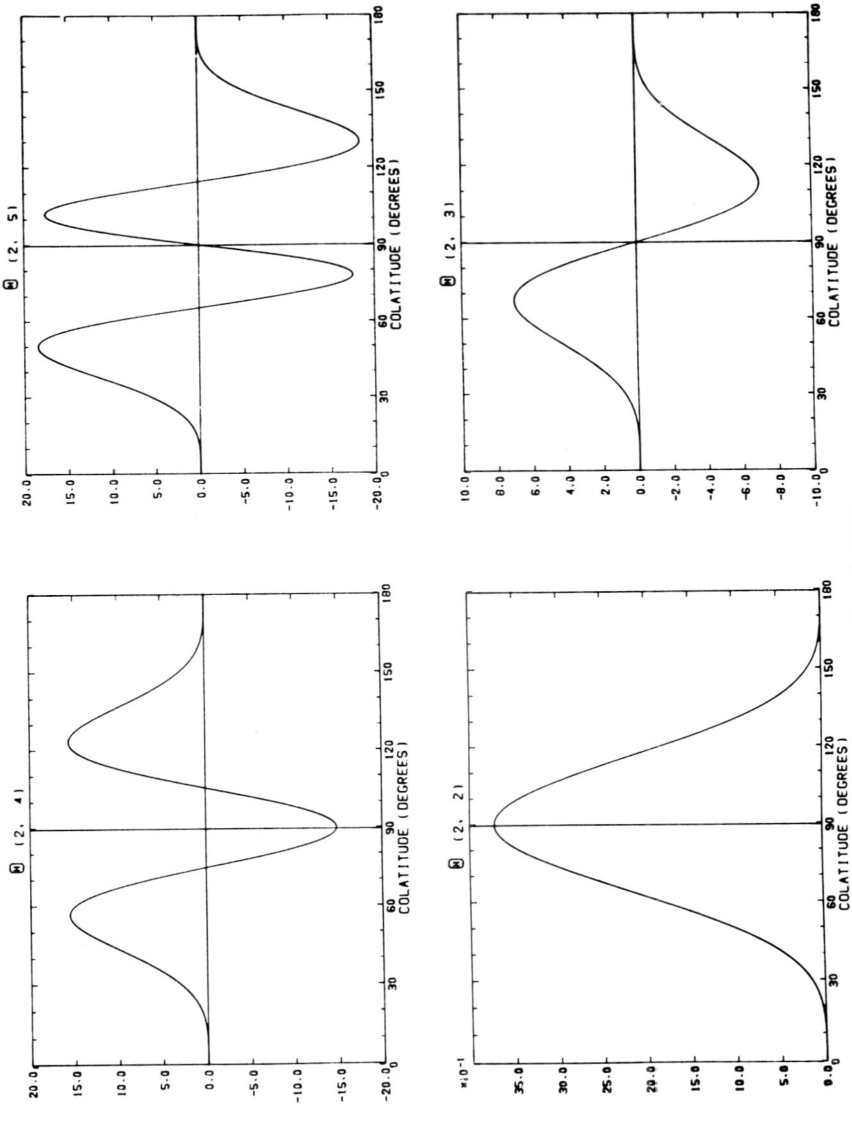
(2, 5)
(2, 4)
(2, 3)
(2, 2)
COLATITUDE (DEGREES)
COLATITUDE (DEGREES)
COLATITUDE (DEGREES)
COLATITUDE (DEGREES)
x10^-1

Fig. 3.2 (A)

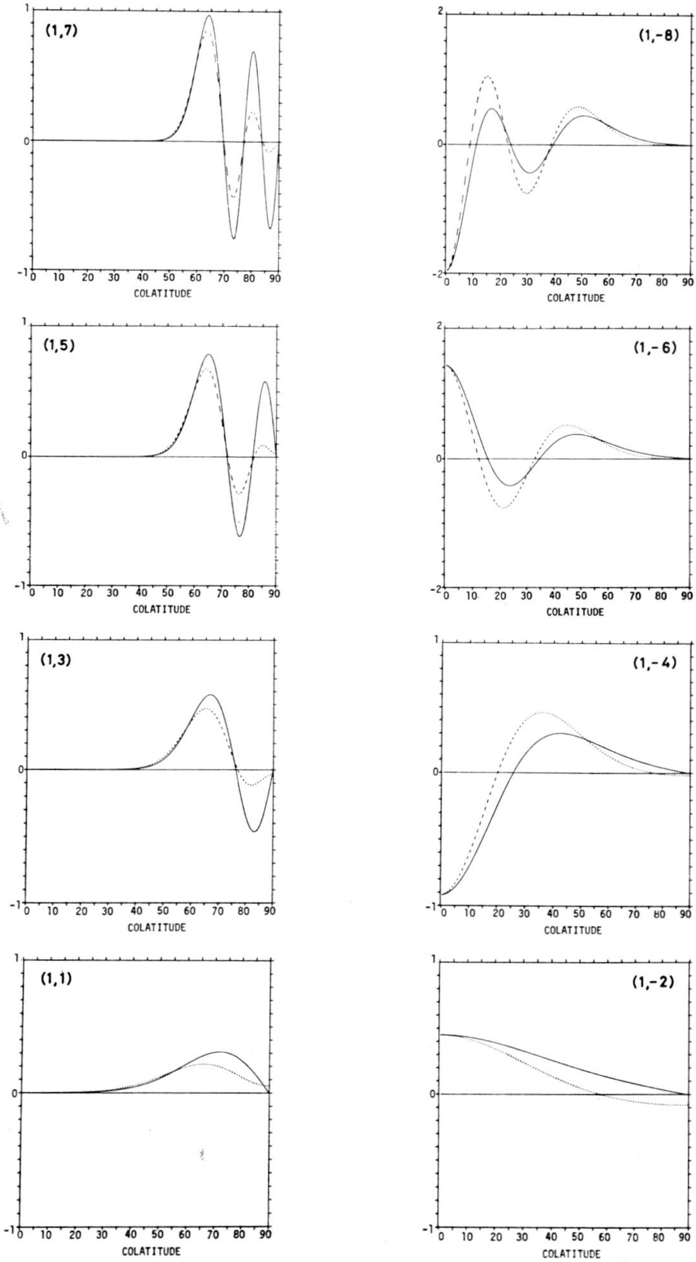

(a)

FIG. 3. 2 (B)

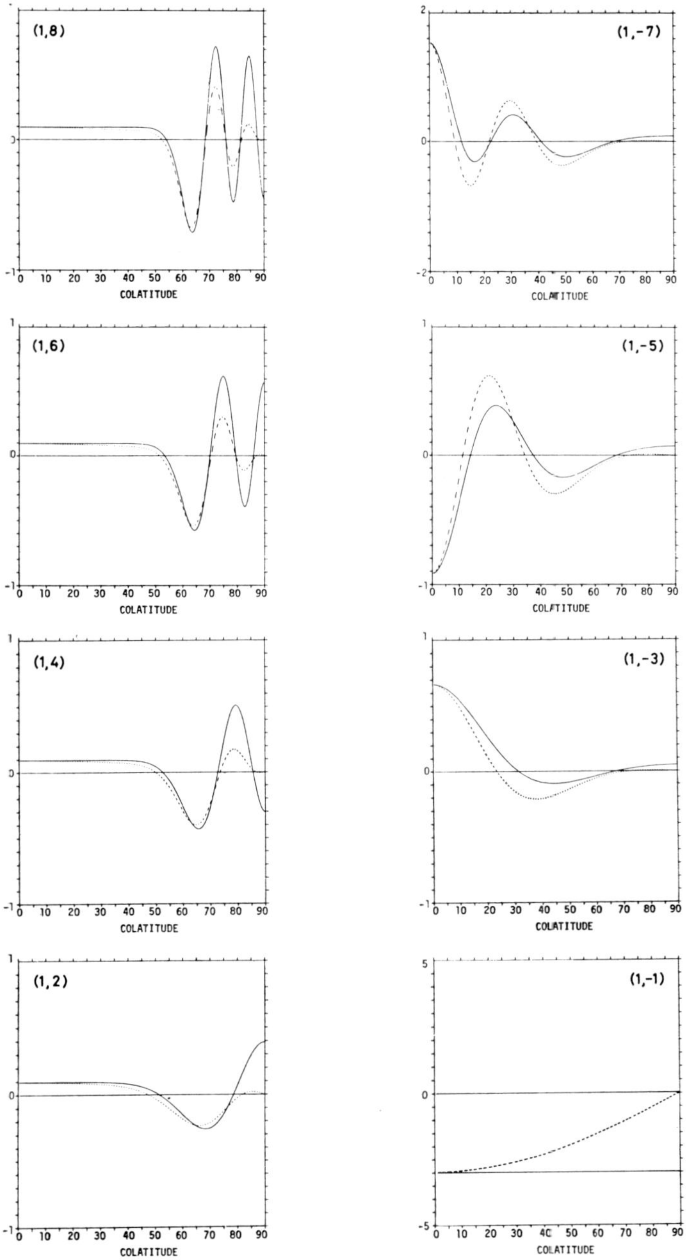

(b)

FIG. 3. 2 (B)

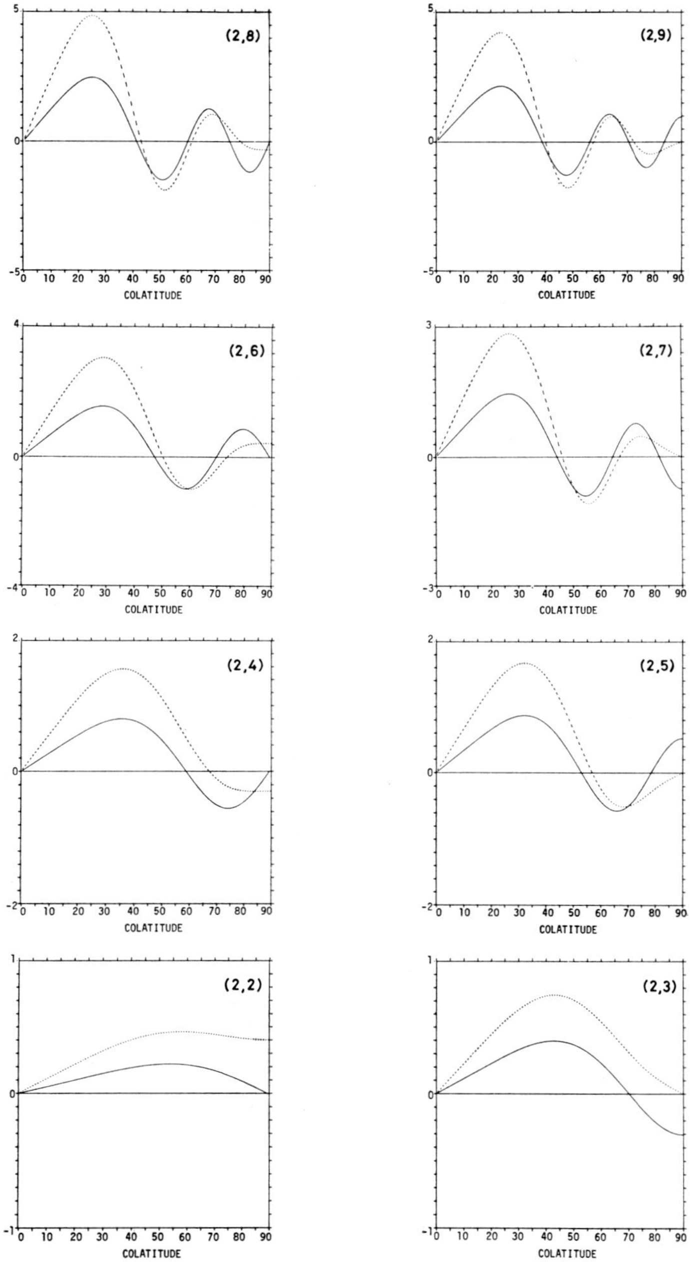

(c)

FIG. 3. 2 (B)

Similarly (see (A. 3.15)–(A. 3.19)),

$$W_{\phi,\omega}=\frac{-\omega}{4a\Omega^2\sin\theta}\sum_{m,n}\frac{m}{f^2}\frac{p_{\omega,m,n}}{\rho_0}\left[\mathscr{D}\Theta_{n,1}{}^{(\omega,m)}-m\left(\frac{\mu}{f}\mathscr{D}-m\right)\Theta_{n,2}{}^{(\omega,m)}\right]\exp(\mathrm{i}m\phi) \tag{3.37}$$

which becomes

$$W_{\phi,\omega}=(a\omega)\sum_{m,n}\sum_{j=m}\left(\frac{1}{m}\right)\left(\frac{p_{\omega,m,n}}{\rho_0 g h_{\omega,m,n}}\right)[J_j{}^{(\omega,m)}\Gamma_{j-2,n}{}^{(\omega,m)}$$

$$+L_j{}^{(\omega,m)}\Gamma_{j,n}{}^{(\omega,m)}+S_j{}^{(\omega,m)}\Gamma_{j+2,n}{}^{(\omega,m)}]\left(\frac{P_{j,m}}{\sin\theta}\right)\exp(\mathrm{i}m\phi) \tag{3.38}$$

where

$$J_j=\frac{(j-m)(j-m-1)}{(2j-1)(2j-3)}\left[\frac{2(m/f)^2(j-1)}{j(j-1)-(m/f)}+\left(\frac{m}{f}\right)(j-2)\right]\frac{1}{[(j-2)(j-1)+(m/f)]}$$

$$L_j=\left\{\frac{2(m/f)^2}{[j(j-1)-(m/f)]}\cdot\frac{(j-1)(j-m)(j+m)}{(2j-1)(2j+1)}-\frac{2(m/f)^2}{[(j+1)(j+2)-(m/f)]}\right.$$

$$\times\frac{(j+2)(j+m+1)(j-m+1)}{(2j+1)(2j+3)}-\left(\frac{m}{f}\right)\left[\frac{(j+1)(j-m)(j+m)}{(2j-1)(2j+1)}\right.$$

$$\left.\left.-\frac{j(j-m+1)(j+m+1)}{(2j+1)(2j+3)}-mf\right]\right\}\frac{1}{[j(j+1)+(m/f)]}$$

$$S_j=-\frac{(j+m+1)(j+m+2)}{(2j+3)(2j+5)}\left[\frac{2(m/f)^2(j+2)}{(j+1)(j+2)-(m/f)}\right.$$

$$\left.+\left(\frac{m}{f}\right)(j+3)\right]\frac{1}{[(j+2)(j+3)+(m/f)]}. \tag{3.39}$$

FIG. 3. 2. (A) Hough functions $\Theta_{\omega,m,n}$. (B) The corresponding velocity $W_{\theta,\omega}$ and $W_{\phi,\omega}$, solid and dotted curves, respectively; $(f/m)^2\sum_{j=m}[A_j{}^{(\omega,m)}\Gamma_{j-1,n}{}^{(\omega,m)}+B_j{}^{(\omega,m)}\Gamma_{j,n}{}^{(\omega,m)}](P_{j,m}/\sin\theta)$ for $W_{\theta,\omega}$ of various diurnal and semidiurnal modes specified by n, and $(f/m)^2\sum_{j=m}[J_j{}^{(\omega,m)}\Gamma_{j-2,n}{}^{(\omega,m)}+L_j{}^{(\omega,m)}\Gamma_{j,n}{}^{(\omega,m)}+S_j{}^{(\omega,m)}\Gamma_{j+2,n}{}^{(\omega,m)}](P_{j,m}/\sin\theta)$ for $W_{\phi,\omega}$ of various modes as $W_{\theta,\omega}$. In (A) and (B), there are (a), (b), and (c) as follows:

	f	m	n	
(a)	1/2	1	−2, −4, −6, −8	diurnal, symmetric
			1, 3, 5, 7	diurnal, symmetric
(b)	1/2	1	−1, −3, −5, −7	diurnal, anti-symmetric
			2, 4, 6, 8	diurnal, anti-symmetric
(c)	1	2	2, 4, 6, 8	semidiurnal, symmetric
			3, 5, 7, 9	semidiurnal, anti-symmetric

Note that no singularity occurs in (3. 35) and (3. 38) by (3. 24′). $(p_{\omega,m,n}/\rho_0 h_{\omega,m,n})$ in $W_{\theta,\phi,\omega}$ depends on the vertical structure of tide by (3. 31).

Since the tidal winds are often measured in observation for the upper atmosphere tides, relations (3. 33)–(3. 39) would be uesful. We have calculated winds $(W_{\theta,\phi})$ of various modes both symmetric and anti-symmetric for $f=1/2$ and $m=1$, and for $f=1$ and $m=2$, representing the global diurnal and semi-diurnal tides, respectively (Fig. 3. 2 (B)). The small difference between the solar and sidereal day is neglected, and a certain care should be paid in calculation of $\boldsymbol{W}$ for (3. 24′). It is found that for the *anti-symmetric diurnal* mode except for the *first mode*, contains *no component proportional* to $P_{2,1}(\theta)$ (see Table 3. 1). A special care must be paid for the (1, −1)-mode which has $h\rightarrow\pm\infty$. For this mode, $\Gamma_{2,-1}\neq 0$. But no singularity occurs in $W_{\theta,\omega}$ in (3. 34) and $W_{\phi,\omega}$ in (3. 38), because $(p_{\omega,m,n}/\rho_0 h_{\omega,m,n})\rightarrow 0$ by (3. 31) and (3. 17). Thus, it turns out that for the (1, −1)-mode $W_{\theta,\omega}$ and $W_{\phi,\omega}$ remain finite, and by (3. 7) $Dp/Dt\equiv 0$. This behavior is associated to that of the global ocean tide with infinite uniform depth (LAMB, 1945). ((3. 24′) has not been considered either in KATO's work (1966d) or in CHAPMAN and LINDZEN's work (1970)).

Recently, the incoherent scatter radars have measured the temperature tide (e.g. WAND, 1976). The following expression would be helpful for comparison with theory. By (1. 3) and (3. 6)

$$\frac{\partial T_1}{\partial t} = -W_z\frac{\mathrm{d}T_0}{\mathrm{d}z}+(\kappa-1)T_0\hat{\chi} \tag{3.40}$$

which reduces by (3.32) to

$$\frac{T_{1,\omega}}{T_0} = \sum_{n,m}\frac{T_{\omega,m,n}}{T_0}\Theta_{\omega,m,n}\exp(\mathrm{i}m\phi) \tag{3.41}$$

where

$$\frac{T_{\omega,m,n}}{T_0} = -\frac{i\kappa\exp(s/2)}{\omega}\left\{\left[\left(\frac{\kappa-1}{\kappa}-\frac{1}{H}\frac{\mathrm{d}H}{\mathrm{d}s}\right)+\frac{1}{2}\left(\frac{h_{\omega,m,n}}{H}\right)\right]\hat{Y}_{\omega,m,n} -\left(\frac{h_{\omega,m,n}}{H}\right)\left(\frac{1}{H}\frac{\mathrm{d}H}{\mathrm{d}s}\right)\frac{\mathrm{d}\hat{Y}_{\omega,m,n}}{\mathrm{d}s}\right\}. \tag{3.42}$$

Thus, the temperature tide has a different vertical structure from the wind tide whose structure is, by (3. 35) and (3. 38), proportional to that of $(p_{1,\omega}/\rho_0)$ in (3. 31).

Note that in treating tides, the gravitational tidal potential must be introduced in the right-hand side of (3. 1) and (3. 2) in addition to $(p_{1,\omega}/\rho_0)$. However, the tidal potential changes so slightly with z that its second derivative is neglected and does not appear in the final equation (3. 12) (e.g. WILKES, 1949). The gravitational tidal force thus affects the atmosphere only at the lower boundary, the ground, as will be discussed in Subsection 3. 1. 3. The effect is readily introduced by replacing $(p_{1,\omega}/\rho_0)$ in (3.9) with $(p_{1,\omega}/\rho_0)+\Lambda_\omega(\theta,\phi,z)$ where Λ_ω is the gravitational tidal potential (in the ω-space). The same

TABLE 3. 1

EIGENVALUES, $h_{\omega,m,n}$, BETA AND COEFFICIENTS OF HOUGH FUNCTIONS, $\Gamma_{j,n}{}^{(\omega,m)}$ FOR EACH $P_{n,m}$ (θ).

$P_{n,m}$ are Associated Legendre functions of un-normalized Neuman-type. Each mode is specified by (M, N) and F corresponding to (m, n) and f in text; BETA is defined as $[4a^2\Omega^2/(gh_{\omega, m, n})]$; $J=j=1, \cdots, 100$. Note that $\Gamma^{(1.1)}{}_{2, n}\equiv 0$ even for n=even number, i.e. the diurnal anti-symmetric mode has no component of $P_{2,1}$ (θ) (see (3. 24′)); superscript ω is normalized by Ω.

TABLE 3.1

M = 1.0 F = 0.5

H (M)	INFINITY	-1.21475D+04	-1.79359D+03	-1.73927D+03	-6.38856D+02	-6.37252D+02	-3.26172D+02	-3.26087D+02
BETA	0.0	-7.20544D+00	-4.88005D+01	-5.03247D+01	-1.37008D+02	-1.37352D+02	-2.68350D+02	-2.68419D+02
J N	-1	-2	-3	-4	-5	-6	-7	-8
1	0.0	1.00000D+00	0.0	1.00000D+00	0.0	1.00000D+00	0.0	1.00000D+00
2	1.00000D+00	0.0	0.0	0.0	0.0	0.0	0.0	0.0
3	0.0	3.06646D-01	0.0	-1.08067D+00	0.0	-1.22756D+00	0.0	-1.26904D+00
4	0.0	0.0	1.00000D+00	0.0	1.00000D+00	0.0	1.00000D+00	0.0
5	0.0	2.50898D-02	0.0	-1.41301D+00	0.0	-4.17886D-01	0.0	-9.71263D-02
6	0.0	0.0	5.60675D-01	0.0	-6.97032D-01	0.0	-1.03760D+00	0.0
7	0.0	9.81950D-04	0.0	-5.13199D-01	0.0	1.91932D+00	0.0	1.94412D+00
8	0.0	0.0	1.35961D-01	0.0	-1.50557D+00	0.0	-8.58221D-01	0.0
9	0.0	2.26707D-05	0.0	-9.63257D-02	0.0	1.80295D+00	0.0	-3.39793D-01
10	0.0	0.0	1.91653D-02	0.0	-8.89494D-01	0.0	1.28688D+00	0.0
11	0.0	3.46033D-07	0.0	-1.13037D-02	0.0	7.70438D-01	0.0	-2.52658D+00
12	0.0	0.0	1.78943D-03	0.0	-2.91425D-01	0.0	1.90474D+00	0.0
13	0.0	3.75472D-09	0.0	-9.15695D-04	0.0	2.03193D-01	0.0	-2.18396D+00
14	0.0	0.0	1.19474D-04	0.0	-6.32167D-02	0.0	1.16815D+00	0.0
15	0.0	3.04657D-11	0.0	-5.45074D-05	0.0	3.72605D-02	0.0	-1.03448D+00
16	0.0	0.0	6.00426D-06	0.0	-9.89955D-03	0.0	4.47449D-01	0.0
17	0.0	1.91885D-13	0.0	-2.49021D-06	0.0	5.07874D-03	0.0	-3.30566D-01
18	0.0	0.0	2.35703D-07	0.0	-1.18036D-03	0.0	1.21772D-01	0.0
19	0.0	9.65518D-16	0.0	-9.01841D-08	0.0	5.37638D-04	0.0	-7.79179D-02
20	0.0	0.0	7.43478D-09	0.0	-1.11182D-04	0.0	2.51594D-02	0.0
21	0.0	3.97092D-18	0.0	-2.65468D-09	0.0	4.56194D-05	0.0	-1.42662D-02
22	0.0	0.0	1.92667D-10	0.0	-8.50231D-06	0.0	4.11450D-03	0.0
23	0.0	1.35992D-20	0.0	-6.48012D-11	0.0	3.17802D-06	0.0	-2.10032D-03
24	0.0	0.0	4.17666D-12	0.0	-5.39204D-07	0.0	5.48407D-04	0.0
25	0.0	3.93849D-23	0.0	-1.33345D-12	0.0	1.85251D-07	0.0	-2.54980D-04
26	0.0	0.0	7.68815D-14	0.0	-2.88467D-08	0.0	6.08962D-05	0.0
27	0.0	9.77240D-26	0.0	-2.34502D-14	0.0	9.17643D-09	0.0	-2.60241D-05
28	0.0	0.0	1.21690D-15	0.0	-1.32032D-09	0.0	5.73141D-06	0.0
29	0.0	2.10071D-28	0.0	-3.55578D-16	0.0	3.91283D-10	0.0	-2.26782D-06
30	0.0	0.0	1.67420D-17	0.0	-5.23157D-11	0.0	4.63647D-07	0.0
31	0.0	3.95010D-31	0.0	-4.73528D-18	0.0	1.45194D-11	0.0	-1.70895D-07
32	0.0	0.0	2.02079D-19	0.0	-1.81276D-12	0.0	3.26146D-08	0.0
33	0.0	6.55219D-34	0.0	-5.53991D-20	0.0	4.73278D-13	0.0	-1.12556D-08
34	0.0	0.0	2.15739D-21	0.0	-5.54107D-14	0.0	2.01470D-09	0.0
35	0.0	9.65880D-37	0.0	-5.75365D-22	0.0	1.36626D-14	0.0	-6.53853D-10
36	0.0	0.0	2.05187D-23	0.0	-1.50558D-15	0.0	1.10218D-10	0.0
37	0.0	1.27374D-39	0.0	-5.34079D-24	0.0	3.51812D-16	0.0	-3.37669D-11
38	0.0	0.0	1.74965D-25	0.0	-3.66095D-17	0.0	5.37950D-12	0.0
39	0.0	1.51153D-42	0.0	-4.45766D-26	0.0	8.13204D-18	0.0	-1.56098D-12
40	0.0	0.0	1.34526D-27	0.0	-8.01421D-19	0.0	2.35771D-13	0.0
41	0.0	1.62267D-45	0.0	-3.36355D-28	0.0	1.69691D-19	0.0	-6.49899D-14
42	0.0	0.0	9.37454D-30	0.0	-1.58794D-20	0.0	9.33245D-15	0.0
43	0.0	1.58345D-48	0.0	-2.30567D-30	0.0	3.21288D-21	0.0	-2.45019D-15
44	0.0	0.0	5.94833D-32	0.0	-2.86156D-22	0.0	3.35346D-16	0.0
45	0.0	1.41069D-51	0.0	-1.44222D-32	0.0	5.54503D-23	0.0	-8.40573D-17
46	0.0	0.0	3.45124D-34	0.0	-4.71065D-24	0.0	1.09899D-17	0.0
47	0.0	1.15194D-54	0.0	-8.26511D-35	0.0	8.75984D-25	0.0	-2.63564D-18
48	0.0	0.0	1.83809D-36	0.0	-7.11191D-26	0.0	3.29849D-19	0.0
49	0.0	8.65341D-58	0.0	-4.35567D-37	0.0	1.27153D-26	0.0	-7.58344D-20
50	0.0	0.0	9.01788D-39	0.0	-9.88332D-28	0.0	9.10161D-21	0.0

J N	-1	-2	-3	-4	-5	-6	-7	-8
51	0.0	6.00003D-61	0.0	-2.11799D-39	0.0	1.70179D-28	0.0	-2.00957D-21
52	0.0	0.0	4.08885D-41	0.0	-1.26848D-29	0.0	2.31696D-22	0.0
53	0.0	3.85187D-64	0.0	-9.53258D-42	0.0	2.10684D-30	0.0	-4.92090D-23
54	0.0	0.0	1.71855D-43	0.0	-1.50822D-31	0.0	5.45891D-24	0.0
55	0.0	2.29605D-67	0.0	-3.98266D-44	0.0	2.41985D-32	0.0	-1.11694D-24
56	0.0	0.0	6.71420D-46	0.0	-1.66603D-33	0.0	1.19389D-25	0.0
57	0.0	1.27420D-70	0.0	-1.54873D-46	0.0	2.58563D-34	0.0	-2.35661D-26
58	0.0	0.0	2.44468D-48	0.0	-1.71431D-35	0.0	2.43041D-27	0.0
59	0.0	6.59953D-74	0.0	-5.61958D-49	0.0	2.57674D-36	0.0	-4.63408D-28
60	0.0	0.0	8.31556D-51	0.0	-1.64719D-37	0.0	4.61691D-29	0.0
61	0.0	3.19748D-77	0.0	-1.90706D-51	0.0	2.40067D-38	0.0	-8.51367D-30
62	0.0	0.0	2.64837D-53	0.0	-1.48132D-39	0.0	8.20361D-31	0.0
63	0.0	0.0	0.0	-6.06604D-54	0.0	2.09561D-40	0.0	-1.46466D-31
64	0.0	0.0	7.91399D-56	0.0	-1.24947D-41	0.0	1.36645D-32	0.0
65	0.0	0.0	0.0	-1.81221D-56	0.0	1.71755D-42	0.0	-2.36455D-33
66	0.0	0.0	2.22331D-58	0.0	-9.90495D-44	0.0	2.13803D-34	0.0
67	0.0	0.0	0.0	-5.09457D-59	0.0	1.32425D-44	0.0	-3.58932D-35
68	0.0	0.0	5.88301D-61	0.0	-7.39338D-46	0.0	3.14846D-36	0.0
69	0.0	0.0	0.0	-1.35015D-61	0.0	9.52238D-47	0.0	-5.13267D-37
70	0.0	0.0	1.45876D-63	0.0	-5.20556D-48	0.0	4.37157D-38	0.0
71	0.0	0.0	0.0	-3.37834D-64	0.0	6.60082D-49	0.0	-6.92632D-39
72	0.0	0.0	3.46550D-66	0.0	-3.46299D-50	0.0	5.73283D-40	0.0
73	0.0	0.0	0.0	-7.99764D-67	0.0	4.28172D-51	0.0	-8.83506D-41
74	0.0	0.0	7.73971D-69	0.0	-2.18010D-52	0.0	7.11201D-42	0.0
75	0.0	0.0	0.0	-1.79316D-69	0.0	2.63033D-53	0.0	-1.06694D-42
76	0.0	0.0	1.63857D-71	0.0	-1.30075D-54	0.0	8.35922D-44	0.0
77	0.0	0.0	0.0	-3.81386D-72	0.0	1.53250D-55	0.0	-1.22163D-44
78	0.0	0.0	3.29300D-74	0.0	-7.36569D-57	0.0	9.32210D-46	0.0
79	0.0	0.0	0.0	-7.70526D-75	0.0	8.47989D-58	0.0	-1.32804D-46
80	0.0	0.0	6.29045D-77	0.0	-3.96385D-59	0.0	9.87704D-48	0.0
81	0.0	0.0	0.0	-1.48063D-77	0.0	4.46213D-60	0.0	-1.37257D-48
82	0.0	0.0	0.0	0.0	-2.02982D-61	0.0	9.95557D-50	0.0
83	0.0	0.0	0.0	0.0	0.0	2.23551D-62	0.0	-1.35036D-50
84	0.0	0.0	0.0	0.0	-9.90284D-64	0.0	9.55795D-52	0.0
85	0.0	0.0	0.0	0.0	0.0	1.06774D-64	0.0	-1.26615D-52
86	0.0	0.0	0.0	0.0	-4.60811D-66	0.0	8.75043D-54	0.0
87	0.0	0.0	0.0	0.0	0.0	4.86672D-67	0.0	-1.13274D-54
88	0.0	0.0	0.0	0.0	-2.04750D-68	0.0	7.64793D-56	0.0
89	0.0	0.0	0.0	0.0	0.0	2.11922D-69	0.0	-9.67955D-57
90	0.0	0.0	0.0	0.0	-8.69600D-71	0.0	6.38806D-58	0.0
91	0.0	0.0	0.0	0.0	0.0	8.82527D-72	0.0	-7.90888D-59
92	0.0	0.0	0.0	0.0	-3.53380D-73	0.0	5.10440D-60	0.0
93	0.0	0.0	0.0	0.0	0.0	3.51818D-74	0.0	-6.18498D-61
94	0.0	0.0	0.0	0.0	-1.37533D-75	0.0	3.90564D-62	0.0
95	0.0	0.0	0.0	0.0	0.0	1.34385D-76	0.0	-4.63379D-63
96	0.0	0.0	0.0	0.0	-5.13116D-78	0.0	2.86426D-64	0.0
97	0.0	0.0	0.0	0.0	0.0	0.0	0.0	-3.32891D-65
98	0.0	0.0	0.0	0.0	0.0	0.0	2.01508D-66	0.0
99	0.0	0.0	0.0	0.0	0.0	0.0	0.0	-2.29517D-67
100	0.0	0.0	0.0	0.0	0.0	0.0	1.36114D-68	0.0

TABLE 3.1 (continued)

M = 1.0 F = 0.5

H (M)	-1.97457D+02	-1.97451D+02	-1.32239D+02	-1.32239D+02	-9.47046D+01	-9.47046D+01	-7.11455D+01	-7.11455D+01
BETA	-4.43277D+02	-4.43290D+02	-6.61893D+02	-6.61896D+02	-9.24222D+02	-9.24223D+02	-1.23027D+03	-1.23027D+03
J N	-9	-10	-11	-12	-13	-14	-15	-16
1	0.0	1.00000D+00	0.0	1.00000D+00	0.0	1.00000D+00	0.0	1.00000D+00
2	0.0	0.0	0.0	0.0	0.0	0.0	0.0	0.0
3	0.0	-1.28618D+00	0.0	-1.29487D+00	0.0	-1.29988D+00	0.0	-1.30302D+00
4	1.00000D+00	0.0	1.00000D+00	0.0	1.00000D+00	0.0	1.00000D+00	0.0
5	0.0	4.05938D-02	0.0	1.11555D-01	0.0	1.52747D-01	0.0	1.78732D-01
6	-1.17779D+00	0.0	-1.24883D+00	0.0	-1.28971D+00	0.0	-1.31537D+00	0.0
7	0.0	1.85664D+00	0.0	1.78953D+00	0.0	1.74384D+00	0.0	1.71249D+00
8	-4.56493D-01	0.0	-2.22843D-01	0.0	-7.91941D-02	0.0	1.43953D-02	0.0
9	0.0	-1.18569D+00	0.0	-1.57322D+00	0.0	-1.77894D+00	0.0	-1.90062D+00
10	1.69563D+00	0.0	1.73869D+00	0.0	1.70570D+00	0.0	1.66186D+00	0.0
11	0.0	-2.01900D+00	0.0	-1.38016D+00	0.0	-9.12404D-01	0.0	-5.85909D-01
12	4.96462D-01	0.0	-4.09365D-01	0.0	-9.21309D-01	0.0	-1.22309D+00	0.0
13	0.0	1.04003D+00	0.0	2.20807D+00	0.0	2.56462D+00	0.0	2.64148D+00
14	-1.78479D+00	0.0	-2.01178D+00	0.0	-1.66720D+00	0.0	-1.28390D+00	0.0
15	0.0	3.04413D+00	0.0	1.80561D+00	0.0	4.38198D-01	0.0	-5.13579D-01
16	-2.24957D+00	0.0	-8.02491D-02	0.0	1.24242D+00	0.0	1.85638D+00	0.0
17	0.0	2.54799D+00	0.0	-1.67179D+00	0.0	-2.90769D+00	0.0	-2.88271D+00
18	-1.42294D+00	0.0	2.21904D+00	0.0	2.10423D+00	0.0	1.20503D+00	0.0
19	0.0	1.29612D+00	0.0	-3.50855D+00	0.0	-1.46095D+00	0.0	6.16904D-01
20	-6.03717D-01	0.0	2.56245D+00	0.0	-3.38605D-01	0.0	-1.93702D+00	0.0
21	0.0	4.70009D-01	0.0	-2.89551D+00	0.0	2.24527D+00	0.0	3.34380D+00
22	-1.90402D-01	0.0	1.66340D+00	0.0	-2.60865D+00	0.0	-2.05830D+00	0.0
23	0.0	1.30657D-01	0.0	-1.55342D+00	0.0	3.93623D+00	0.0	1.05803D+00
24	-4.71428D-02	0.0	7.60165D-01	0.0	-2.85395D+00	0.0	7.43393D-01	0.0
25	0.0	2.90639D-02	0.0	-6.17258D-01	0.0	3.22895D+00	0.0	-2.77137D+00
26	-9.48835D-03	0.0	2.66587D-01	0.0	-1.89394D+00	0.0	2.96549D+00	0.0
27	0.0	5.32550D-03	0.0	-1.93253D-01	0.0	1.80642D+00	0.0	-4.33657D+00
28	-1.59106D-03	0.0	7.51927D-02	0.0	-9.16657D-01	0.0	3.12977D+00	0.0
29	0.0	8.21069D-04	0.0	-4.94744D-02	0.0	7.70018D-01	0.0	-3.55067D+00
30	-2.26449D-04	0.0	1.75813D-02	0.0	-3.48687D-01	0.0	2.11698D+00	0.0
31	0.0	1.08279D-04	0.0	-1.06217D-02	0.0	2.64044D-01	0.0	-2.05558D+00
32	-2.77594D-05	0.0	3.48175D-03	0.0	-1.08660D-01	0.0	1.07308D+00	0.0
33	0.0	1.23759D-05	0.0	-1.94819D-03	0.0	7.52758D-02	0.0	-9.26930D-01
34	-2.96612D-06	0.0	5.93580D-04	0.0	-2.84979D-02	0.0	4.35573D-01	0.0
35	0.0	1.23940D-06	0.0	-3.09706D-04	0.0	1.82486D-02	0.0	-3.41787D-01
36	-2.79022D-07	0.0	8.82440D-05	0.0	-5.41224D-03	0.0	1.46971D-01	0.0
37	0.0	1.09755D-07	0.0	-4.31703D-05	0.0	3.82493D-03	0.0	-1.06171D-01
38	-2.33048D-08	0.0	1.15607D-05	0.0	-1.25602D-03	0.0	4.22440D-02	0.0
39	0.0	8.66250D-09	0.0	-5.32723D-06	0.0	7.02188D-04	0.0	-2.83598D-02
40	-1.74095D-09	0.0	1.34646D-06	0.0	-2.16668D-04	0.0	1.05259D-02	0.0
41	0.0	6.13526D-10	0.0	-5.86707D-07	0.0	1.14089D-04	0.0	-6.61348D-03
42	-1.17071D-10	0.0	1.40467D-07	0.0	-3.32284D-05	0.0	2.30405D-03	0.0
43	0.0	3.92289D-11	0.0	-5.80727D-08	0.0	1.65479D-05	0.0	-1.36241D-03
44	-7.12665D-12	0.0	1.32114D-08	0.0	-4.56650D-06	0.0	4.47761D-04	0.0
45	0.0	2.27657D-12	0.0	-5.19747D-09	0.0	2.15836D-06	0.0	-2.50308D-04
46	-3.94718D-13	0.0	1.12666D-09	0.0	-5.66179D-07	0.0	7.79250D-05	0.0
47	0.0	1.20484D-13	0.0	-4.22869D-10	0.0	2.54753D-07	0.0	-4.13385D-05
48	-1.99811D-14	0.0	8.75568D-11	0.0	-6.37045D-08	0.0	1.22329D-05	0.0
49	0.0	5.84004D-15	0.0	-3.14254D-11	0.0	2.73599D-08	0.0	-6.17803D-06
50	-9.28236D-16	0.0	6.22876D-12	0.0	-6.53836D-09	0.0	1.74298D-06	0.0

J N	-9	-10	-11	-12	-13	-14	-15	-16
51	0.0	2.60277D-16	0.0	-2.14226D-12	0.0	2.68665D-09	0.0	-8.40356D-07
52	-3.97207D-17	0.0	4.07275D-13	0.0	-6.14941D-10	0.0	2.26621D-07	0.0
53	0.0	1.07035D-17	0.0	-1.34479D-13	0.0	2.42263D-10	0.0	-1.04563D-07
54	-1.57097D-18	0.0	2.45664D-14	0.0	-5.32157D-11	0.0	2.70149D-08	0.0
55	0.0	4.07475D-19	0.0	-7.80093D-15	0.0	2.01384D-11	0.0	-1.19543D-08
56	-5.76053D-20	0.0	1.37156D-15	0.0	-4.25292D-12	0.0	2.96498D-09	0.0
57	0.0	1.44031D-20	0.0	-4.19507D-16	0.0	1.54862D-12	0.0	-1.26073D-09
58	-1.96402D-21	0.0	7.10953D-17	0.0	-3.14940D-13	0.0	3.00737D-10	0.0
59	0.0	4.74005D-22	0.0	-2.09751D-17	0.0	1.10519D-13	0.0	-1.23089D-10
60	-6.24266D-23	0.0	3.43113D-18	0.0	-2.16765D-14	0.0	2.82858D-11	0.0
61	0.0	1.45611D-23	0.0	-9.77726D-19	0.0	7.34115D-15	0.0	-1.11614D-11
62	-1.85438D-24	0.0	1.54573D-19	0.0	-1.39051D-15	0.0	2.47458D-12	0.0
63	0.0	4.18520D-25	0.0	-4.25953D-20	0.0	4.55086D-16	0.0	-9.42743D-13
64	-5.15972D-26	0.0	6.51591D-21	0.0	-8.33511D-17	0.0	2.01934D-13	0.0
65	0.0	1.12799D-26	0.0	-1.73838D-21	0.0	2.63935D-17	0.0	-7.43725D-14
66	-1.34765D-27	0.0	2.57592D-22	0.0	-4.67984D-18	0.0	1.54102D-14	0.0
67	0.0	2.85668D-28	0.0	-6.66047D-23	0.0	1.43540D-18	0.0	-5.49351D-15
68	-3.31070D-29	0.0	9.56994D-24	0.0	-2.46656D-19	0.0	1.10238D-15	0.0
69	0.0	6.81112D-30	0.0	-2.40057D-24	0.0	7.33571D-20	0.0	-3.80802D-16
70	-7.66410D-31	0.0	3.34775D-25	0.0	-1.22288D-20	0.0	7.40862D-17	0.0
71	0.0	1.53166D-31	0.0	-8.15443D-26	0.0	3.52995D-21	0.0	-2.48246D-17
72	-1.67481D-32	0.0	1.10473D-26	0.0	-5.71407D-22	0.0	4.68716D-18	0.0
73	0.0	3.25413D-33	0.0	-2.61522D-27	0.0	1.60235D-22	0.0	-1.52494D-18
74	-3.46061D-34	0.0	3.44477D-28	0.0	-2.52088D-23	0.0	2.79691D-19	0.0
75	0.0	6.54229D-35	0.0	-7.93196D-29	0.0	6.87329D-24	0.0	-8.84323D-20
76	-6.77166D-36	0.0	1.01664D-29	0.0	-1.05181D-24	0.0	1.57694D-20	0.0
77	0.0	1.24654D-36	0.0	-2.27872D-30	0.0	2.79060D-25	0.0	-4.84959D-21
78	-1.25671D-37	0.0	2.84404D-31	0.0	-4.15704D-26	0.0	8.41477D-22	0.0
79	0.0	2.25416D-38	0.0	-6.20982D-32	0.0	1.07405D-26	0.0	-2.51905D-22
80	-2.21500D-39	0.0	7.55252D-33	0.0	-1.55864D-27	0.0	4.25640D-23	0.0
81	0.0	3.87396D-40	0.0	-1.60751D-33	0.0	3.92440D-28	0.0	-1.24127D-23
82	-3.71268D-41	0.0	1.90644D-34	0.0	-5.55178D-29	0.0	2.04388D-24	0.0
83	0.0	6.33541D-42	0.0	-3.95809D-35	0.0	1.36314D-29	0.0	-5.81049D-25
84	-5.92544D-43	0.0	4.58027D-36	0.0	-1.88114D-30	0.0	9.33005D-26	0.0
85	0.0	9.87132D-44	0.0	-9.28159D-37	0.0	4.50704D-31	0.0	-2.58742D-26
86	-9.01548D-45	0.0	1.04863D-37	0.0	-6.07106D-32	0.0	4.05415D-27	0.0
87	0.0	1.46710D-45	0.0	-2.07530D-38	0.0	1.42022D-32	0.0	-1.09744D-27
88	-1.30914D-46	0.0	2.29048D-39	0.0	-1.86844D-33	0.0	1.67097D-20	0.0
89	0.0	2.08215D-47	0.0	-4.42948D-40	0.0	4.27017D-34	0.0	-4.43897D-29
90	-1.81627D-48	0.0	4.77839D-41	0.0	-5.48985D-35	0.0	6.63481D-30	0.0
91	0.0	2.82479D-49	0.0	-9.03455D-42	0.0	1.22642D-35	0.0	-1.71424D-30
92	-2.41001D-50	0.0	9.53112D-43	0.0	-1.54163D-36	0.0	2.50461D-31	0.0
93	0.0	3.66710D-51	0.0	-1.76275D-43	0.0	3.36817D-37	0.0	-6.32735D-32
94	-3.06147D-52	0.0	1.81950D-44	0.0	-4.14174D-38	0.0	9.04154D-33	0.0
95	0.0	4.55972D-53	0.0	-3.29328D-45	0.0	8.85429D-39	0.0	-2.23453D-33
96	-3.72668D-54	0.0	3.32753D-46	0.0	-1.06562D-39	0.0	3.12447D-34	0.0
97	0.0	5.43541D-55	0.0	-5.89702D-47	0.0	2.23014D-40	0.0	-7.55779D-35
98	-4.35099D-56	0.0	5.83516D-48	0.0	-2.62810D-41	0.0	1.03458D-35	0.0
99	0.0	6.21712D-57	0.0	-1.01295D-48	0.0	5.38681D-42	0.0	-2.45052D-36
100	-4.87643D-58	0.0	9.82034D-50	0.0	-6.21862D-43	0.0	3.28551D-37	0.0

TABLE 3.1 (continued)

M = 1.0 F = 0.5

H (M)	6.94260D+02	2.39598D+02	1.20885D+02	7.27965D+01	4.86201D+01	3.47652D+01	2.60911D+01	2.03020D+01
BETA	1.26074D+02	3.65313D+02	7.24060D+02	1.20237D+03	1.80024D+03	2.51769D+03	3.35472D+03	4.31131D+03
J N	1	2	3	4	5	6	7	8
1	1.00000D+00	0.0	1.00000D+00	0.0	1.00000D+00	0.0	1.00000D+00	0.0
2	0.0	0.0	0.0	0.0	0.0	0.0	0.0	0.0
3	-1.40504D+00	0.0	-1.32861D+00	0.0	-1.31898D+00	0.0	-1.31598D+00	0.0
4	0.0	1.00000D+00	0.0	1.00000D+00	0.0	1.00000D+00	0.0	1.00000D+00
5	1.07758D+00	0.0	3.94281D-01	0.0	3.12382D-01	0.0	2.87043D-01	0.0
6	0.0	-1.65382D+00	0.0	-1.47215D-00	0.0	-1.43072D+00	0.0	-1.41497D+00
7	-4.95951D-01	0.0	1.37882D+00	0.0	1.52091D+00	0.0	1.56110D+00	0.0
8	0.0	1.49606D+00	0.0	6.43589D-01	0.0	4.67762D-01	0.0	4.02702D-01
9	1.51952D-01	0.0	-2.62014D+00	0.0	-2.41324D+00	0.0	-2.33180D+00	0.0
10	0.0	-9.21639D-01	0.0	9.63490D-01	0.0	1.22445D+00	0.0	1.30869D+00
11	-3.31335D-02	0.0	2.68138D+00	0.0	1.37114D+00	0.0	9.74226D-01	0.0
12	0.0	4.20594D-01	0.0	-2.23937D+00	0.0	-2.16598D+00	0.0	-2.09334D+00
13	5.39005D-03	0.0	-1.94723D+00	0.0	9.31310D-01	0.0	1.54112D+00	0.0
14	0.0	-1.48939D-01	0.0	2.56558D+00	0.0	1.50592D+00	0.0	1.07594D+00
15	-6.77809D-04	0.0	1.09451D+00	0.0	-3.00636D+00	0.0	-3.16208D+00	0.0
16	0.0	4.22052D-02	0.0	-2.10357D+00	0.0	3.29644D-01	0.0	1.02195D+00
17	6.77525D-05	0.0	-4.97163D-01	0.0	3.86671D+00	0.0	2.59800D+00	0.0
18	0.0	-9.79046D-03	0.0	1.36175D+00	0.0	-2.20412D+00	0.0	-2.57981D+00
19	-5.50597D-06	0.0	1.87420D-01	0.0	-3.51304D+00	0.0	-2.15686D-01	0.0
20	0.0	1.89273D-03	0.0	-7.27906D-01	0.0	3.22825D+00	0.0	2.45586D+00
21	3.70604D-07	0.0	-5.97410D-02	0.0	2.53740D+00	0.0	-2.54963D+00	0.0
22	0.0	-3.09456D-04	0.0	3.29769D-01	0.0	-3.22510D+00	0.0	-7.43595D-01
23	-2.09861D-08	0.0	1.63314D-02	0.0	-1.53085D+00	0.0	4.38329D+00	0.0
24	0.0	4.33212D-05	0.0	-1.28844D-01	0.0	2.55540D+00	0.0	-1.54145D+00
25	1.01313D-09	0.0	-3.87267D-03	0.0	7.92738D-01	0.0	-4.81204D+00	0.0
26	0.0	-5.24822D-06	0.0	4.39721D-02	0.0	-1.69995D+00	0.0	3.30838D+00
27	-4.21766D-11	0.0	8.04183D-04	0.0	-3.58553D-01	0.0	4.14437D+00	0.0
28	0.0	5.55331D-07	0.0	-1.32394D-02	0.0	9.77948D-01	0.0	-4.01337D-00
29	1.52920D-12	0.0	-1.47429D-04	0.0	1.43406D-01	0.0	-2.99751D+00	0.0
30	0.0	-5.17439D-08	0.0	3.54533D-03	0.0	-4.95469D-01	0.0	3.74095D+00
31	-4.87111D-14	0.0	2.40306D-05	0.0	-5.11960D-02	0.0	1.88211D+00	0.0
32	0.0	4.27641D-09	0.0	-8.50178D-04	0.0	2.23855D-01	0.0	-2.91637D+00
33	1.37373D-15	0.0	-3.50448D-06	0.0	1.64360D-02	0.0	-1.04624D+00	0.0
34	0.0	-3.15516D-10	0.0	1.83648D-04	0.0	-9.10293D-02	0.0	1.97568D+00
35	-3.45352D-17	0.0	4.59819D-07	0.0	-4.77450D-03	0.0	5.21664D-01	0.0
36	0.0	2.09035D-11	0.0	-3.59198D-05	0.0	3.35570D-02	0.0	-1.18849D+00
37	7.78697D-19	0.0	-5.45568D-08	0.0	1.26155D-03	0.0	-2.35511D-01	0.0
38	0.0	-1.25014D-12	0.0	6.39082D-06	0.0	-1.12800D-02	0.0	6.43780D-01
39	-1.58351D-20	0.0	5.88027D-09	0.0	-3.04578D-04	0.0	9.69637D-02	0.0
40	0.0	6.78145D-14	0.0	-1.03863D-06	0.0	3.47444D-03	0.0	-3.17105D-01
41	2.91866D-22	0.0	-5.78162D-10	0.0	6.74623D-05	0.0	-3.66163D-02	0.0
42	0.0	-3.35128D-15	0.0	1.54769D-07	0.0	-9.84749D-04	0.0	1.43080D-01
43	-4.89813D-24	0.0	5.20561D-11	0.0	-1.37579D-05	0.0	1.27425D-02	0.0
44	0.0	1.51483D-16	0.0	-2.12192D-08	0.0	2.57765D-04	0.0	-5.94792D-02
45	7.51554D-26	0.0	-4.30721D-12	0.0	2.59173D-06	0.0	-4.10285D-03	0.0
46	0.0	-6.28612D-18	0.0	2.68518D-09	0.0	-6.25157D-05	0.0	2.28871D-02
47	-1.05833D-27	0.0	3.28582D-13	0.0	-4.52335D-07	0.0	1.22649D-03	0.0
48	0.0	2.40295D-19	0.0	-3.14556D-10	0.0	1.40893D-05	0.0	-8.18375D-03
49	1.37256D-29	0.0	-2.31807D-14	0.0	7.33418D-08	0.0	-3.41433D-04	0.0
50	0.0	-8.48816D-21	0.0	3.42045D-11	0.0	-2.95855D-06	0.0	2.72833D-03

J N	1	2	3	4	5	6	7	8
51	-1.64470D-31	0.0	1.51659D-15	0.0	-1.10754D-08	0.0	8.87536D-05	0.0
52	0.0	2.77882D-22	0.0	-3.46125D-12	0.0	5.80240D-07	0.0	-8.50522D-04
53	1.82634D-33	0.0	-9.22594D-17	0.0	1.56136D-09	0.0	-2.15955D-05	0.0
54	0.0	-8.45399D-24	0.0	3.26724D-13	0.0	-1.06524D-07	0.0	2.48560D-04
55	-1.88460D-35	0.0	5.23137D-18	0.0	-2.05937D-10	0.0	4.92944D-06	0.0
56	0.0	2.39618D-25	0.0	-2.88330D-14	0.0	1.83442D-08	0.0	-6.82547D-05
57	1.81182D-37	0.0	-2.77129D-19	0.0	2.54646D-11	0.0	-1.05773D-06	0.0
58	0.0	-6.34248D-27	0.0	2.38378D-15	0.0	-2.96894D-09	0.0	1.76478D-05
59	-1.62672D-39	0.0	1.37450D-20	0.0	-2.95767D-12	0.0	2.13752D-07	0.0
60	0.0	1.57125D-28	0.0	-1.84997D-16	0.0	4.52415D-10	0.0	-4.30453D-06
61	1.36706D-41	0.0	-6.39562D-22	0.0	3.23262D-13	0.0	-4.07533D-08	0.0
62	0.0	-3.65071D-30	0.0	1.35017D-17	0.0	-6.50197D-11	0.0	9.92193D-07
63	-1.07758D-43	0.0	2.79723D-23	0.0	-3.33039D-14	0.0	7.34242D-09	0.0
64	0.0	7.97092D-32	0.0	-9.28317D-19	0.0	8.82708D-12	0.0	-2.16472D-07
65	7.98285D-46	0.0	-1.15202D-24	0.0	3.23945D-15	0.0	-1.25200D-09	0.0
66	0.0	-1.63846D-33	0.0	6.02292D-20	0.0	-1.13373D-12	0.0	4.47707D-08
67	-5.56825D-48	0.0	4.47524D-26	0.0	-2.97951D-16	0.0	2.02339D-10	0.0
68	0.0	3.17625D-35	0.0	-3.69319D-21	0.0	1.37956D-13	0.0	-8.78984D-09
69	3.66346D-50	0.0	-1.64247D-27	0.0	2.59506D-17	0.0	-3.10354D-11	0.0
70	0.0	-5.81644D-37	0.0	2.14350D-22	0.0	-1.59257D-14	0.0	1.64034D-09
71	-2.27716D-52	0.0	5.70383D-29	0.0	-2.14326D-18	0.0	4.52369D-12	0.0
72	0.0	1.00772D-38	0.0	-1.17920D-23	0.0	1.74639D-15	0.0	-2.91336D-10
73	1.33937D-54	0.0	-1.87694D-30	0.0	1.68073D-19	0.0	-6.27359D-13	0.0
74	0.0	-1.65423D-40	0.0	6.15715D-25	0.0	-1.82139D-16	0.0	4.93025D-11
75	-7.46549D-57	0.0	5.86069D-32	0.0	-1.25301D-20	0.0	8.28760D-14	0.0
76	0.0	2.57655D-42	0.0	-3.05528D-26	0.0	1.80879D-17	0.0	-7.95868D-12
77	3.94886D-59	0.0	-1.73871D-33	0.0	8.89138D-22	0.0	-1.04402D-14	0.0
78	0.0	-3.81276D-44	0.0	1.44256D-27	0.0	-1.71231D-18	0.0	1.22679D-12
79	-1.98481D-61	0.0	4.90713D-35	0.0	-6.01216D-23	0.0	1.25550D-15	0.0
80	0.0	5.36722D-46	0.0	-6.48833D-29	0.0	1.54684D-19	0.0	-1.80754D-13
81	9.49183D-64	0.0	-1.31906D-36	0.0	3.87803D-24	0.0	-1.44273D-16	0.0
82	0.0	-7.19598D-48	0.0	2.78312D-30	0.0	-1.33482D-20	0.0	2.54811D-14
83	-4.32402D-66	0.0	3.38090D-38	0.0	-2.38872D-25	0.0	1.58574D-17	0.0
84	0.0	9.19938D-50	0.0	-1.13970D-31	0.0	1.10138D-21	0.0	-3.43997D-15
85	1.87858D-68	0.0	-8.27181D-40	0.0	1.40644D-26	0.0	-1.66862D-18	0.0
86	0.0	-1.12261D-51	0.0	4.46020D-33	0.0	-8.69747D-23	0.0	4.45121D-16
87	-7.79202D-71	0.0	1.93383D-41	0.0	-7.92318D-28	0.0	1.68247D-19	0.0
88	0.0	1.30906D-53	0.0	-1.66972D-34	0.0	6.37936D-24	0.0	-5.52530D-17
89	3.08890D-73	0.0	-4.32431D-43	0.0	4.27459D-29	0.0	-1.62693D-20	0.0
90	0.0	-1.46009D-55	0.0	5.98505D-36	0.0	-4.77179D-25	0.0	6.58471D-18
91	-1.17146D-75	0.0	9.25781D-45	0.0	-2.21050D-30	0.0	1.50999D-21	0.0
92	0.0	1.55922D-57	0.0	-2.05596D-37	0.0	3.32084D-26	0.0	-7.53977D-19
93	4.25433D-78	0.0	-1.89928D-46	0.0	1.09662D-31	0.0	-1.34619D-22	0.0
94	0.0	-1.59566D-59	0.0	6.77421D-39	0.0	-2.21937D-27	0.0	8.30125D-20
95	0.0	0.0	3.73711D-48	0.0	-5.22324D-33	0.0	1.15370D-23	0.0
96	0.0	1.56626D-61	0.0	-2.14269D-40	0.0	1.42548D-28	0.0	-8.79438D-21
97	0.0	0.0	-7.05857D-50	0.0	2.39049D-34	0.0	-9.51155D-25	0.0
98	0.0	-1.47584D-63	0.0	6.51126D-42	0.0	-8.80570D-30	0.0	8.97106D-22
99	0.0	0.0	1.28080D-51	0.0	-1.05202D-35	0.0	7.54890D-26	0.0
100	0.0	1.33605D-65	0.0	-1.90242D-43	0.0	5.23537D-31	0.0	-8.81757D-23

TABLE 3.1 (continued)

M = 1.0 F = 0.5

H (M)	1.62466D+01	1.32956D+01	1.10815D+01	9.37790D+00	8.03902D+00	6.96772D+00	6.09715D+00	5.38013D+00
BETA	5.38749D+03	6.58323D+03	7.89855D+03	9.33344D+03	1.08879D+04	1.25620D+04	1.43556D+04	1.62688D+04
J N	9	10	11	12	13	14	15	16
1	1.00000D+00	0.0	1.00000D+00	0.0	1.00000D+00	0.0	1.00000D+00	0.0
2	0.0	0.0	0.0	0.0	0.0	0.0	0.0	0.0
3	-1.31467D+00	0.0	-1.31398D+00	0.0	-1.31357D+00	0.0	-1.31331D+00	0.0
4	0.0	1.00000D+00	0.0	1.00000D+00	0.0	1.00000D+00	0.0	1.00000D+00
5	2.76000D-01	0.0	2.70212D-01	0.0	2.66806D-01	0.0	2.64632D-01	0.0
6	0.0	-1.40734D+00	0.0	-1.40307D+00	0.0	-1.40045D+00	0.0	-1.39872D+00
7	1.57805D+00	0.0	1.58680D+00	0.0	1.59190D+00	0.0	1.59514D+00	0.0
8	0.0	3.71544D-01	0.0	3.54223D-01	0.0	3.43603D-01	0.0	3.36623D-01
9	-2.29388D+00	0.0	-2.27343D+00	0.0	-2.26121D+00	0.0	-2.25334D+00	0.0
10	0.0	1.34659D+00	0.0	1.36697D+00	0.0	1.37922D+00	0.0	1.38717D+00
11	8.03930D-01	0.0	7.15462D-01	0.0	6.63649D-01	0.0	6.30700D-01	0.0
12	0.0	-2.05035D+00	0.0	-2.02422D+00	0.0	-2.00742D+00	0.0	-1.99606D+00
13	1.76436D+00	0.0	1.87104D+00	0.0	1.93053D+00	0.0	1.96721D+00	0.0
14	0.0	8.67524D-01	0.0	7.51515D-01	0.0	6.80436D-01	0.0	6.33762D-01
15	-3.10383D+00	0.0	-3.04627D+00	0.0	-3.00425D+00	0.0	-2.97441D+00	0.0
16	0.0	1.30278D+00	0.0	1.44375D+00	0.0	1.52470D+00	0.0	1.57560D+00
17	1.90778D+00	0.0	1.53045D+00	0.0	1.30576D+00	0.0	1.16194D+00	0.0
18	0.0	-2.59917D+00	0.0	-2.56892D+00	0.0	-2.53672D+00	0.0	-2.51011D+00
19	9.95727D-01	0.0	1.53951D+00	0.0	1.82667D+00	0.0	1.99651D+00	0.0
20	0.0	1.87036D+00	0.0	1.50905D+00	0.0	1.27919D+00	0.0	1.12568D+00
21	-3.46159D+00	0.0	-3.63384D+00	0.0	-3.64483D+00	0.0	-3.61903D+00	0.0
22	0.0	3.96771D-01	0.0	9.59366D-01	0.0	1.27126D+00	0.0	1.46103D+00
23	3.78458D+00	0.0	3.01619D+00	0.0	2.47049D+00	0.0	2.09730D+00	0.0
24	0.0	-2.58803D+00	0.0	-2.87250D+00	0.0	-2.94477D+00	0.0	-2.95181D+00
25	-1.80358D+00	0.0	-7.51367D-02	0.0	8.63917D-01	0.0	1.41196D+00	0.0
26	0.0	3.25212D+00	0.0	2.72551D+00	0.0	2.27945D+00	0.0	1.94865D+00
27	-1.31112D+00	0.0	-3.15903D+00	0.0	-3.78712D+00	0.0	-4.00583D+00	0.0
28	0.0	-2.03018D+00	0.0	-5.87468D-01	0.0	2.76417D-01	0.0	8.07738D-01
29	4.06404D+00	0.0	4.61485D+00	0.0	4.09408D+00	0.0	3.51643D+00	0.0
30	0.0	-3.47088D-01	0.0	-2.10324D+00	0.0	-2.80622D+00	0.0	-3.09536D+00
31	-5.51918D+00	0.0	-3.50058D+00	0.0	-1.54013D+00	0.0	-2.36374D-01	0.0
32	0.0	2.72962D+00	0.0	3.66890D+00	0.0	3.46628D+00	0.0	3.07091D+00
33	5.56488D+00	0.0	5.13481D-01	0.0	-2.19235D+00	0.0	-3.43569D+00	0.0
34	0.0	-4.25335D+00	0.0	-3.26164D+00	0.0	-1.79500D+00	0.0	-6.99936D-01
35	-4.65206D+00	0.0	2.88636D+00	0.0	4.82654D+00	0.0	4.90035D+00	0.0
36	0.0	4.65310D+00	0.0	1.18365D+00	0.0	-1.12421D+00	0.0	-2.32257D+00
37	3.37413D+00	0.0	-5.40777D+00	0.0	-4.94582D+00	0.0	-3.17837D+00	0.0
38	0.0	-4.16078D+00	0.0	1.53406D+00	0.0	3.54243D+00	0.0	3.90990D+00
39	-2.17565D+00	0.0	6.46554D+00	0.0	2.61440D+00	0.0	-6.24136D-01	0.0
40	0.0	3.21463D+00	0.0	-3.82182D+00	0.0	-4.16889D+00	0.0	-3.03073D+00
41	1.26624D+00	0.0	-6.18270D+00	0.0	9.83764D-01	0.0	4.26235D+00	0.0
42	0.0	-2.20689D+00	0.0	5.05994D+00	0.0	2.78435D+00	0.0	2.52604D-01
43	-6.72152D-01	0.0	5.07245D+00	0.0	-4.38922D+00	0.0	-5.80815D+00	0.0
44	0.0	1.36908D+00	0.0	-5.18488D+00	0.0	-1.20943D-01	0.0	2.80335D+00
45	3.27915D-01	0.0	-3.69010D+00	0.0	6.58410D+00	0.0	4.61536D+00	0.0
46	0.0	-7.76147D-01	0.0	4.51035D+00	0.0	-2.71076D+00	0.0	-4.52806D+00
47	-1.47895D-01	0.0	2.42582D+00	0.0	-7.26330D+00	0.0	-1.37109D+00	0.0
48	0.0	4.05354D-01	0.0	-3.46576D+00	0.0	4.80326D+00	0.0	4.16052D+00
49	6.19574D-02	0.0	-1.45883D+00	0.0	6.69926D+00	0.0	-2.53321D+00	0.0
50	0.0	-1.96225D-01	0.0	2.40391D+00	0.0	-5.75139D+00	0.0	-1.98552D+00

J N	9	10	11	12	13	14	15	16
51	-2.42031D-02	0.0	8.09523D-01	0.0	-5.42646D+00	0.0	5.77767D+00	0.0
52	0.0	8.84718D-02	0.0	-1.52572D+00	0.0	5.63305D+00	0.0	-1.03332D+00
53	8.84537D-03	0.0	-4.17179D-01	0.0	3.95895D+00	0.0	-7.61132D+00	0.0
54	0.0	-3.72987D-02	0.0	8.94382D-01	0.0	-4.80543D+00	0.0	3.82970D+00
55	-3.03290D-03	0.0	2.00660D-01	0.0	-2.64151D+00	0.0	7.94215D+00	0.0
56	0.0	1.47521D-02	0.0	-4.87564D-01	0.0	3.67961D+00	0.0	-5.67754D+00
57	9.78085D-04	0.0	-9.04476D-02	0.0	1.62844D+00	0.0	-7.13631D+00	0.0
58	0.0	-5.48914D-03	0.0	2.48474D-01	0.0	-2.57343D+00	0.0	6.34713D+00
59	-2.97324D-04	0.0	3.83338D-02	0.0	-9.34390D-01	0.0	5.72756D+00	0.0
60	0.0	1.92623D-03	0.0	-1.18873D-01	0.0	1.66257D+00	0.0	-6.01428D+00
61	8.53644D-05	0.0	-1.53194D-02	0.0	5.01798D-01	0.0	-4.18966D+00	0.0
62	0.0	-6.38855D-04	0.0	5.35709D-02	0.0	-1.00013D+00	0.0	5.05661D+00
63	-2.31900D-05	0.0	5.78678D-03	0.0	-2.53320D-01	0.0	2.82882D+00	0.0
64	0.0	2.00641D-04	0.0	-2.28066D-02	0.0	5.63533D-01	0.0	-3.86290D+00
65	5.97053D-06	0.0	-2.07058D-03	0.0	1.20639D-01	0.0	-1.77841D+00	0.0
66	0.0	-5.97740D-05	0.0	9.19483D-03	0.0	-2.98783D-01	0.0	2.72013D+00
67	-1.45906D-06	0.0	7.03114D-04	0.0	-5.43569D-02	0.0	1.04767D+00	0.0
68	0.0	1.69187D-05	0.0	-3.51808D-03	0.0	1.49612D-01	0.0	-1.78266D+00
69	3.38916D-07	0.0	-2.26974D-04	0.0	2.32305D-02	0.0	-5.81154D-01	0.0
70	0.0	-4.55630D-06	0.0	1.27985D-03	0.0	-7.09696D-02	0.0	1.09483D+00
71	-7.49265D-08	0.0	6.97600D-05	0.0	-9.43692D-03	0.0	3.04730D-01	0.0
72	0.0	1.16903D-06	0.0	-4.43440D-04	0.0	3.19726D-02	0.0	-6.33402D-01
73	1.57847D-08	0.0	-2.04421D-05	0.0	3.65080D-03	0.0	-1.51518D-01	0.0
74	0.0	-2.86119D-07	0.0	1.46549D-04	0.0	-1.37099D-02	0.0	3.46611D-01
75	-3.17247D-09	0.0	5.71863D-06	0.0	-1.34727D-03	0.0	7.16275D-02	0.0
76	0.0	6.68771D-08	0.0	-4.62591D-05	0.0	5.60616D-03	0.0	-1.79996D-01
77	6.08959D-10	0.0	-1.52905D-06	0.0	4.74986D-04	0.0	-3.22650D-02	0.0
78	0.0	-1.49447D-08	0.0	1.39644D-05	0.0	-2.18978D-03	0.0	8.89462D-02
79	-1.11752D-10	0.0	3.91193D-07	0.0	-1.60195D-04	0.0	1.38759D-02	0.0
80	0.0	3.19612D-09	0.0	-4.03606D-06	0.0	8.18247D-04	0.0	-4.19212D-02
81	1.96255D-11	0.0	-9.58634D-08	0.0	5.17483D-05	0.0	-5.70693D-03	0.0
82	0.0	-6.54784D-10	0.0	1.11806D-06	0.0	-2.92888D-04	0.0	1.88817D-02
83	-3.30135D-12	0.0	2.25230D-08	0.0	-1.60291D-05	0.0	2.24806D-03	0.0
84	0.0	1.28621D-10	0.0	-2.97154D-07	0.0	1.00549D-04	0.0	-8.14134D-03
85	5.32413D-13	0.0	-5.07817D-09	0.0	4.76583D-06	0.0	-8.49295D-04	0.0
86	0.0	-2.42460D-11	0.0	7.58412D-08	0.0	-3.31437D-05	0.0	3.36557D-03
87	-8.23874D-14	0.0	1.09969D-09	0.0	-1.36146D-06	0.0	3.08093D-04	0.0
88	0.0	4.38976D-12	0.0	-1.86045D-08	0.0	1.05005D-06	0.0	-1.33572D-03
89	1.22427D-14	0.0	-2.28914D-10	0.0	3.74026D-07	0.0	-1.07438D-04	0.0
90	0.0	-7.63934D-13	0.0	4.39012D-09	0.0	-3.20048D-06	0.0	5.09559D-04
91	-1.74839D-15	0.0	4.58406D-11	0.0	-9.88992D-08	0.0	3.60516D-05	0.0
92	0.0	1.27883D-13	0.0	-9.97285D-10	0.0	9.39291D-07	0.0	-1.87057D-04
93	2.40140D-16	0.0	-8.83739D-12	0.0	2.51897D-08	0.0	-1.16516D-05	0.0
94	0.0	-2.06075D-14	0.0	2.18257D-10	0.0	-2.65657D-07	0.0	6.61439D-05
95	-3.17442D-17	0.0	1.64136D-12	0.0	-6.18466D-09	0.0	3.63009D-06	0.0
96	0.0	3.19883D-15	0.0	-4.60499D-11	0.0	7.24622D-08	0.0	-2.25499D-05
97	4.04148D-18	0.0	-2.93887D-13	0.0	1.46481D-09	0.0	-1.09111D-06	0.0
98	0.0	-4.78633D-16	0.0	9.37326D-12	0.0	-1.90760D-08	0.0	7.41837D-06
99	-4.95882D-19	0.0	5.07618D-14	0.0	-3.34898D-10	0.0	3.16641D-07	0.0
100	0.0	6.90772D-17	0.0	-1.84176D-12	0.0	4.85002D-09	0.0	-2.35681D-06

TABLE 3.1 (continued)

M = 2.0 F = 1.0

H (M)	7.86490D+03	3.67705D+03	2.11764D+03	1.37307D+03	9.61212D+02	7.09971D+02	5.45604D+02	4.32270D+02
BETA	1.11290D+01	2.38039D+01	4.13329D+01	6.37463D+01	9.10602D+01	1.23284D+02	1.60424D+02	2.02485D+02
J N	2	3	4	5	6	7	8	9
2	1.00000D+00	0.0	1.00000D+00	0.0	1.00000D+00	0.0	1.00000D+00	0.0
3	0.0	1.00000D+00	0.0	1.00000D-00	0.0	1.00000D+00	0.0	1.00000D+00
4	-8.69705D-02	0.0	1.29193D+00	0.0	1.56938D+00	0.0	1.66910D+00	0.0
5	0.0	-2.10994D-01	0.0	9.05843D-01	0.0	1.22728D+00	0.0	1.36189D+00
6	4.82147D-03	0.0	-4.84058D-01	0.0	9.82268D-01	0.0	1.63730D+00	0.0
7	0.0	2.23769D-02	0.0	-5.46969D-01	0.0	4.69583D-01	0.0	1.05205D+00
8	-1.65606D-04	0.0	8.11605D-02	0.0	-9.54961D-01	0.0	2.24822D-01	0.0
9	0.0	-1.44641D-03	0.0	1.31542D-01	0.0	-8.22649D-01	0.0	-1.59658D-01
10	3.82399D-06	0.0	-8.18587D-03	0.0	3.12106D-01	0.0	-1.20557D+00	0.0
11	0.0	6.32748D-05	0.0	-1.87208D-02	0.0	3.54991D-01	0.0	-8.56523D-01
12	-6.32520D-08	0.0	5.60605D-04	0.0	-5.89814D-02	0.0	6.79642D-01	0.0
13	0.0	-2.00277D-06	0.0	1.80892D-03	0.0	-8.56738D-02	0.0	6.35242D-01
14	7.85559D-10	0.0	-2.79648D-05	0.0	7.54104D-03	0.0	-2.04199D-01	0.0
15	0.0	4.80590D-08	0.0	-1.27824D-04	0.0	1.38884D-02	0.0	-2.33804D-01
16	-7.59171D-12	0.0	1.06503D-06	0.0	-7.06261D-04	0.0	4.07444D-02	0.0
17	0.0	-9.05105D-10	0.0	6.93024D-06	0.0	-1.64842D-03	0.0	5.62220D-02
18	5.87045D-14	0.0	-3.20432D-08	0.0	5.09163D-05	0.0	-5.93698D-03	0.0
19	0.0	1.37414D-11	0.0	-2.98309D-07	0.0	1.50870D-04	0.0	-9.82595D-03
20	-3.71432D-16	0.0	7.81769D-10	0.0	-2.92494D-06	0.0	6.67363D-04	0.0
21	0.0	-1.71810D-13	0.0	1.04609D-08	0.0	-1.10310D-05	0.0	1.32358D-03
22	1.95843D-18	0.0	-1.57898D-11	0.0	1.37384D-07	0.0	-6.00219D-05	0.0
23	0.0	1.80021D-15	0.0	-3.04979D-10	0.0	6.61276D-07	0.0	-1.42746D-04
24	-8.73731D-21	0.0	2.68502D-13	0.0	-5.38328D-09	0.0	4.43496D-06	0.0
25	0.0	-1.60387D-17	0.0	7.51481D-12	0.0	-3.31615D-08	0.0	1.26645D-05
26	3.34107D-23	0.0	-3.89834D-15	0.0	1.78849D-10	0.0	-2.74715D-07	0.0
27	0.0	1.23006D-19	0.0	-1.58648D-13	0.0	1.41369D-09	0.0	-9.43469D-07
28	-1.10721D-25	0.0	4.89015D-17	0.0	-5.10587D-12	0.0	1.44968D-08	0.0
29	0.0	-8.20694D-22	0.0	2.90287D-15	0.0	-5.19147D-11	0.0	5.99798D-08
30	3.21045D-28	0.0	-5.35412D-19	0.0	1.26676D-13	0.0	-6.60353D-10	0.0
31	0.0	4.80734D-24	0.0	-4.64919D-17	0.0	1.66060D-12	0.0	-3.29712D-09
32	-8.21349D-31	0.0	5.16189D-21	0.0	-2.75770D-15	0.0	2.62524D-11	0.0
33	0.0	-2.49217D-26	0.0	6.57362D-19	0.0	-4.67086D-14	0.0	1.58445D-10
34	1.86777D-33	0.0	-4.41612D-23	0.0	5.31199D-17	0.0	-9.19434D-13	0.0
35	0.0	1.15153D-28	0.0	-8.26730D-21	0.0	1.16478D-15	0.0	-6.71844D-12
36	-3.80014D-36	0.0	3.37559D-25	0.0	-9.12015D-19	0.0	2.85983D-14	0.0
37	0.0	-4.77237D-31	0.0	9.30967D-23	0.0	-2.59372D-17	0.0	2.53387D-13
38	6.95837D-39	0.0	-2.31941D-27	0.0	1.40471D-20	0.0	-7.95597D-16	0.0
39	0.0	1.78403D-33	0.0	-9.44235D-25	0.0	5.19016D-19	0.0	-8.55961D-15
40	-1.15276D-41	0.0	1.44043D-29	0.0	-1.95215D-22	0.0	1.99197D-17	0.0
41	0.0	-6.04625D-36	0.0	8.67161D-27	0.0	-9.38567D-21	0.0	2.60583D-16
42	1.73606D-44	0.0	-8.12513D-32	0.0	2.46050D-24	0.0	-4.51352D-19	0.0
43	0.0	1.86632D-38	0.0	-7.24549D-29	0.0	1.54159D-22	0.0	-7.18841D-18
44	-2.38712D-47	0.0	4.18142D-34	0.0	-2.82576D-26	0.0	9.30131D-21	0.0
45	0.0	-5.26893D-41	0.0	5.53176D-31	0.0	-2.31033D-24	0.0	1.80565D-19
46	3.00876D-50	0.0	-1.97123D-36	0.0	2.96953D-28	0.0	-1.75110D-22	0.0
47	0.0	1.36570D-43	0.0	-3.87438D-33	0.0	3.17228D-26	0.0	-4.14823D-21
48	-3.48883D-53	0.0	8.54440D-39	0.0	-2.86649D-30	0.0	3.02396D-24	0.0
49	0.0	-3.26147D-46	0.0	2.49833D-35	0.0	-4.00589D-28	0.0	8.75087D-23
50	3.73425D-56	0.0	-3.41694D-41	0.0	2.55066D-32	0.0	-4.80780D-26	0.0

J N	2	3	4	5	6	7	8	9
51	0.0	7.19940D-49	0.0	-1.48816D-37	0.0	4.66825D-30	0.0	-1.70129D-24
52	-3.70081D-59	0.0	1.26464D-43	0.0	-2.09895D-34	0.0	7.06139D-28	0.0
53	0.0	-1.47335D-51	0.0	8.21351D-40	0.0	-5.03630D-32	0.0	3.05838D-26
54	3.40565D-62	0.0	-4.34444D-46	0.0	1.60212D-36	0.0	-9.61077D-30	0.0
55	0.0	2.80311D-54	0.0	-4.21226D-42	0.0	5.04478D-34	0.0	-5.09935D-28
56	-2.91786D-65	0.0	1.38901D-48	0.0	-1.13745D-38	0.0	1.21561D-31	0.0
57	0.0	-4.97072D-57	0.0	2.01256D-44	0.0	-4.70458D-36	0.0	7.90815D-30
58	2.33324D-68	0.0	-4.14351D-51	0.0	7.53061D-41	0.0	-1.43270D-33	0.0
59	0.0	8.23537D-60	0.0	-8.98033D-47	0.0	4.09487D-38	0.0	-1.14368D-31
60	-1.74537D-71	0.0	1.15595D-53	0.0	-4.66038D-43	0.0	1.57728D-35	0.0
61	0.0	-1.27763D-62	0.0	3.75088D-49	0.0	-3.33437D-40	0.0	1.54618D-33
62	1.22400D-74	0.0	-3.02246D-56	0.0	2.70194D-45	0.0	-1.62575D-37	0.0
63	0.0	1.85991D-65	0.0	-1.46959D-51	0.0	2.54560D-42	0.0	-1.95847D-35
64	-8.06341D-78	0.0	7.42206D-59	0.0	-1.47061D-47	0.0	1.57224D-39	0.0
65	0.0	-2.54567D-68	0.0	5.41188D-54	0.0	-1.82582D-44	0.0	2.32912D-37
66	0.0	0.0	-1.71499D-61	0.0	7.52895D-50	0.0	-1.42948D-41	0.0
67	0.0	3.28197D-71	0.0	-1.87674D-56	0.0	1.23267D-46	0.0	-2.60581D-39
68	0.0	0.0	3.73555D-64	0.0	-3.63233D-52	0.0	1.22418D-43	0.0
69	0.0	-3.99250D-74	0.0	6.13945D-59	0.0	-7.84764D-49	0.0	2.74771D-41
70	0.0	0.0	-7.68319D-67	0.0	1.65424D-54	0.0	-9.89209D-46	0.0
71	0.0	4.59037D-77	0.0	-1.89778D-61	0.0	4.71923D-51	0.0	-2.73548D-43
72	0.0	0.0	1.49459D-69	0.0	-7.12335D-57	0.0	7.55495D-48	0.0
73	0.0	0.0	0.0	5.55183D-64	0.0	-2.68496D-53	0.0	2.57537D-45
74	0.0	0.0	-2.75391D-72	0.0	2.90475D-59	0.0	-5.46206D-50	0.0
75	0.0	0.0	0.0	-1.53937D-66	0.0	1.44742D-55	0.0	-2.29648D-47
76	0.0	0.0	4.81340D-75	0.0	-1.12332D-61	0.0	3.74376D-52	0.0
77	0.0	0.0	0.0	4.05117D-69	0.0	-7.40390D-58	0.0	1.94239D-49
78	0.0	0.0	-7.99131D-78	0.0	4.12542D-64	0.0	-2.43611D-54	0.0
79	0.0	0.0	0.0	-1.01327D-71	0.0	3.59853D-60	0.0	-1.56049D-51
80	0.0	0.0	0.0	0.0	-1.44070D-66	0.0	1.50696D-56	0.0
81	0.0	0.0	0.0	2.41173D-74	0.0	-1.66398D-62	0.0	1.19235D-53
82	0.0	0.0	0.0	0.0	4.79022D-69	0.0	-8.87297D-59	0.0
83	0.0	0.0	0.0	-5.46909D-77	0.0	7.32924D-65	0.0	-8.67582D-56
84	0.0	0.0	0.0	0.0	-1.51821D-71	0.0	4.97879D-61	0.0
85	0.0	0.0	0.0	0.0	0.0	-3.07867D-67	0.0	6.01855D-58
86	0.0	0.0	0.0	0.0	4.59188D-74	0.0	-2.66540D-63	0.0
87	0.0	0.0	0.0	0.0	0.0	1.23465D-69	0.0	-3.98609D-60
88	0.0	0.0	0.0	0.0	-1.32678D-76	0.0	1.36289D-65	0.0
89	0.0	0.0	0.0	0.0	0.0	-4.73211D-72	0.0	2.52125D-62
90	0.0	0.0	0.0	0.0	0.0	0.0	-6.66293D-68	0.0
91	0.0	0.0	0.0	0.0	0.0	1.73516D-74	0.0	-1.52570D-64
92	0.0	0.0	0.0	0.0	0.0	0.0	3.11754D-70	0.0
93	0.0	0.0	0.0	0.0	0.0	-6.09280D-77	0.0	8.83948D-67
94	0.0	0.0	0.0	0.0	0.0	0.0	-1.39738D-72	0.0
95	0.0	0.0	0.0	0.0	0.0	0.0	0.0	-4.90787D-69
96	0.0	0.0	0.0	0.0	0.0	0.0	6.00571D-75	0.0
97	0.0	0.0	0.0	0.0	0.0	0.0	0.0	2.61372D-71
98	0.0	0.0	0.0	0.0	0.0	0.0	-2.47710D-77	0.0
99	0.0	0.0	0.0	0.0	0.0	0.0	0.0	-1.33628D-73
100	0.0	0.0	0.0	0.0	0.0	0.0	0.0	0.0

TABLE 3.1 (continued)

M = 2.0 F = 1.0

H (M)	3.50858D+02	2.90427D+02	2.44346D+02	2.08412D+02	1.79851D+02	1.56778D+02	1.37873D+02	1.22189D+02
BETA	2.49469D+02	3.01378D+02	3.58213D+02	4.19977D+02	4.86670D+02	5.58293D+02	6.34847D+02	7.16331D+02
J N	10	11	12	13	14	15	16	17
2	1.00000D+00	0.0	1.00000D+00	0.0	1.00000D+00	0.0	1.00000D+00	0.0
3	0.0	1.00000D+00	0.0	1.00000D+00	0.0	1.00000D+00	0.0	1.00000D+00
4	1.71582D+00	0.0	1.74138D+00	0.0	1.75685D+00	0.0	1.76692D+00	0.0
5	0.0	1.43065D+00	0.0	1.47040D+00	0.0	1.49543D+00	0.0	1.51220D+00
6	1.96753D+00	0.0	2.15443D+00	0.0	2.26977D+00	0.0	2.34571D+00	0.0
7	0.0	1.38527D+00	0.0	1.58895D+00	0.0	1.72133D+00	0.0	1.81179D+00
8	1.13197D+00	0.0	1.73165D+00	0.0	2.13174D+00	0.0	2.40745D+00	0.0
9	0.0	5.11793D-01	0.0	1.01569D+00	0.0	1.37906D+00	0.0	1.64314D+00
10	-6.43728D-01	0.0	2.24044D-01	0.0	9.83457D-01	0.0	1.58270D+00	0.0
11	0.0	-7.20319D-01	0.0	-1.98449D-01	0.0	3.45701D-01	0.0	8.17179D-01
12	-1.04417D+00	0.0	-1.24461D+00	0.0	-7.39924D-01	0.0	-5.10714D-02	0.0
13	0.0	-5.77929D-01	0.0	-9.74690D-01	0.0	-8.05072D-01	0.0	-4.06055D-01
14	1.05030D+00	0.0	-4.71947D-01	0.0	-1.31645D+00	0.0	-1.38032D+00	0.0
15	0.0	8.49482D-01	0.0	-6.31902D-02	0.0	-8.13898D-01	0.0	-1.06978D+00
16	-4.69421D-01	0.0	1.23543D+00	0.0	3.12858D-01	0.0	-8.25189D-01	0.0
17	0.0	-4.60102D-01	0.0	8.69089D-01	0.0	4.94412D-01	0.0	-3.05203D-01
18	1.34004D-01	0.0	-8.15205D-01	0.0	1.08717D+00	0.0	9.98756D-01	0.0
19	0.0	1.54297D-01	0.0	-7.09136D-01	0.0	6.28200D-01	0.0	8.67646D-01
20	-2.76043D-02	0.0	3.19016D-01	0.0	-1.13052D+00	0.0	5.83014D-01	0.0
21	0.0	-3.69721D-02	0.0	3.22770D-01	0.0	-8.83721D-01	0.0	1.67752D-01
22	4.37309D-03	0.0	-8.80356D-02	0.0	5.98596D-01	0.0	-1.26849D+00	0.0
23	0.0	6.78890D-03	0.0	-1.01828D-01	0.0	5.46556D-01	0.0	-8.83120D-01
24	-5.54567D-04	0.0	1.85237D-02	0.0	-2.14782D-01	0.0	9.24802D-01	0.0
25	0.0	-9.96841D-04	0.0	2.43314D-02	0.0	-2.22338D-01	0.0	7.71638D-01
26	5.78944D-05	0.0	-3.11083D-03	0.0	5.78692D-02	0.0	-4.26009D-01	0.0
27	0.0	1.20509D-04	0.0	-4.62730D-03	0.0	6.71803D-02	0.0	-4.02856D-01
28	-5.08070D-06	0.0	4.29920D-04	0.0	-1.23630D-02	0.0	1.43767D-01	0.0
29	0.0	-1.22549D-05	0.0	7.23434D-04	0.0	-1.60175D-02	0.0	1.51425D-01
30	3.80997D-07	0.0	-4.99929D-05	0.0	2.16770D-03	0.0	-3.80532D-02	0.0
31	0.0	1.06594D-06	0.0	-9.51497D-05	0.0	3.12724D-03	0.0	-4.43072D-02
32	-2.47386D-08	0.0	4.97541D-06	0.0	-3.19567D-04	0.0	8.22352D-03	0.0
33	0.0	-8.03678D-08	0.0	1.07140D-05	0.0	-5.12840D-04	0.0	1.05460D-02
34	1.40619D-09	0.0	-4.29574D-07	0.0	4.03329D-05	0.0	-1.49060D-03	0.0
35	0.0	5.31064D-09	0.0	-1.04720D-06	0.0	7.19789D-05	0.0	-2.10147D-03
36	-7.06243D-11	0.0	3.25358D-08	0.0	-4.42010D-06	0.0	2.31101D-04	0.0
37	0.0	-3.10423D-10	0.0	8.98525D-08	0.0	-8.77293D-06	0.0	3.57846D-04
38	3.15907D-12	0.0	-2.18195D-09	0.0	4.25454D-07	0.0	-3.11098D-05	0.0
39	0.0	1.61789D-11	0.0	-6.83177D-09	0.0	9.39445D-07	0.0	-5.28884D-05
40	-1.26726D-13	0.0	1.30598D-10	0.0	-3.63114D-08	0.0	3.67993D-06	0.0
41	0.0	-7.57047D-13	0.0	4.63984D-10	0.0	-8.92421D-08	0.0	6.86841D-06
42	4.58684D-15	0.0	-7.02470D-12	0.0	2.77010D-09	0.0	-3.86269D-07	0.0
43	0.0	3.19964D-14	0.0	-2.83419D-11	0.0	7.58168D-09	0.0	-7.91645D-07
44	-1.50609D-16	0.0	3.41608D-13	0.0	-1.90199D-10	0.0	3.62760D-08	0.0
45	0.0	-1.22803D-15	0.0	1.56647D-12	0.0	-5.80070D-10	0.0	8.16605D-08
46	4.50785D-18	0.0	-1.50991D-14	0.0	1.18250D-11	0.0	-3.06956D-09	0.0
47	0.0	4.30051D-17	0.0	-7.87564D-14	0.0	4.02105D-11	0.0	-7.59251D-09
48	-1.23526D-19	0.0	6.09479D-16	0.0	-6.69226D-13	0.0	2.35452D-10	0.0
49	0.0	-1.38008D-18	0.0	3.61889D-15	0.0	-2.53892D-12	0.0	6.40199D-10
50	3.11122D-21	0.0	-2.25633D-17	0.0	3.46396D-14	0.0	-1.64594D-11	0.0

J N	10	11	12	13	14	15	16	17
51	0.0	4.07431D-20	0.0	-1.52627D-16	0.0	1.46708D-13	0.0	-4.92184D-11
52	-7.22833D-23	0.0	7.69060D-19	0.0	-1.64677D-15	0.0	1.05356D-12	0.0
53	0.0	-1.11047D-21	0.0	5.93081D-18	0.0	-7.79089D-15	0.0	3.46642D-12
54	1.55420D-24	0.0	-2.42188D-20	0.0	7.21792D-17	0.0	-6.20106D-14	0.0
55	0.0	2.80326D-23	0.0	-2.13078D-19	0.0	3.81678D-16	0.0	-2.24599D-13
56	-3.10203D-26	0.0	7.06920D-22	0.0	-2.92690D-18	0.0	3.36888D-15	0.0
57	0.0	-6.57388D-25	0.0	7.10038D-21	0.0	-1.73092D-17	0.0	1.34386D-14
58	5.76317D-28	0.0	-1.91820D-23	0.0	1.10152D-19	0.0	-1.69514D-16	0.0
59	0.0	1.43605D-26	0.0	-2.20095D-22	0.0	7.28943D-19	0.0	-7.45096D-16
60	-9.99238D-30	0.0	4.03175D-25	0.0	-3.85853D-21	0.0	7.92476D-18	0.0
61	0.0	-2.92963D-28	0.0	6.36344D-24	0.0	-2.85886D-20	0.0	3.84002D-17
62	1.62072D-31	0.0	-1.14677D-26	0.0	1.26140D-22	0.0	-3.45198D-19	0.0
63	0.0	5.59461D-30	0.0	-1.72029D-25	0.0	1.04694D-21	0.0	-1.84483D-18
64	-2.46456D-33	0.0	2.53882D-28	0.0	-3.85790D-24	0.0	1.40473D-20	0.0
65	0.0	-1.00229D-31	0.0	4.35854D-27	0.0	-3.58876D-23	0.0	8.28363D-20
66	3.52103D-35	0.0	-5.27609D-30	0.0	1.10640D-25	0.0	-5.35325D-22	0.0
67	0.0	1.68801D-33	0.0	-1.03715D-28	0.0	1.15409D-24	0.0	-3.48477D-21
68	-4.73521D-37	0.0	1.03132D-31	0.0	-2.98162D-27	0.0	1.91477D-23	0.0
69	0.0	-2.67760D-35	0.0	2.32257D-30	0.0	-3.48921D-26	0.0	1.37653D-22
70	6.00537D-39	0.0	-1.89974D-33	0.0	7.56546D-29	0.0	-6.44167D-25	0.0
71	0.0	4.00761D-37	0.0	-4.90385D-32	0.0	9.93699D-28	0.0	-5.11639D-24
72	-7.19474D-41	0.0	3.30361D-35	0.0	-1.81078D-30	0.0	2.04225D-26	0.0
73	0.0	-5.66927D-39	0.0	9.77930D-34	0.0	-2.67068D-29	0.0	1.79285D-25
74	8.15581D-43	0.0	-5.43249D-37	0.0	4.09538D-32	0.0	-6.11278D-28	0.0
75	0.0	7.59210D-41	0.0	-1.84501D-35	0.0	6.78538D-31	0.0	-5.93352D-27
76	-8.76104D-45	0.0	8.46076D-39	0.0	-8.76663D-34	0.0	1.73033D-29	0.0
77	0.0	-9.63924D-43	0.0	3.29822D-37	0.0	-1.63236D-32	0.0	1.85783D-28
78	8.93112D-47	0.0	-1.24986D-40	0.0	1.77888D-35	0.0	-4.63957D-31	0.0
79	0.0	1.16195D-44	0.0	-5.59490D-39	0.0	3.72395D-34	0.0	-5.51209D-30
80	-8.65187D-49	0.0	1.75375D-42	0.0	-3.42663D-37	0.0	1.18015D-32	0.0
81	0.0	-1.33160D-46	0.0	9.01852D-41	0.0	-8.06801D-36	0.0	1.55200D-31
82	7.97497D-51	0.0	-2.34045D-44	0.0	6.27464D-39	0.0	-2.85185D-34	0.0
83	0.0	1.45266D-48	0.0	-1.38318D-42	0.0	1.66223D-37	0.0	-4.15287D-33
84	-7.00319D-53	0.0	2.97445D-46	0.0	-1.09364D-40	0.0	6.55589D-36	0.0
85	0.0	-1.51035D-50	0.0	2.02099D-44	0.0	-3.26088D-39	0.0	1.05746D-34
86	5.86567D-55	0.0	-3.60419D-48	0.0	1.81660D-42	0.0	-1.43550D-37	0.0
87	0.0	1.49836D-52	0.0	-2.81646D-46	0.0	6.09857D-41	0.0	-2.56558D-36
88	-4.69112D-57	0.0	4.16866D-50	0.0	2.07700D-44	0.0	2.99756D-39	0.0
89	0.0	-1.41989D-54	0.0	3.74787D-48	0.0	-1.08861D-42	0.0	5.93793D-38
90	3.58619D-59	0.0	-4.60727D-52	0.0	4.35845D-46	0.0	-5.97615D-41	0.0
91	0.0	1.28662D-56	0.0	-4.76730D-50	0.0	1.85673D-44	0.0	-1.31252D-39
92	-2.62317D-61	0.0	4.87076D-54	0.0	-6.30902D-48	0.0	1.13877D-42	0.0
93	0.0	-1.11593D-58	0.0	5.80244D-52	0.0	-3.02905D-46	0.0	2.77374D-41
94	1.83770D-63	0.0	-4.93042D-56	0.0	8.74137D-50	0.0	-2.07620D-44	0.0
95	0.0	9.27309D-61	0.0	-6.76430D-54	0.0	4.73135D-48	0.0	-5.61002D-43
96	-1.23419D-65	0.0	4.78317D-58	0.0	-1.16039D-51	0.0	3.62529D-46	0.0
97	0.0	-7.38945D-63	0.0	7.55985D-56	0.0	-7.08266D-50	0.0	1.08699D-44
98	7.95300D-68	0.0	-4.45125D-60	0.0	1.47718D-53	0.0	-6.06838D-48	0.0
99	0.0	5.65170D-65	0.0	-8.10714D-58	0.0	1.01704D-51	0.0	-2.01957D-46
100	-4.92142D-70	0.0	3.97703D-62	0.0	-1.80490D-55	0.0	9.74650D-50	0.0

replacement occurs in the left-hand side of (3. 30), and an additional term $(-i\omega/g)\Lambda_\omega$ appears in (3. 33) where

$$\Lambda_\omega = \sum_{m,n} \Lambda_{\omega,m,n}\Theta_{\omega,m,n} \exp(\mathrm{i}m\phi) \tag{3.43}$$

$$\Lambda_{\omega,m,n} = \text{function of only } z.$$

A different mathematical formulation of the tidal theory has been attempted by JONES (1970, 1971) who used matrices in the Hilbert space.

3. 1. 3 *Mode dispersion due to horizontal structure*

Figures 3.1 and 3.2 show that $h_{\omega,m,n}$ can be negative as well as positive for $\omega<2\Omega$; $\beta_e=4a^2\Omega^2/gh$. There are two distinct classes of modes, one consisting of those which remain in the absence of the earth's rotation, and the other of those which do not. The former can be called GW class or positive modes and the other rotational wave class or negative modes. The GW class was called oscillation of the first kind by Hough (LAMB, 1945). Of the rotational class, the modes for $f>0$ i.e. westward traveling modes were called oscillation of the second kind; modes for $f<0$ in this class were unknown. The situation will be clarified by putting $f\to\infty\,(\Omega\to 0)$. Then, (3. 14) tends to an associated Legendre equation as

$$\frac{\mathrm{d}^2\Theta}{\mathrm{d}\theta^2}+\cot\theta\,\frac{\mathrm{d}\Theta}{\mathrm{d}\theta}-\frac{m^2}{\sin^2\theta}\,\Theta+\frac{a^2\omega^2}{gh}\,\Theta=0 \tag{3.44}$$

which requires

$$\frac{a^2\omega^2}{gh}=n(n+1) \text{ as } n=\text{positive integers} \tag{3.44$'$}$$

or

$$f=\pm\sqrt{\frac{n(n+1)}{\beta_e}} \text{ for } \beta_e>0 \tag{3.44$''$}$$

$$\Theta\propto P_{n,m}(\theta) \tag{3.45}$$

by the boundary conditions same as those mentioned in Subsection 3. 1. 2. We find that (3. 44″) gives with $f\to\infty$ those curves which belong to GW mode in Fig. 3. 1. Further, we transform (3. 14) by using local Cartesian coordinates x and y, positive towards the south and the east, respectively, for $|x|$, $|y|\ll a$ as (with $f\to\infty$)

$$\mathrm{d}x=a\,\mathrm{d}\theta,\quad \mathrm{d}y=a\sin\theta\,\mathrm{d}\phi\,.$$

Then, (3.14) reduces to

$$\frac{\partial^2\Theta}{\partial x^2}+\frac{\partial^2\Theta}{\partial y^2}+\frac{\omega^2}{gh}\,\Theta=0 \tag{3.44$'''$}$$

which gives $gh=\omega^2/(\alpha^2+\beta^2)$ for $\Theta\propto\exp[-i(\alpha x+\beta y)]$. Then (3. 17) gives the solution for isothermal adiabatic case $\mathrm{d}H/\mathrm{d}s=q_{\omega,m,n}=0$ as

$$\gamma^2 = \left(\frac{\omega_g'}{\omega}\right)^2(\alpha^2+\beta^2)-\frac{1}{4H^2} \tag{3.45'}$$

which is identical to (2. 17) in Chapter 2 for $\alpha^2 \rightarrow (\alpha^2+\beta^2)$ and $\omega \rightarrow 0$.

Oscillation of the second kind is also called the Rossby wave and the characteristic frequencies for $h \rightarrow \infty$ i.e. $\beta_e \rightarrow 0$ are obtained analytically as follows (ECKART, 1960): Define $\hat{\Theta}$ as

$$\hat{\Theta} = \frac{1}{(f^2-\mu^2)}\left[(1-\mu^2)\frac{d\Theta}{d\mu}-\frac{m}{f}\mu\Theta\right] \tag{3.46}$$

or

$$\frac{d\Theta}{d\mu} = \frac{(f^2-\mu^2)\hat{\Theta}+(m/f)\mu\Theta}{1-\mu^2}$$

which gives after being substituted into (3.14)

$$\Theta = \frac{f^2}{m^2}\left[(1-\mu^2)\frac{d\hat{\Theta}}{d\mu}+\frac{m}{f}\mu\hat{\Theta}\right]. \tag{3.46'}$$

Then, substitution of Θ into (3.46) leads us to

$$\frac{d}{d\mu}\left[(1-\mu^2)\frac{d\hat{\Theta}}{d\mu}\right]+\left(\frac{m}{f}-\frac{m^2}{1-\mu^2}\right)\hat{\Theta} = 0 \tag{3.47}$$

which gives also an associated Legendre equation with its solution as (3.44) only if

$$\frac{m}{f} = n(n+1)$$

or

$$\omega = \frac{2m\Omega}{n(n+1)} \tag{3.48}$$

which shows readily that $\omega \rightarrow 0$ with $\Omega \rightarrow 0$, i.e. that the mode can exist only with the Coriolis force. These situations can be understood simply by looking at Fig. 3. 1; $\Omega \rightarrow 0$ implies $\beta_e \rightarrow 0$ by definition. If the mode exists without the Coriolis force, ω should remain finite even with $\Omega \rightarrow 0$, implying that $f \rightarrow \infty$ with $\Omega \rightarrow 0$. The GW mode behaves this way in Fig. 3. 1. On the other hand, the disappearing mode implies that with $\Omega \rightarrow 0\,(\beta_e \rightarrow 0)$, either $f(=\omega/2\Omega)$ is finite i.e. $\omega=0$ (as in $f>0$) or ω never approaches $\beta_e=0$ (as in $f<0$).

The tide is a special case where $\omega = m\Omega$ and $\hat{\chi} \propto e^{im(\Omega t+\phi)}$, the variation depending on only local time $(\Omega t+\phi)$. Table 3. 1 shows h's and Θ's for the diurnal tide $(f=1/2)$ and those for the semidiurnal $(f=1)$. There are two possible conventions for the mode designations (m, n) where m gives the wave number and is common in the both ways; $m=1, 2$ are diurnal and semidiurnal, respectively. As to n, the designation is fairly similar to that of the associated Legendre function $P_{n,m}(\theta)$ which is a limiting form of $\Theta_{n,m}$ with $\Omega \rightarrow 0$ by (3. 44). However, the eigenvalue is no longer as simple as in (3. 44') where the eigenvalue (h) is always positive, decreasing with n, and the number of the zeros of $P_{n,m}$ in $0<\theta<\pi$ is $(n-m)=0$. Now, the eigenvalue can be negative as well as positive for the diurnal tide. However, we arrange the modes in the order of decreasing magnitude of h, separately for the positive and negative modes. Then, for diurnal negative mode, $n=-1, -2, -3$, etc. correspond to $h=-\infty, -12.2, -1.79$ km,... respectively; for diurnal positive mode $n=1, 2, 3,\ldots$ correspond to $h=0.698$, 0.240, 0.121 km, ... respectively. The number of the zeros is different in expression between the negative and the positive modes, a situation which is more complicated than for $P_{n,m}(\theta)$ with $(n-m)$ zeros in both symmetric and anti-symmetric cases. The negative and positive symmetric modes have $(|n|-m-1)$ and $(n-m+2)$ zeros, respectively, whereas the negative and positive anti-symmetric modes have $(|n|-m+1)$ and $(n-m+2)$ zeros, respectively. A confusion may arise, when some people call $(1, -2)$-modes the *first* negative mode! As to the semidiurnal tide, the situation is as simple as in $P_{n,m}$ and (m, n)-mode has $(n-m)$ zeros in $0<\theta<\pi$ and n is numbered according to the order of decreasing h starting $n=m$ i.e. (2, 2), (2, 3), etc. Table 3.1 shows h in this designation. To distinguish solar tide from lunar one can write as $S\,(m, n)$*.

In the other designation, we observe the principle that $(|n|-m)$ gives the number of the zero of $\Theta_{n,m}$. Then, the first negative mode is $(1, -1)$, which may look more suitable than the other. This time, the first positive mode is (1, 3), which may look less suitable. Also, the $(1, -2)$-mode has larger h than the $(1, -1)$. This designation seems less in fashion now.

The negative modes of tide and of other atmospheric waves with $f<1$ and $h_{\omega,m,n}<0$ do not travel in the vertical direction, because $N^2<0$ is always true for any distribution of H. The negative modes of tides belong to external waves in the classification in Chapter 2.

It is found that the atmospheric wave in general tends to behave very distinctly with the frequency lower than 2Ω, for the rotational modes appear. Note that the first four diurnal modes are negative. HAURWITZ (1965) tried to express the observed diurnal pressure tide distribution in a few terms of the Hough functions of only positive modes. His failure to have a good convergence (Fig. 3. 3) seems now to be only natural, because the negative modes $(1, -2)$ and $(1, -4)$ are the first and second fundamentals. Improvement was remarkable when the negative modes were introduced by KATO (1966d).

The finding of the diurnal first negative mode (KATO, 1966a; LINDZEN, 1967) was of vital significance in the history of the atmospheric dynamo theory which attempts to explain the solar daily geomagnetic variation in terms of electric current driven by the upper atmospheric tidal motion in the presence of geomagnetic field. This will be discussed in the next chapter.

* $S_m{}^n$ or $S_n{}^m$ is used also.

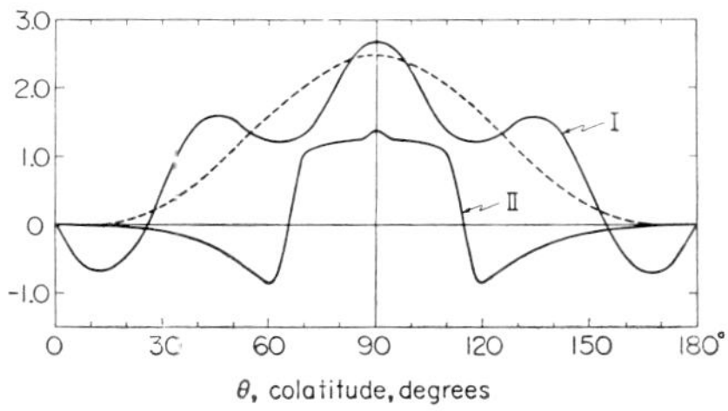

FIG. 3. 3. Observed diurnal pressure variation and the approximation by a combination of the Hough function. Curve I is a combination of (1, −2), (1, −4), (1, 1) and Curve II is a combination by (1, 1), (1, 3), (1, 5) i.e. only positive modes (KATO, 1966d). The broken curve corresponds to observation by HAURWITZ (1965).

3. 1. 4 *Vertical structure of forced waves and boundary conditions*

The vertical structure of $\hat{\chi}$ i.e. $\hat{Y}_{\omega,m,n}$ is determined by (3.17) for each mode by $h_{\omega,m,n}$, the eigenvalues of the horizontal structure function. Generally, the heat input $q_{\omega,n,n}$ must be specified in (3. 17) to find the vertical distribution of the waves. The waves consist of those which are directly excited by heat, and those which are produced in the presence of boundaries. Mathematically, the former crresponds to a particular solution of (3. 17), and the latter to the general solution of (3, 17) with the vanishing right-hand side. These two types of solutions are combined so as to satisfy the two boundary conditions, both at lower and upper levels. The boundary condition at lower level is

$$W_{z,\omega} = 0 \quad \text{at} \quad z = 0\ (s = 0) \tag{3.49}$$

which is generally true for invicid fluid at a rigid boundary. By (3. 33), (3. 49) is given as

$$\frac{d\hat{Y}_{\omega,m,n}}{ds} + \left(\frac{H}{h_{\omega,m,n}} - \frac{1}{2}\right)\hat{Y}_{\omega,m,n} = 0 \quad \text{at} \quad s = 0. \tag{3.50}$$

The boundary condition at the upper level varies with the behavior of $\hat{Y}_{\omega,m,n}$ for large s i.e. with $N^2 \gtreqless 0$ in (3. 17′). If $N^2 < 0$ in the upper region ($s > s_0 \gg 1$) where $q_{\omega,n,n} = 0$, the evanescent wave $e^{-|N|s}$ instead of $e^{|N|s}$ is adopted. If $N^2 > 0$, however, no such choice is available and $\hat{Y}_{\omega,m,n}$ is a traveling wave. We adopt the radiation condition which says that the energy flows upwards in the region where $q_{\omega,m,n} = 0$. Thus, the solution e^{iNs} instead of e^{-iNs} is adopted, because, then, by (2. 5) $\hat{\chi}_\omega \propto e^{i(\omega t + Ns)}$ which gives downward phase progression and the upward energy flow, as is explained in Chapter 2. The vertical enregy flow $\mathscr{F}_z$ is obtained by (2. 18), (3.30) and (3.32) as (WILKES, 1949).

$$\mathscr{F}_{z,\omega} = \frac{1}{2}\,\mathscr{R}(W_{z,\omega}p_{1,}{}^{c}{}_{\omega})$$

$$= \frac{p_{00}}{2}\sum_{m,n} h_{\omega,m,n}(\kappa^2/\omega)\,[\Theta_{\omega,m,n}]^2\,\mathscr{I}\left[\hat{Y}_{\omega}{}^{c}{}_{,m,n}\left(\frac{\mathrm{d}\hat{Y}_{\omega,m,n}}{\mathrm{d}s}\right)\right] \qquad (3.51)$$

where $\mathscr{R}$ and $\mathscr{I}$ denote, respectively, to take real and imaginary parts.

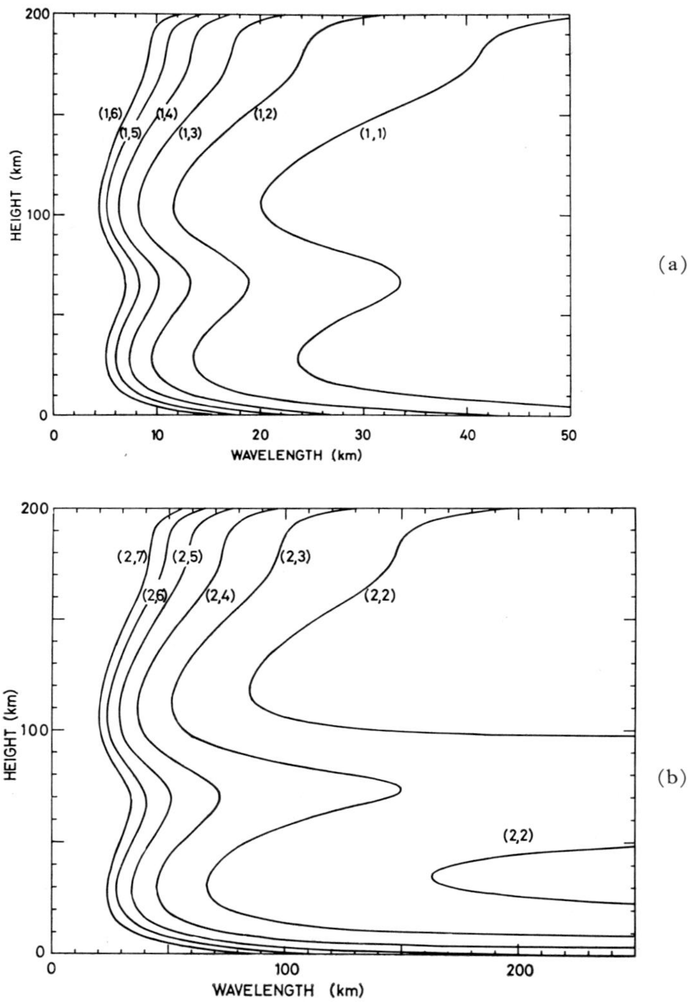

FIG. 3.4. Vertical wavelength of each tidal modes (m, n). Calculation is based on $T_0\ (z)$ in Fig. 1.1. (a) diurnal, (b) semidiurnal.

(3. 51) shows that the case for $N^2<0$ gives $\mathscr{F}_{z,\omega}=0$, whilst $(+N)$ for $N^2>0$ gives $\mathscr{F}_{z,\omega}>0$. Figure 3.4 shows N in terms of wavelength $(2\pi H/N)$ for a model $T_0(z)$ in Fig. 1. 1. It is found that any (positive) mode of the diurnal tide has relatively small wavelength in association to small eigenvalues; the (2. 2)-mode of the semidiurnal tide is exceptionally large wavelength; $\hat{Y}_{\omega,m,n}$ of all modes have the smallest wavelength above the mesopause around 100 km.

Note that the vertical wavelength of tides in p_1, T_1, and $\boldsymbol{W}$ depends not only on N in (3. 17′) but also on " the vertical wavelength of Q." While the former can be determined only from h (the eigenvalue of each mode) and $T_0(z)$ (the vertical temperature profile), the vertical wavelength of Q needs an additional information of the source distribution. Thus, it is not always relevant to identify the observed tides with particular modes by the vertical wavelength alone. Also, the phase of the vertical structure differs among p_1, T_1, and $\boldsymbol{W}$, as is understood in the respective equations above. On the contrary, the horizontal structure depends only on h. In principle, the mode identification by the horizontal structure should be less ambiguous. However, the global observation is very scarce still now except for the pressure tide at the surface. And the mode identification is based usually on the vertical structure observation of the wind and temperature tide at meteor heights and in the thermosphere at particular locations (Section 3. 5).

The thermal excitation is mainly produced by radiative heating of H_2O, O_3 etc. which are distributed in the vertical direction over certain height range. But the thermal conduction near the surface can contribute to the excitation to some degree. An approximate treatment of the excitation is at first to obtain a daily temperature variation, as a function of height, $T(t,z)$ due to thermal conduction from the ground in the absence of tides. Then, the tidal excitation is calculated for the corresponding Q where $Q=C_p[\partial T(t, z)/\partial t]$. It was found, however, that such an excitation is negligible (e. g. SIEBERT, 1961).

3. 1. 5 *Free oscillation*

Here we shall consider a free oscillation in the earth's atmosphere. The subject, as is stated briefly at the beginning of the present chapter, has long been the main concern of the atmospheric tidal theory.

In the preceding subsection it is found that we can determine $Y_{\omega,m,n}$ for $h_{\omega,m,n}$ which is determined from the structure equation (3.14) for specified(ω,m). Note that (3. 17) depends only on $h_{\omega,m,n}$, not on ω,m and n, explicitly. Remove here the subscript (ω,m,n) from (3. 17), putting $q_{\omega,m,n}=0$ for a while. Now, the question arises as to whether we can find those solutions of the equation which satisfy the boundary conditions both at lower and upper levels. Such solutions are available only for special h's; generally, only when a particular solution of (3.17) with $q_{\omega,m,n}\neq 0$ is included, can the solution satisfy the boundary conditions. It was expected that for the earth's atmosphere the vertical temperature distribution would allow such a special h as an eigenvalue and the h

would coincide with that of semidiurnal oscillation i.e. $\omega=2\Omega, m=2$ in (3. 14). This famous resonance theory was to explain the observed predominant semidiurnal pressure tide, for then no energy input is required to have the oscillation ($q_{\omega,m,n}=0$) and, therefore, any continuous input would build up the oscillation indefinitely with time. It is necessary that no energy should leak for the existence of the resonance. The resonance, thus, cannot occur with $N^2>0$ at the upper levels for large s, because then the vertical energy flow $\mathscr{F}_z \propto \mathscr{J}(Y^c \, \mathrm{d}Y/\mathrm{d}s)$ gives a leak. Accordingly, by (3. 17′) h must be large enough to have $N^2<0$, for which $\mathscr{F}_z=0$ (Note that $h>0$ for semidiurnal tide). In the earth's thermosphere, however, $N^2<0$ cannot be true for any possible h (see Fig. 3. 4), because the temperature increases monotonously with altitude (see Fig. 1. 1). However, in the mesopause $N^2<0$ could be possible due to the decreasing temperature, providing a fairly good insulator against the energy leakage from below. Thus, the lower atmosphere could trap energy to a certain extent and an incomplete resonance may arise. It would be interesting to know the magnification of the dynamical tide to the equilibrium tide as a function of h (GIWA, 1968). The magnification ζ is defined as the ratio of $(p_{1,\omega}/\rho_0)$ at $s=0$ between the dynamical and the equilibrium tide, both of which are assumed to be generated by gravity mainly near the ground where the density is highest. In the equilibrium tide, the tidal force balances the resulting pressure force $p_{1,\omega}$ as

$$\nabla\left(\frac{p_{1,\omega}}{\rho_0}+\Lambda_\omega\right)=0$$

or

$$\frac{p_{1,\omega}}{\rho_0}+\Lambda_\omega(0)=0 \tag{3. 52}$$

considering $p_{1,\omega}=0$ without Λ_ω.

In the dynamical tide, $(p_{1,\omega}/\rho_0)$ is given by (3. 11′), where $W_{z,\omega}=0$ at $z=0$, as

$$\frac{p_{1,\omega}}{\rho_0}(\text{dynamical tide})=i\frac{C^2}{\omega}\chi_\omega. \tag{3. 53}$$

The left-hand side of (3.31) is now replaced by $(p_{\omega,m,n}/\rho_0)+\Lambda_{\omega,m,n}$ as is remarked in the last paragraph of Subsection 3. 1. 2. Then, substitution of (3. 53) into $p_{\omega,m,n}/\rho_0$ in (3.31) gives

$$\Lambda_{\omega,m,n}=-i\left(\frac{\kappa gh}{\omega}\right)\left[\left(-\frac{1}{2}+\frac{H}{h}\right)\hat{Y}+\frac{\mathrm{d}\hat{Y}}{\mathrm{d}s}\right]. \tag{3. 54}$$

Thus, substitution of (3. 54) into (3. 52) gives

$$\zeta=\left|\frac{p_1(\text{dynamical})}{p_1(\text{equilibrium})}\right|=\left(\frac{H}{h}\right)\left|\frac{\hat{Y}}{[-(1/2)+(H/h)]\hat{Y}+\mathrm{d}\hat{Y}/\mathrm{d}s}\right| \tag{3. 55}$$

by (3. 53).

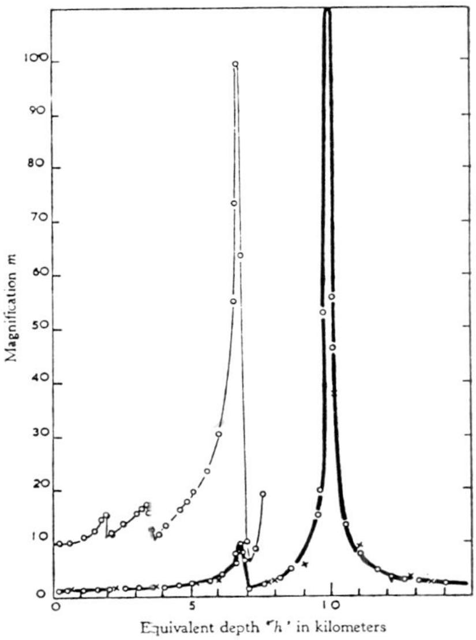

FIG. 3. 5. Magnification of the dynamical tide, to the equilibrium tide, versus h (GIWA, 1968). The thin curve enlarges the solid curve by 10 times for restricted parts. ζ is denoted by m on the diagram ordinate.

ζ is calculated for $\hat{Y}$ with various given h's and $q=0$, the lower boundary conditions being (3. 49), which gives (3. 50) with $(i\omega A/\kappa gh)$ in the right-hand side; the upper boundary condition is the radiation condition in Subsection 3. 1. 3. Figure 3. 5 shows ζ versus h; ζ is very large (9,220) at $h=9.94$ km and is almost at the resonance (the true resonance has $\zeta=\infty$). The second peak ($\zeta=10.0$) is at $h=6.72$ and would be slightly leaky. It is found that these h's, especially $h=9.94$ km, are not close to those h's which are eigenvalues of the semidiurnal tide. ζ becomes very small (~ 2) at $h=7.9$ km, which equals the eigenvalue for the (2, 2)-mode. If the temperature maximum around 60 km exceeded 320°K, the second peak of ζ would be at $h=7.9$ km and could explain ζ as large as 70 (WILKES, 1949). However, such high temperature is unrealistic, and thus, so is the resonance theory.

3.2 Recent Developments in Tidal Theory

The tidal theory in Section 3. 1 was established in the 1960's and is now called classical tidal theory. The theory is elegant and successful to find all possible *implications* of the atmospheric waves in a lossless hydrostatic shallow atmosphere surrounding a spherical rotating planet. Then, the question arises as to how well the theory is able to explain tides in a dissipative non-uniform and non-stationary atmosphere. A model of the heat input Q was given by a series of Hough functions for each frequency. It was found that Q due to the ozone in the middle atmosphere (BUTLER and SMALL, 1963) and the water vapor in the troposphere (SIEBERT, 1961) could explain the observed semidiurnal global pressure tide near the ground. Lindzen has attempted to explain further the observed tide above the ground. The calculated result of diurnal tide by LINDZEN (1967, Fig. 3. 6) was far less satisfactory to explain the surface pressure tide which was based on HAURWITZ's analysis of the observed data (1965). Nevertheless, he could show that the diurnal heating of twice the semidiurnal produce only a diurnal surface pressure tide of (1/6) of the semidiurnal at the equator. This demonstrated clearly an inefficient generation of the diurnal tide.

There are two factors to produce small diurnal tides. One is a destructive interference of $S(1, 1)$ generated by the ozone of the middle atmosphere on the surface because of its short vertical wavelength (30 km between the surface and 90 km). The other factor is the fact that fundamental mode of diurnal tide is $S(1, -2)$ and more heating goes into this mode than $S(1, 1)$

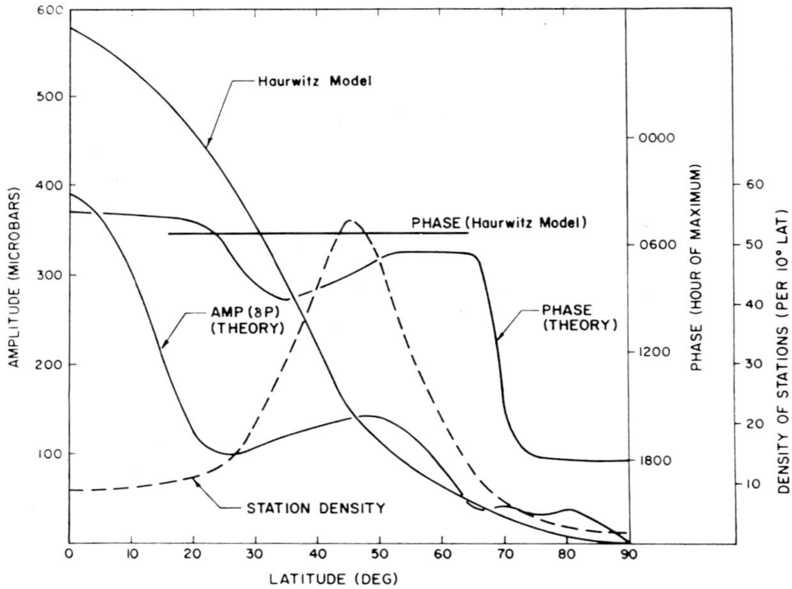

FIG. 3. 6. Amplitude and phase of the solar diurnal tide as was calculated by Lindzen on the classical tidal theory. The result of analysis by Haurwitz for observed data is also shown. The distribution of the density of stations is given (CHAPMAN and LINDZEN, 1970).

(see Fig. 3. 3). But $S(1, -2)$-mode generated in the middle atmosphere by heating due to ozone cannot reach the surface. As far as the diurnal tides due to water vapor near the surface are concerned, $S(1, -2)$ should be larger than $S(1, 1)$ on the ground. In summary, the observed diurnal tides on the ground should consist mainly of $S(1, -2)$ due to water vaper and $S(1, 1)$ due to ozone and water vapor. Which is more effective, water vapor or ozone, depends on the generation of $S(1, 1)$ on the surface and seems to depend on the model of the water vapor and ozone heating. MURATA (1968) has shown that $S(1, -2)$ due to water vapor and $S(1, 1)$ due to only ozone could explain well the observation on the ground, though his model assumed an ozone distribution to be concentrated between 55 and 65 km. Lindzen's model assumed a wider distribution of ozone with a smaller peak. Details of the calculation by the classical theory and comparison to observation will be presented in the next section.

The tidal phenomena are not confined near the surface but occur at any height of the atmosphere. The propagating positive tidal modes which are excited in the lower atmosphere travel upwards and eventually reach the thermosphere. On the way the propagation of the wave is affected by the steady zonal wind. The wave is also modified by the electrical and viscous forces. The wave itself would give rise to the electrical field $\boldsymbol{E}$ and current $\boldsymbol{J}$ as is discussed in the next chapter. There is also *in situ* excitation of the tidal wave in the thermosphere. KATO (1966b) has pointed out a possible thermal excitation of the diurnal first negative mode $(1, -2)$ in the lower thermosphere by the dissociation and ionization of molecular oxygen (Fig. 3. 7). He found

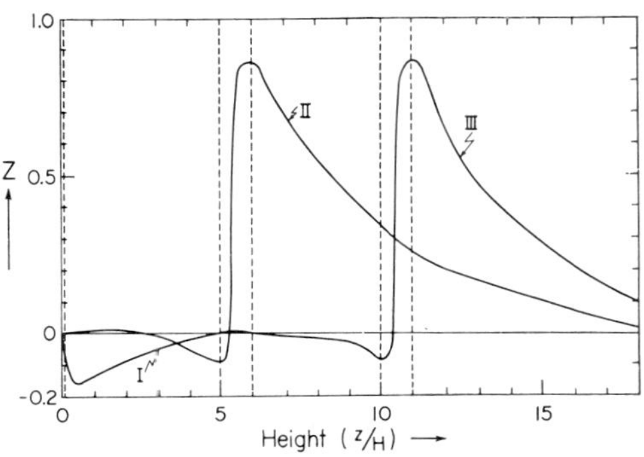

FIG. 3. 7. Vertical distribution of tidal wind the of $(1, -2)$-mode. Z is proportional to $W_{\theta,\phi}$. The vertical broken lines bound the region where energy is supplied; the lower region represents the ozone layer and the upper the lower ionosphere. The numbers of each curve I, II, III correspond to the excitation near the ground, in the ozone layer and the ionosphere E region, respectively (KATO, 1966b).

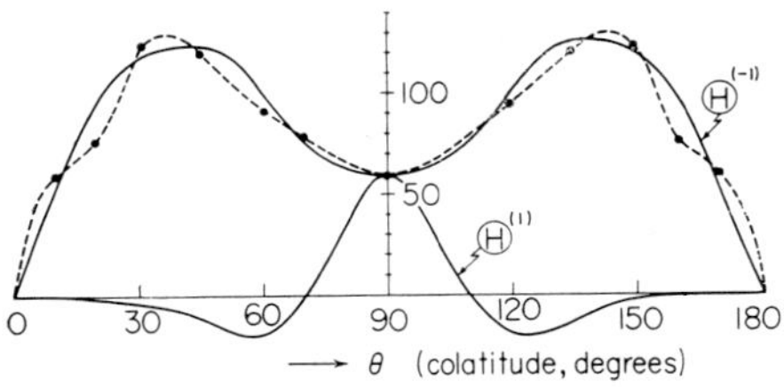

FIG. 3. 8. (p_1/ρ_0) for (1, −2) and (1, 1). Note that $(p_1/\rho_0) \propto \Theta_{\omega, m, n}$. The broken curve is (p_1/ρ_0) which was obtained from geomagnetic S_q. $\Theta^{(-1)}$ and $\Theta^{(1)}$ correspond to the present (1, −2) and (1, 1) modes, respectively (KATO, 1957a).

that the ionospheric pressure tide which had been deduced from the observed geomagnetic tide (S_q) is consistent with that of (1, −2)-mode (Fig. 3. 8.). Since the mode cannot travel vertically, his finding meant in situ excitation of the wave in the dynamo region, the lower thermosphere. This original idea has survived, with modification, various observational and theoretical trials to date (e.g. RICHMOND *et al.*, 1976; VOLLAND, 1971).

3. 2. 1 *Hydromagnetic tide*

Of various effects which are not taken into consideration in the classical theory, the hydromagnetic force would be most complicated for rigorous treatment. It requires additional electromagnetic equations along with the fluid dynamic ones including the hydromagnetic force. Valid solution has not yet been obtained except in a fairly rudimental form. A simple way is to separate the electromagnetic equations simply by neglecting the electric field which connects the electromagnetic equations to the fluid dynamic equations of motion through the Amper force ($\boldsymbol{J}\times\boldsymbol{B}$) where $\boldsymbol{J}=[\sigma](\boldsymbol{E}+\boldsymbol{W}\times\boldsymbol{B}_0)$ and $[\sigma]$ is the electrical conductivity (for details, see Section 4. 2 in Chapter 4). KATO and MATSUSHITA (1969) discussed the attenuation and reflection of a positive mode which travels from below, using this approximation. A simpler approach is to replace $(\boldsymbol{J}\times\boldsymbol{B}_0/\rho_0)$ by $(-\nu\boldsymbol{W})\mathscr{F}_0(\theta,\phi,t)$ which is usually called "the ion-drag-force"; $\mathscr{F}_0(\theta,\phi,t)$ is a function of (θ,ϕ,t) and ν the collision frequency of ions with neutrals. If we pick up in $(\boldsymbol{J}\times\boldsymbol{B}_0)/\rho_0$ only the term proportional to $\boldsymbol{W}$, we have $\tilde{\alpha}\boldsymbol{W}_\perp$ where $\tilde{\alpha}=(\sigma_1 B_0^2/\rho_0)$; $\tilde{\alpha}\boldsymbol{W}_\perp$ approximates $[\sqrt{1+\cos^2\theta}\,\nu\boldsymbol{W}_\perp]$. The negelcted terms in $(\boldsymbol{J}\times\boldsymbol{B})/\rho_0$ in this approach are $\tilde{\beta}(\boldsymbol{W}\times\boldsymbol{B}_0)/B_0$ where $\tilde{\beta}=(\sigma_2 B_0^2/\rho_0)$ and these are important in the upper dynamo region (Fig. 3. 9). Thus, the ion-drag-force approach to neglect σ_2 is too primitive.

Looking at equations (3. 1) and (3. 2) where $-(\tilde{\alpha}W_\theta-2\cos\theta\,\tilde{\beta}W_\phi)$ and $(-2\cos\theta\tilde{\beta}W_\theta-\tilde{\alpha}W_\phi)$ are now included in the right-hand side, respectively, we notice that $W_{\theta,\phi}$ are given also by (3. 8) where ω is replaced by $(\omega+i\tilde{\alpha})$, and Ω by $(\Omega-\tilde{\beta})$, f being modified accordingly. If $\tilde{\alpha}$ and $\tilde{\beta}$ were constant, the

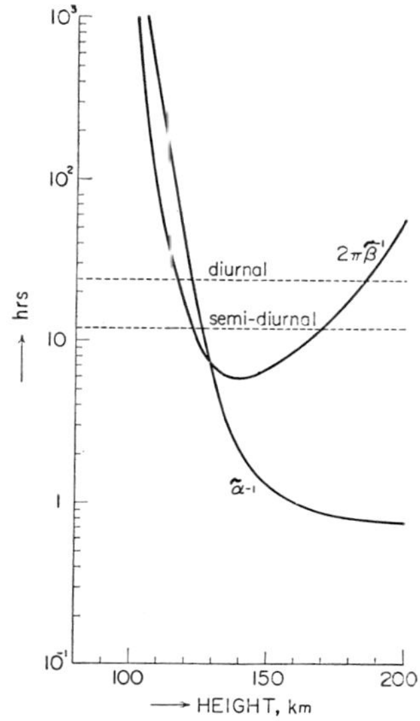

FIG. 3. 9. $\tilde{\alpha}$ and $\tilde{\beta}$ versus height (KATO and MATSUSHITA, 1969). $\tilde{\alpha}=\sigma_1 B_0{}^2/\rho_0$, $\tilde{\beta}=\sigma_2 B_0{}^2/\rho_0$.

basic equation as (3. 12) remains separable. We can get solution for the hydromagnetic tide only by these replacements in the Laplace tidal equation (3. 14). VOLLAND (1974) started with $\tilde{\beta}=0$ and $\tilde{\alpha}=\nu=$const. As a limiting case, $\nu\gg\omega$ was considered. Then, (3. 10) and (3. 14) show that by $f\rightarrow i\nu/(2\Omega)$ and $(f^2-\cos^2\theta)\rightarrow f^2$, (3. 14) tends to be an associated Legendre equation as (3. 44) and (3. 47) for

$$\frac{a^2\nu^2}{gh}=-n'(n'+1) \tag{3. 56}$$

where n' is positive integers as in (3. 44′) and (3. 48); the minus sign in (3. 56) comes from the relation $f^2\rightarrow(-\nu^2/4\Omega^2)<0$. And $\Theta\rightarrow P_{n\ ,m}(\theta)$, which shows that, under strong ion drag, any tidal mode tends to be GW-mode or oscillation of the first kind. But the situation becomes a little complicated by the fact that h tends to be negative by (3. 56). Physically, the negative h is understood as attenuation in a lossy medium, when the losses are characterized by ν, the ion collision. Generally, however, the above limiting case is not realistic, and any mode has a complex h and is a propagating mode with attenuation. Some important modes in the thermospheric tide are thus modified as illustrated in Fig. 3. 10. Note that the modified (1, −2)-mode which is defined as (1, −1) in Fig. 3. 10(a) (see Subsection 3. 1. 3) is less attenuated above 100 km, and the modified (2, 2)-mode is now partly evanescent (with $H_z>0$), remarkably

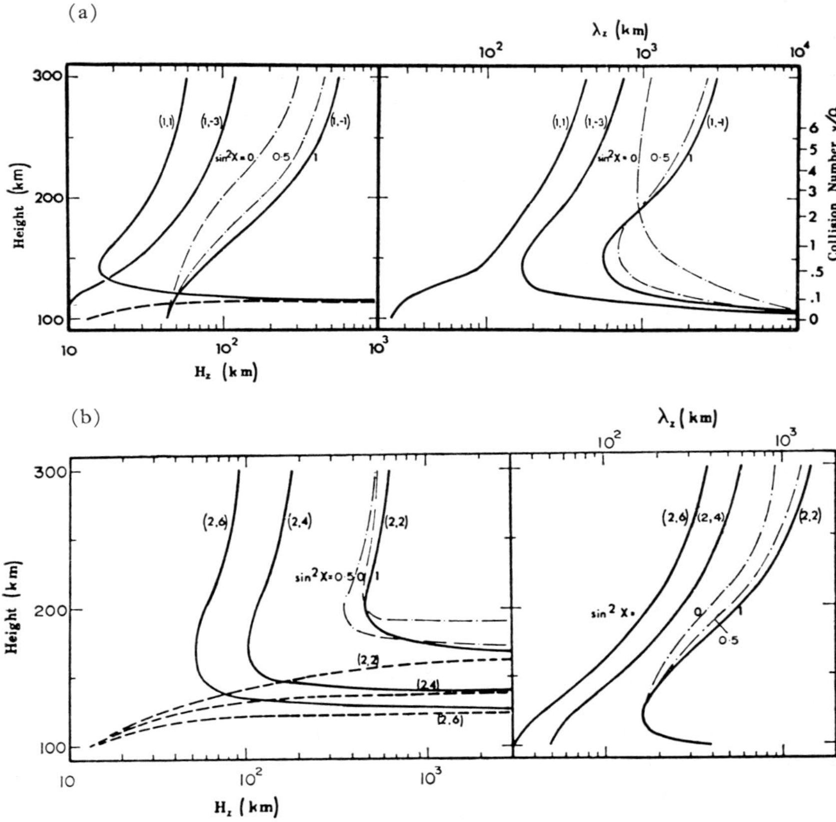

FIG. 3. 10. Vertical attenuation rate H_z. H_z is defined as $Y \propto [e^{-(z/H_z)} \exp (i\, 2\pi \lambda_z^{-1}\, z)]$ where λ_z is real, and the isothermal atmosphere is assumed. The dashed line corresponds to negative value of H_z. The chain curves are related to the (1, −1)-mode in (a), and (2, 2)-mode in (b); the dip angle (χ in the diagram) equals 0°, 45° (chain), and 90° (solid) (VOLLAND, 1974). The mode designation is different; see Subsection 3. 1. 3.

above 165 km. The Laplace tidal equation for a complex effective frequency $(\omega - i\alpha)$ needs certain care in the treatment, because no completeness of the eigenfunctions exists any more, implying that the modal analysis may lose its significance.

A self-consistent approach has been attempted by VOLLAND and GRELLMANN (1978) by introducing, in addition to constant $\tilde{\alpha}$ and $\tilde{\beta}$ as above, the *height-averaged* electric current. They discussed diurnal and semidiurnal tides in various situations which are distinguished by δ as

$$\Omega\delta = \tilde{\alpha}+\frac{\tilde{\beta}^2}{\tilde{\alpha}} = \frac{\sigma_3 B_0{}^2}{\rho_0}. \tag{3.57}$$

$\delta=0$ corresponds to the lower neutral atmosphere, $\delta=1$ the dynamo region and $\delta=10$ the F region. An important finding in their work was that the electric field had an important influence on the structure of tide. The influence is stronger on the classical positive modes than on the negative ones. A negative mode is also produced for the semidiurnal tide, even though no negative mode exists in the classical theory as is discussed in the preceding Subsection. This is because $(\boldsymbol{J}\times\boldsymbol{B}_0)$, especially its part proportional to σ_2, behaves similarly as the Coriolis force which causes the occurrence of negative modes in the classical theory. Table 3.2 shows the properties of the novel tidal mode. Of course, the theory is still fairly incomplete. In addition to the above mentioned problem of significance of modal analysis, an averaged current was adopted throughout the paper. However, $\mathrm{div_H}\langle\boldsymbol{J}\rangle=0$ is valid only for $J_z=0$ at $z=z_{1,2}$, where $\langle\ \rangle$ denotes average, $\mathrm{div_H}$ the divergence in the horizontal plane and $z_{1,2}$ the lower and upper boundary. Note that the *approximation* $J_z=0$ in $z_1<z<z_2$ cannot lead to $\mathrm{div_H}\boldsymbol{J}=0$, because $\partial J_\theta/\partial\theta$, $\partial J_\phi/\partial\phi$ can be of similar magnitude as $\partial J_z/\partial z$. Thus, in $\mathrm{div_H}\langle\boldsymbol{J}\rangle=0$ the average must be done for the *entire alti-*

TABLE 3.2

VERTICAL WAVELENGTH λ_z, ATTENUATION SCALE HEIGHT H_z, FOR DIURNAL AND SEMIDIURNAL TIDAL WAVES FOR $\delta=0$, 1, 10 (VOLLAND and GRELLMANN, 1978)

The parameters used are as follows: for $\delta=0$, $H=6$ km, for $\delta=1$, $H=11.3$ km, $\rho_0=3.64\times10^{-8}$ kg m^{-3}, $\mu_0=5\times10^{-5}$ kg m^{-1} sec^{-1}; for $\delta=10$, $H=45$ km, $\rho_0=5\times10^{-11}$ kg m^{-3}, $\mu_0=1\times10^{-4}$ kg m^{-1} sec^{-1}; μ_0 is viscosity coefficient, $\mu_v=\mu_0\rho_0{}^{-1}$. H_z, λ_z, and modes are defined in the caption of Fig. 3.10. Unlike in the text, $\tilde{\alpha}+i\tilde{\beta}=2N$ for complex h's in the table (see (3.17′)) defining the complex refractive index.

	$\delta=0$				$\delta=1$				$\delta=10$			
Mode	$\tilde{\alpha}$	$\tilde{\beta}$	λ_z, km	H_z, km	$\tilde{\alpha}$	$\tilde{\beta}$	λ_z, km	H_z, km	$\tilde{\alpha}$	$\tilde{\beta}$	λ_z, km	H_z, km
(1, −1)	0	1	∞	∞	0.19	0.45	730	−41	8.2	10.7	69	9
(1, −2)	0	1.9	∞	42	0.05	1.3	2810	69	0.29	0.74	1940	−341
(1, −3)	0	2.3	∞	9	0.50	2.1	280	21	0.51	0.99	1110	−6140
(1, −4)	0	2.4	∞	9	0.55	2.2	260	19	0.85	1.3	670	350
(1, 1)	3.2	0	23	−12	4.9	0.57	29	−52	12.8	6.1	44	18
(2, 2)	0.11	0	660	−12	1.1	0.07	130	−25	4.5	2.6	126	60
(2, 3)	1.1	0	69	−12	2.2	0.21	66	−28	13	11	45	9
(2, 4)	1.7	0	45	−12	3.3	0.50	43	−46	22	20	25	5
(2, 5)	2.2	0	35	−12	4.4	0.99	32	−1510	38	36	15	3
(2, −2)					0.88	2.2	161	19	0.53	0.97	1060	−3350
(2, −3)					0.88	2.2	161	19	0.41	0.86	1370	660
(2, −4)					1.9	2.6	76	15	1.0	1.2	550	370
(2, −5)					1.9	2.6	76	15	0.84	1.2	680	580

tude range over which the electric current flows. So, separate discussion for the E and F region which was done by Volland and Grellmann is not convincing. In particular, the non-static electric field, which is generally given by a vector potential, may vary widely between z_1 and z_2. It is likely that the hydromagnetic waves, which contain non-electrostatic field, cannot be neglected in the F region where the magnetic field lines tend to be fairly frozen. TAFFE's work (1969) was similar in appearance to the above work by Volland and Grellmann. But the latters are much advanced, because Volland and Grellmann introduced the boundary condition $J_z=0$ at $z=z_{1,2}$, the condition which is essential to produce a large scale horizontal electrostatic field. Taffe discussed only each possible mode in open space without consideration on the boundary.

3. 2. 2 *Numerical solution*

A full consideration of the effects of height-dependent dissipation on thermospheric tide is possible only in a numerical approach. Though each elementary process may often be obscured due to complex numerical calculations, the result of this approach can be more realistic than the modal analysis. The works along the line are those e.g. by LINDZEN (1971), VOLLAND and MAYR (1972), LINDZEN and HONG (1974), HARRIS and MAYR (1975), HONG and LINDZEN (1976), FORBES and GARRETT (1976) and FORBES and GARRETT (1978). Of these works, those for diurnal tide at equinox and solstice by Forbes and Garrett will be introduced below. Though linear, the model is fairly sophisticated, taking into account molecular viscosity and conductivity, the Coriolis force and anisotropic ion drag $\nu_{\theta,\phi}$ where $\nu_\theta=\nu_0(z,\theta)\sin^2 I$ and $\nu_\phi=\nu_0(z,\theta)$, correct only for $\tilde{\alpha}$ and not for $\tilde{\beta}$; the effect of eddy viscosity and Newtonian cooling were found negligibly small. The heat source is due to absorption upon dissociation and ionization of molecular oxygen in UV (ultraviolet) and XUV (extreme ultraviolet) range of solar radiation. Figure 3. 11 illustrates heating rate distribution, its latitudinal efficiency for the thermal excitation of the diurnal tide, and $T_0(z)$. The additional terms to the right-hand side of (3. 1) and (3. 2) are, respectively,

$$-\nu_\theta W_\theta+\frac{F_\theta}{\rho_0}, \quad -\nu_\phi W_\phi+\frac{F_\phi}{\rho_0} \tag{3.58}$$

where $F_{\theta,\phi}$ are the molecular viscosity term i.e. the divergence of momentum flux due to molecular diffusion

$$F_{\theta,\phi}=\frac{\mu_0(s)}{H^2}\left[\frac{d^2}{ds^2}+\left(1+\frac{1}{2T_0}\frac{dT_0}{ds}-\frac{1}{H}\frac{dH}{ds}\right)\frac{d}{ds}\right.$$
$$\left.+\frac{1}{4}\left(1+\frac{1}{T_0}\frac{dT_0}{ds}-\frac{2}{H}\frac{dH}{ds}\right)\right](e^{-s/2}W_{\theta,\phi}) \tag{3.59}$$

where μ_0 is the viscosity coefficient. The thermal conduction term is now included in the right-hand side of (3.5) as K' where

$$K' = \frac{\kappa_t}{H^2}\left[\frac{d^2}{ds^2}+\left(1-\frac{1}{H}\frac{dH}{ds}+\frac{1}{T_0}\frac{dT_0}{ds}+\frac{1}{R}\frac{dR}{ds}\right)\frac{d}{ds}\right.$$
$$+\frac{1}{2}\left\{\frac{1}{2}-\frac{1}{H}\frac{dH}{ds}+\frac{1}{T_0}\frac{dT_0}{ds}+\frac{1}{R}\frac{dR}{dz}+\frac{1}{T_0}\frac{d^2T_0}{ds^2}\right.$$
$$-\left(\frac{1}{H}\frac{dH}{ds}\right)\left(\frac{1}{T_0}\frac{dT_0}{ds}\right)-\frac{1}{2T_0^2}\left(\frac{dT_0}{ds}\right)^2$$
$$\left.\left.+\left(\frac{1}{R}\frac{dR}{ds}\right)\left(\frac{1}{T_0}\frac{dT_0}{ds}\right)\right\}\right]e^{-s/2}T_1 \qquad (3.60)$$

where $\kappa_t(s)$ is the thermal conductivity coefficient. The boundary condition is that $W_z=0$ at $z=s=0$, and $\partial W_{\theta,\phi}/\partial z=\partial T_1/\partial z=0$ at $z=z_\infty$ where z_∞ corresponds to $s=25.0$, the base of the diffusion dominating region. The mixing ratio of composing gases i.e. O and N_2 was assumed constant. Though neglecting mean zonal wind, temperature variation which depends on z and θ, and excitation in the lower atmosphere, Forbes and Garrett's result seems fairly realistic, explaining thermospheric wind and termperature tide, which vary with season, latitude and sunspot number. Figure 3. 12 shows $W_{\theta,\phi,z}$ and T_1. It is found that (1) $W_{\theta,\phi,z}$ in amplitude and phase, tends to be constant in the upper thermosphere because of increasing molecular diffusion and ion drag. This occurs at higher altitude at sunspot maximum than sunspot minimum because of increasing level of diffusion dominance with the mean exospheric temperature with sunspot; (2) horizontal structure varies with z due to height-dependent diffusion and ion drag; (3) while Q and T_1 vary, respectively, by a factor of 2 and 3 between sunspot maximum and minimum periods, no such large variation occurs in $W_{\theta,\phi,z}$. This may be attributed to the variation in heating and drag largely cancelling each other in these epochs. Besides results (1)–(3), which are in common at both equinox and solstice, there are various solstitial asymmetries dependent on sunspot number, or more exactly, solar activity: The sunspot maximum and minimum correspond to T_∞ (exospheric temperature)$=1{,}400$ and $800\,°$K, respectively. The dependency is interpreted mainly in terms of varying heat input Q and the F region electron density. Density tide of O and N_2 has also been calculated for a posteriori wind and temperature tide as in Fig. 3. 13. The calculation seems to give results consistent with those observations which are now available e.g. seasonal variation of northerly wind. Precise comparison between the theoretical result and observations by incoherent scatter radar and other facilities has yet to be done, because of lack of data due to the difficulties of making measurements in the nighttime E and F1 regions.

In the case of semidiurnal tide, the situation is improved now because of continuous daytime observation possible by incoherent scatter radar at St. Santin, Millstone Hill and Arecibo. The fundamental semidiurnal mode is $S\,(2, 2)$, which is important not only in observed pressure tide at the ground

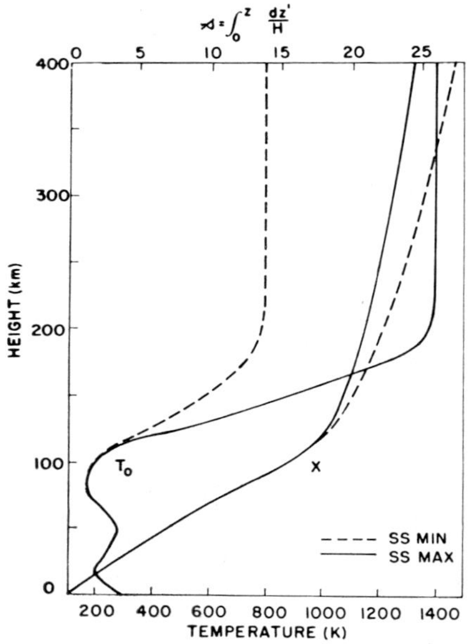

$\mathcal{A} = \int_0^z \frac{dz'}{H}$
0 5 10 15 20 25
HEIGHT (km)
400
300
200
100
0
T_0
X
SS MIN
SS MAX
200 400 600 800 1000 1200 1400
TEMPERATURE (K)

(a)

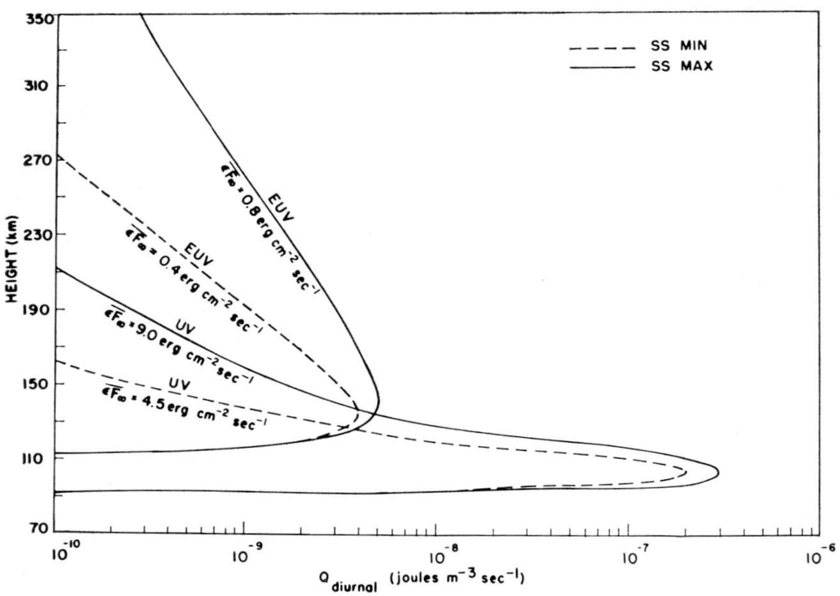

SS MIN
SS MAX
HEIGHT (km)
350
310
270
230
190
150
110
70
EUV
$\epsilon \overline{F_\infty}$ = 0.8 erg cm^{-2} sec^{-1}
EUV
$\epsilon \overline{F_\infty}$ = 0.4 erg cm^{-2} sec^{-1}
UV
$\epsilon \overline{F_\infty}$ = 9.0 erg cm^{-2} sec^{-1}
UV
$\epsilon \overline{F_\infty}$ = 4.5 erg cm^{-2} sec^{-1}
10^{-10}
10^{-9}
10^{-8}
10^{-7}
10^{-6}
$Q_{diurnal}$ (joules m^{-3} sec^{-1})

(b)

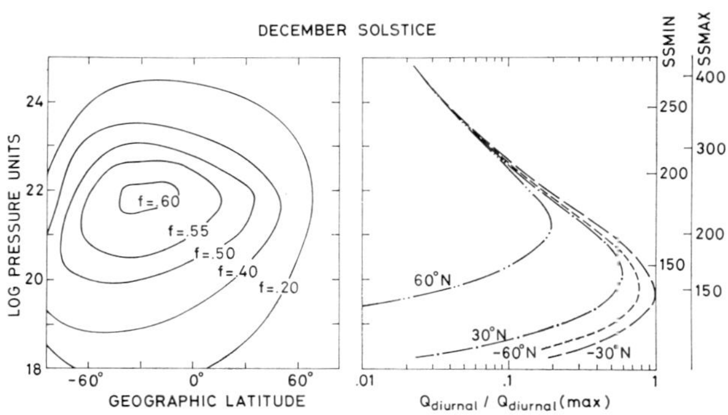

FIG. 3. 11. Atmosphere model by FORBES and GARRETT (1976, 1978). (a) T_0 and s (X in the diagram): at SSMIN (sunspot minimum) and SSMAX (sunspot maximum). The same distribution of T_0 was used both for equinox and solstice. (b) Diurnal component of EUV and UV heat input at SSMIN and SSMAX. Q_{diurnal} is for an overhead sun; generally Q_{diurnal} varies with solar zenith distance χ_s as proportional to exp $[-Ch\ (\chi_s, z)\ e^{-z}]$ where Ch is the Chapman function i.e. $Ch\ (\chi_s, z) = (\text{absorption coefficient}) \times \sec\ \chi_s \int_z^\infty n dz$. This Q_{diurnal} was used only for equinox. (c) Diurnal energy input in December solstice, Q_{diurnal}, at various latitudes. Q_{diurnal} (max) is the maximum Q_{diurnal} for $\chi_s = 0$, and f is the fraction of total heating energy going into diurnal excitation. Vertical distribution of Q_{diurnal} is similar to that in (b) except for 25% increase in the absolute magnitude. FORBES and GARRETT's figure (1978) is redrawn.

but also in observed wind and temperature tide in the lower thermospere (e.g. HARPER, 1977b). Higher order modes as $S\,(2, 4)$ also seem important in the lower themosphere especially in the dynamo region where tidal motion in the presence of geomagnetic field generates electric current (e.g. RICHMOND *et al.*, 1976). Since almost all of semidiurnal modes are propagating and can be excited in the lower atmosphere, an understanding of the transmission of a tidal wave through the lower atmosphere is essential. It is found (LINDZEN and HONG, 1974) that the transmission is seriously affected by mean zonal winds which vary with height and latitude. The simple classical theory predicted a phase reversal by 180° around 30 km for the semidiurnal tide but the reversal does not occur in reality. Also classical theory predicts that the fundamental mode $S\,(2, 2)$ should dominate throughout the atmosphere, but dominant short vertical wavelength semidiurnal tides are frequently observed, especially in winter in the mesosphere. Lindzen and Hong have given a mechanism by which these discrepancies can be explained. The zonal winds cause mode coupling to occur, transferring energy from the thermally excited $S\,(2, 2)$-mode to higher order modes. The zonal wind

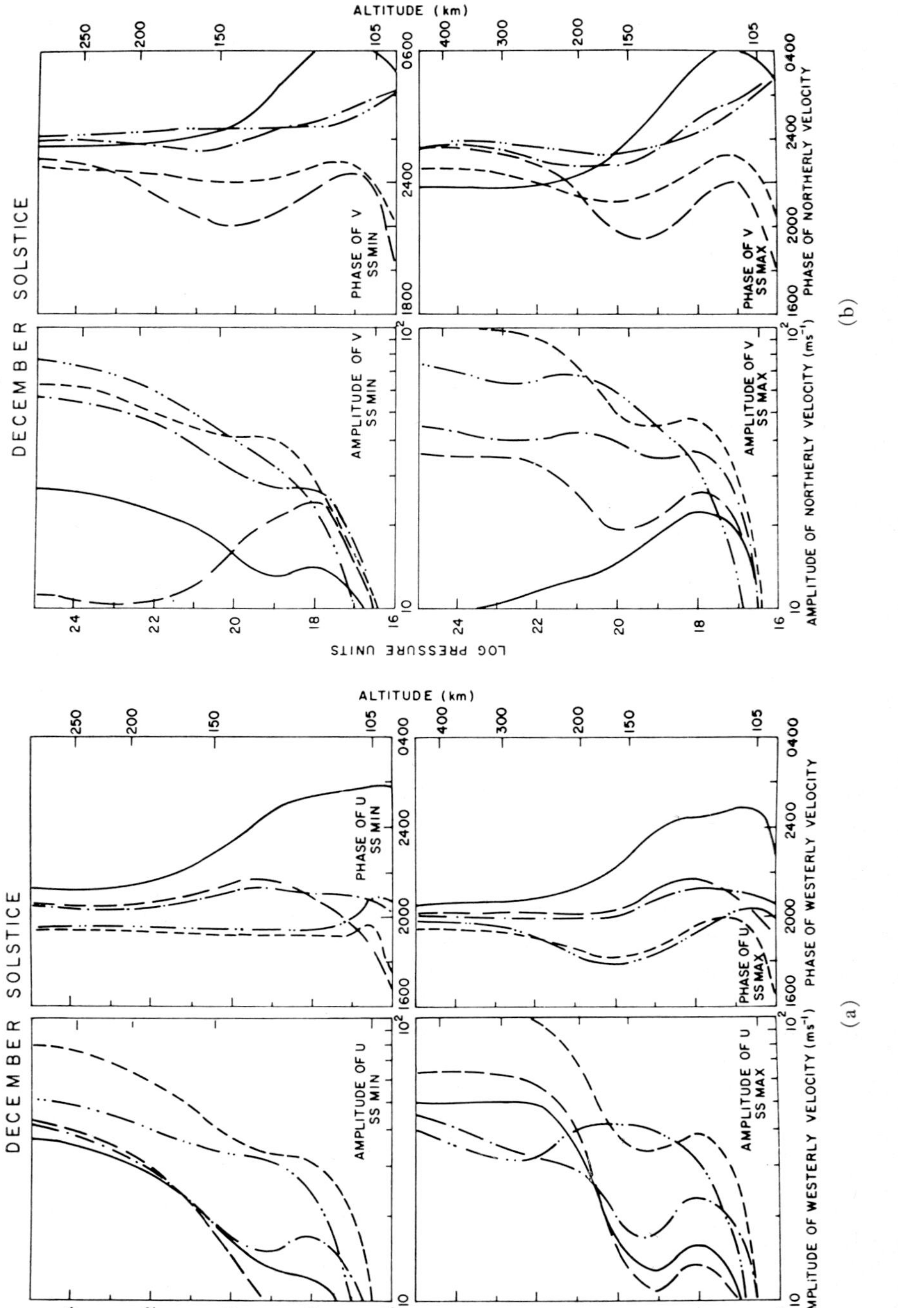
DECEMBER SOLSTICE
AMPLITUDE OF U SS MIN
PHASE OF U SS MIN
AMPLITUDE OF U SS MAX
PHASE OF U SS MAX
AMPLITUDE OF WESTERLY VELOCITY (ms⁻¹)
PHASE OF WESTERLY VELOCITY
LOG PRESSURE UNITS
ALTITUDE (km)
(a)
DECEMBER SOLSTICE
AMPLITUDE OF V SS MIN
PHASE OF V SS MIN
AMPLITUDE OF V SS MAX
PHASE OF V SS MAX
AMPLITUDE OF NORTHERLY VELOCITY (ms⁻¹)
PHASE OF NORTHERLY VELOCITY
LOG PRESSURE UNITS
ALTITUDE (km)
(b)

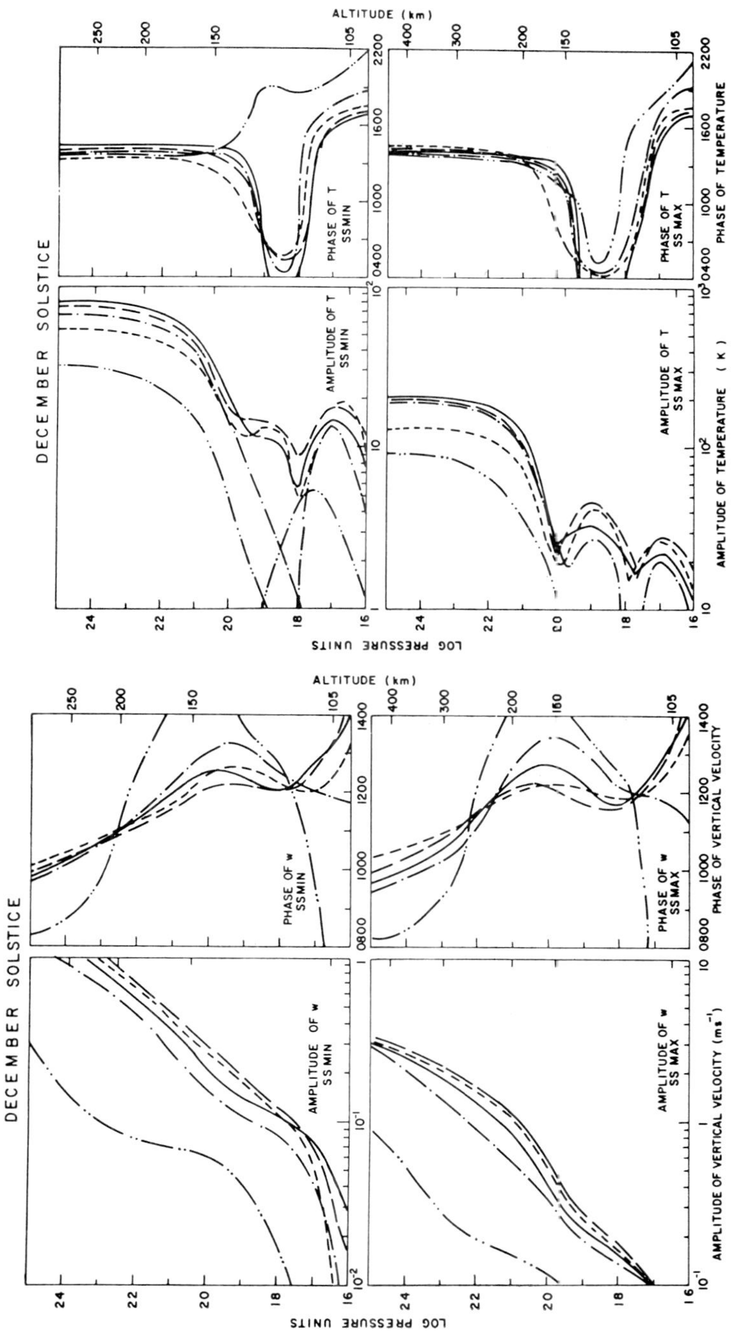
DECEMBER SOLSTICE
AMPLITUDE OF w SS MIN
PHASE OF w SSMIN
AMPLITUDE OF w SS MAX
PHASE OF w SS MAX
AMPLITUDE OF VERTICAL VELOCITY (ms⁻¹)
PHASE OF VERTICAL VELOCITY
LOG PRESSURE UNITS
ALTITUDE (km)
DECEMBER SOLSTICE
AMPLITUDE OF T SS MIN
PHASE OF T SS MIN
AMPLITUDE OF T SS MAX
PHASE OF T SS MAX
AMPLITUDE OF TEMPERATURE (K)
PHASE OF TEMPERATURE
LOG PRESSURE UNITS
ALTITUDE (km)

(c)

(d)

Fig. 3.12

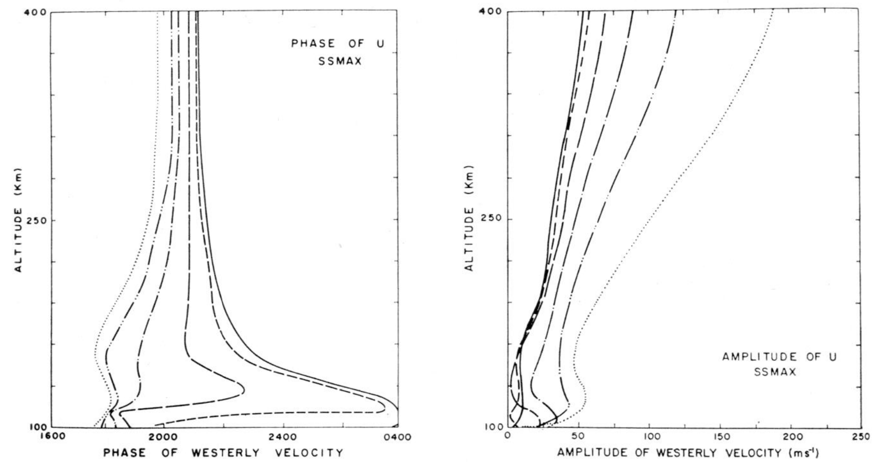
PHASE OF U
SSMAX
ALTITUDE (Km)
400
250
100
1600
2000
2400
0400
PHASE OF WESTERLY VELOCITY
AMPLITUDE OF U
SSMAX
0
50
100
150
200
250
AMPLITUDE OF WESTERLY VELOCITY (ms⁻¹)

(a′)

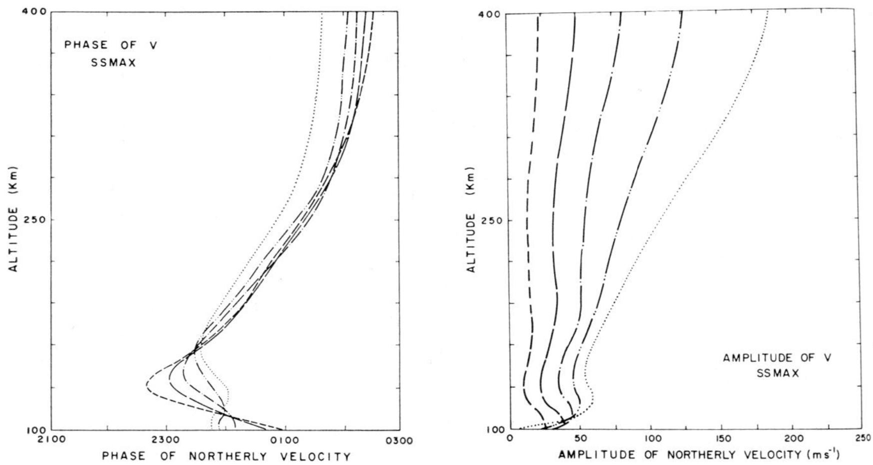
PHASE OF V
SSMAX
ALTITUDE (Km)
400
250
100
2100
2300
0100
0300
PHASE OF NORTHERLY VELOCITY
AMPLITUDE OF V
SSMAX
0
50
100
150
200
250
AMPLITUDE OF NORTHERLY VELOCITY (ms⁻¹)

(b′)

Fig. 3.12

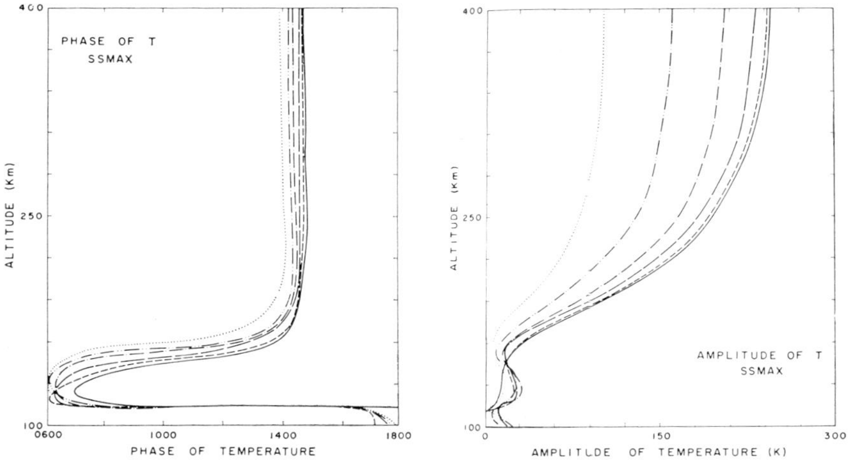

(d′)

FIG. 3. 12. $W_{\theta,\phi}$ and T_1 in equinox and December solstice and in SSMAX and SSMIN (FORBES and GARRETT, 1976, 1978). In these figures $W_\theta \rightarrow V$, $W_\phi \rightarrow U$, $W_z \rightarrow W$, $T_1 \rightarrow T$ according to the symbols in the original figures. In December solstice, the southern hemisphere phase is shifted by 12 hr for convenience of illustration. (a) U (solstice), (b) V (solstice), (c) W (solstice), (d) T (solstice), (a′) U (equinox), (b′) V (eqinox), (d′) T (equinox). Lines (— —), (- - -), (—•—) and (—••—) as in Fig. 3. 11c (right) denote $-30°$, $-60°$, $30°$ and $60°$ of latitude, respectively, in (a)-(c), and $30°$, $15°$, $45°$, $60°$, respectively, in (a′)-(d′); (——) and (······) are at $0°$ and $75°$, respectively. The phase is given by the time of maximum.

W_0 (Figs. 1. 3(b) and 3. 15) which is geostrophic as in (1. 34) and determined by the thermal wind relation (1. 35) is given, consistent with the mesospheric circulation model by LEOVY (1964) for $T_0(\theta, z)$ as

$$T_0 = A_0 + A_2 \cos^2\theta + A_3 \cos^3\theta + A_4 \cos^4\theta \tag{3.61}$$

where $A_{0,2,3,4}$ are function of z: If the mean wind exists, the dynamic equation is generally given by (1. 1) as

$$\frac{\partial \boldsymbol{W}}{\partial t} + (W \cdot \nabla)\boldsymbol{W} - 2(\boldsymbol{W} \times \boldsymbol{\Omega}) = -\frac{1}{\rho} \nabla p + \boldsymbol{g} - \nu_v \boldsymbol{W} \tag{3.62}$$

where ν_v here is the effective kinematic viscosity (Rayleigh friction). The first order perturbation method is applied as $W = W_0(\theta, z) - W'(\theta, \phi, z, t)$ where $\boldsymbol{W}_0$ is given by (1. 35). Liniarization of (3. 62) in spherical coordinates (θ, ϕ, z) is

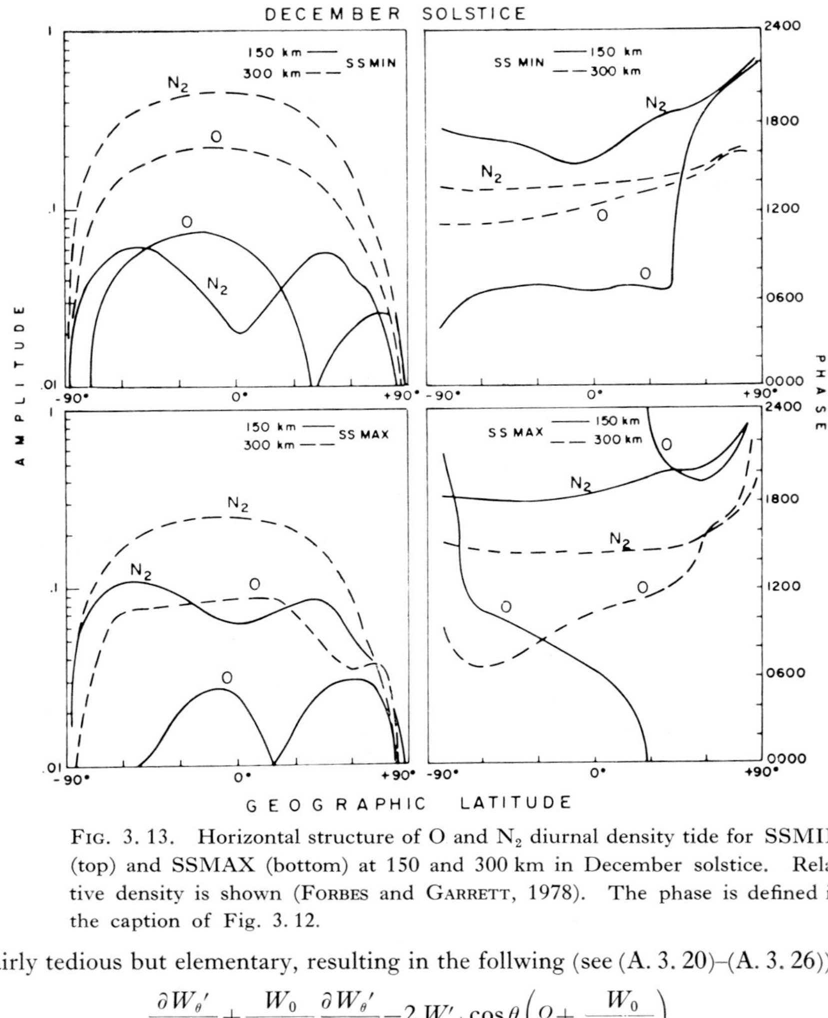

FIG. 3. 13. Horizontal structure of O and N_2 diurnal density tide for SSMIN (top) and SSMAX (bottom) at 150 and 300 km in December solstice. Relative density is shown (FORBES and GARRETT, 1978). The phase is defined in the caption of Fig. 3. 12.

fairly tedious but elementary, resulting in the follwing (see (A. 3. 20)–(A. 3. 26)):

$$\frac{\partial W_\theta'}{\partial t}+\frac{W_0}{a\sin\theta}\frac{\partial W_\theta'}{\partial\phi}-2\,W'_\phi\cos\theta\left(\Omega+\frac{W_0}{a\sin\theta}\right)$$

$$=-\frac{1}{\rho_0}\frac{\partial p_1}{a\partial\theta}+\frac{\rho_1}{{\rho_0}^2}\frac{\partial p_0}{a\partial\theta}-\frac{1}{a}\frac{\partial\Lambda}{\partial\theta}-\nu_v W_\theta' \qquad (3.63)$$

$$\frac{\partial W_\phi'}{\partial t}+\frac{W_0}{a\sin\theta}\frac{\partial W_\phi'}{\partial\phi}+\frac{W_\theta'}{a}\frac{\partial W_0}{\partial\theta}+W_z'\frac{\partial W_0}{\partial z}$$

$$+W_\theta'\cos\theta\left(2\Omega+\frac{W_0}{a\sin\theta}\right)=-\frac{1}{\rho_0}\frac{1}{a\sin\theta}\frac{\partial p_1}{\partial\phi}$$

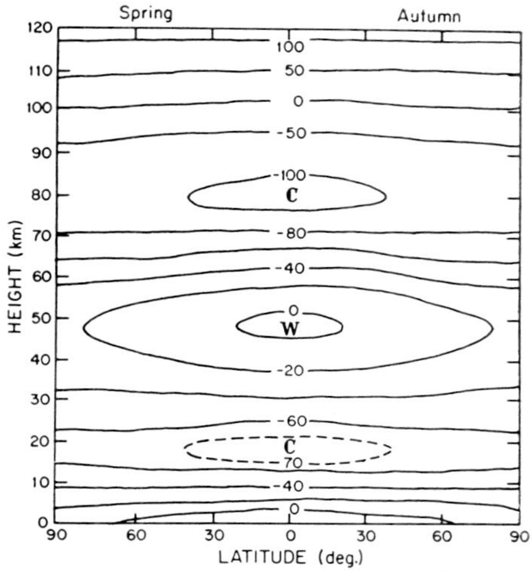

FIG. 3. 14. Meridional cross section of temperature (°) for equinox model atmosphere (LINDZEN and HONG, 1974). See Fig. 1. 3(a) for solstice.

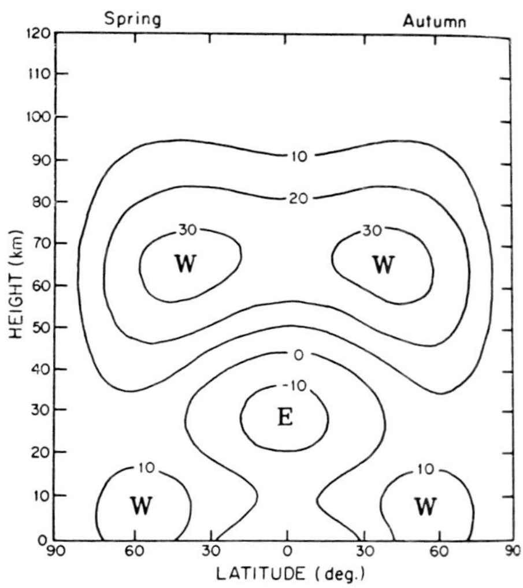

FIG. 3. 15. Meridional cross section of zonal wind (m sec^{-1}) in equinox for model atmosphere (LINDZEN and HONG, 1974). See Fig. 1. 3(b) for solstice.

$$-\frac{1}{a\sin\theta}\frac{\partial \Lambda}{\partial \phi}-\nu_v W_\phi{}' \tag{3.64}$$

$$\frac{\partial p_1}{\partial z} = -\rho_1 g - \rho_0 \frac{\partial \Lambda}{\partial z} \tag{3.65}$$

where $(W_0/a\sin\theta)\,\partial W'_z/\partial\phi$ is neglected in the vertical direction, presumably it being small. As to the equation of continuity, no complication is introduced by W_0 except for one added term as $(W_0/a\sin\theta)\,\partial\rho_1/\partial\phi$ on the left-hand side of (3. 4). The same is true in χ in the left-hand side of the energy equation in (3. 5), though a Newtonian cooling term proportional to T_1 is also included in the right-hand side. The set of equations thus obtained has been numerically solved by finite difference method to get a tidal solution as

$$\frac{p_{1,\omega}}{\rho_0} \propto \left[\frac{p_{1,\omega}(\theta, z)}{\rho_0(\theta, z)}\right] e^{i(\omega t+m\phi)}$$

as by (3. 13) where χ_ω is the dependent variable instead of $(p_{1,\omega}/\rho_0)$. The numerical integration methods adopted is similar to that of LINDZEN and KUO (1969). The boundary conditions are similar to those which are mentioned above in relation to the works by FORBES and GARRETT (1976, 1978) i.e. $W_z=0$ at $z=0$ and $(\partial W_{\theta,\phi}/\partial z)=0$ at $z=300$ km. Generally, a wave which is originally excited due to thermal forcing itself excites other waves due to the mean zonal winds convecting temperature and density perturbations, as well as the velocity perturbations convecting background temperature and density variations. These new perturbations propagate vertically as a new set of excited tidal modes.

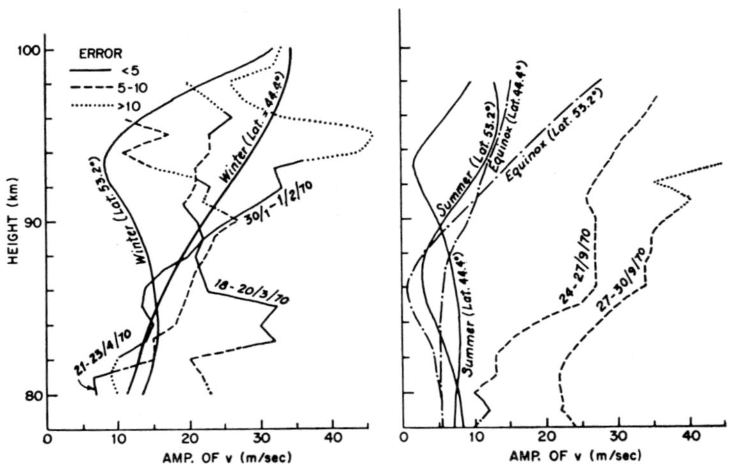

FIG. 3. 16. Comparison between theory and observation (meteor radar at Garchy lat. 40 °N) (LINDZEN and HONG, 1974). V is the eastward velocity.

Thus, the fundamental mode $S\,(2,\,2)$ changes to contain higher order modes during its transmission through the upper atmosphere. The direct comparison between the calculated result and the observed radio meteor wind (Fig. 3. 16) is not satisfying, except certain basic points. Discrepancies would come from various simplifications in the theoretical model. Day-to-day variability of the mean zonal wind and other parameters should be averaged out over fairly long period before comparison. Or, more realistic mean zonal wind and other parameters should be used in the theoretical calculation. The result by Lindzen and Hong, in principle, is applicable up to 115 km, beyond which the assumed Rayleigh friction coefficient becomes unsuitable. Nevertheless, the overall success of the present numerical approach by Lindzen and Hong is encouraging us to improve the models in future for better agreement with observation. The lunar tide, which is not the subject in this book, seems to be far better modeled by introducing the present $\boldsymbol{W}_0$. It tends to improve the unrealistic high sensitivity of the calculated tide, at sea level, to a small temperature variation with height in the conventional models without $\boldsymbol{W}_0$.

3.3 Tidal Observation

There are various methods of observation of atmospheric tides. Some of them are common to other atmospheric waves such as AGW's, planetary waves and synoptic scale waves. The last two waves are important to tropospheric disturbances in meteorology.

The outstanding characteristic of tides among atmospheric waves is its fixed period, a day and its submultiples. Since the present concern is on solar tides, the day is solar. This fixed frequency makes it possible to remove noise by applying statistical methods, usually taking means or differences (e.g. Justus and Woodrum, 1973). Thus, even if the tidal components are small, as is usually the case below the mesosphere, the tidal "signal" can be extracted from the "noisy" record. A difficulty is that we need data from at least a full period for unambiguous detection, one day for the diurnal tide and half a day for the semidiurnal tide. Some observations are possible only in nighttime and others only in daytime. Great care must be taken in the interpretation of observation extending over a fraction of the tidal period, especially as a combination of diurnal and semidiurnal tides usually dominate in the observations (e.g. Hines, 1966).

On the earth's surface, barometers record the pressure tide which is, at low latitudes, of the order of a few millimeters of mercury i.e. a few millibars where one bar is 10^5 N m^{-2}, 750 mm of mercury (Fig. 3. 17). Magnetometers also record the geomagnetic tides which are occurring in the ionosphere (Section 5. 1). The amplitude is of the order of a few gamma; one gamma (γ) is 10^{-9} Wbm^{-2} and the geomagnetic field is about $3\times10^4\,\gamma$ in midlatitudes.

Meteorological balloons, carrying sondes, measure wind velocity above the earth's surface up to 30 km. The frequency of the measurement per

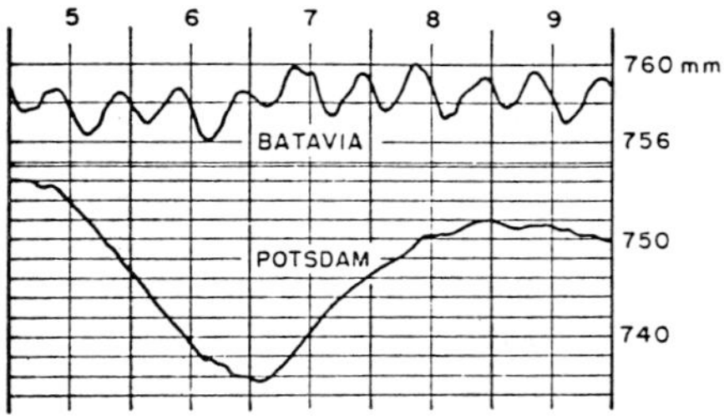

FIG. 3. 17. Barometric variations at Batavia (6 °S) and Potsdam (52 °N). The pressure tide is clear at Batavia, whilst some meteorological disturbance obscures the variation at Potsdam (WILKES, 1949).

day is very limited (usually twice, at 000UT and 1200UT), but the accumulation of data for long periods and at many locations makes it possible to analyse tidal components. The wind velocity is of the order of 0.1 m sec^{-1} near the surface and a few times that value at the 10-mb level (about 30 km) (e.g. WALLACE and TADD, 1974). Beyond 30 to 60 km, meteorological rockets are available for observation. The methods consist of grenades and falling sphere observation. The former is based on the analysis of ground measurements of sound waves produced by the grenade explosion at high altitude to determine wind velocity and temperature. For falling sphere method, the motion of the sphere, which moves horizontally with the wind, is tracked by ground radar (NASTROM and BELMONT, 1975). Though the measurement is done fairly infrequently, some data are available for tidal velocities, because the tidal amplitudes are of the order of 1 m sec^{-1} near 30 km to 10 m sec^{-1} at 60 km. Almost no observations have been made between 60–80 km on a regular basis. Above these levels the atmosphere becomes weakly ionized, and various radio techniques have been developed to study winds. Chemicals such as sodium and TMA (trimethyl aluminium) released from rockets form luminous clouds in the upper atmosphere. Taking photos frequently one can trace the motion of the clouds moving with local winds. This observation is limited in time, i.e. in twilights for sodium and in nighttime for TMA. The height is limited below 150 km above which the clouds diffuse very rapidly. The lower limit is not clear but usually 90 km. Tidal velocity reaches more than a few to several ten m sec^{-1} in the thermosphere.

The meteor radar technique is very suitable for continuous observation of wind velocity. Meteor radars detect the motion of targets by the Doppler shifts, upon reflection, in frequency of radio signals transmitted by the radar. The distance to the targets is determined from the time needed for the radio signals to make the round trip. By measuring the direction of the radio echo we can find height and azimuth of the target. The targets are meteor trails which are always present (several times per hour in the evening and a few hundreds per hour near dawns (Aso *et al.*, 1979)). Meteor trails are detected from heights between 80 and 100 km, and the velocity observed is that along the radar line of sight. Statistical methods can reduce this velocity to horizontal velocities for waves with periods much longer than the mean time of arrival between meteors.

Continuous observation in space is important in addition to continuous observation in time. In this respect, the powerful radars called incoherent scatter radars have contributed much by measuring vertical profiles of tides in electron density, temperature and velocity. Some preliminaries would be helpful for readers to understand the results by these radars. A complete review of the incoherent scatter technique has been presented by Evans (1969).

3. 3. 1 *Principles of wind observation by incoherent scatter radars*

There are five facilities of this kind now in operation at Chatanika (64.9°N, 147.7°W), Millstone Hill (42.6°N, 71.5°W), St. Santin (44.6°N, 2.2°E), Arecibo (18.3°N, 66.5°W) and Jicamarca (11.9°S, 76.0°W). New facilities of this kind have recently been built (SOUSY in Germany) or are now under construction (EISCAT in Scandinavia). The original incoherent scatter radars were constructed at the beginning of the 1960's in order to observe the ionosphere, where free electrons, in thermal motion under the influence of ions, scatter incoherently incident radio waves. The incoherently scattered echoes, which are extremely weak, contain information relative to the ionospheric parameters. Besides simple information about the electron density, which is proportional to the echo intensity, the echo spectrum, which spreads widely because of moving electrons, has been found to tell us much about the state of the ionosphere, e.g. the ion and electron temperature, the composition, the drift velocity etc. Scattering occurs from all heights as the radar pulse passes through the ionosphere. The width of the transmitted pulse determines the height resolution for a monostatic radar. The St. Santin radar uses continuous waves which are then detected at other stations than the transmitting one, i.e. a multistatic radar.

The ion drift along $\boldsymbol{B}_0$ is a combination of $\boldsymbol{W}$ and ambi-polar diffusion, as in (4. 1) and (4. 21). Thus, the measured ion velocity gives the component of $\boldsymbol{W}$ in the geomagnetic meridian. Since the geomagnetic declination at Arecibo, Millstone Hill and St. Santin is (at most) 10° and $|W_\lambda| \ll |W_\theta|$, the component of $\boldsymbol{W}$ in the geomagnetic meridian equals, within an error of 10% at most, W_θ, the north-south velocity component. The diffusion velocity is calculated by using the measured $T_{\mathrm{e,i}}$ and n_{e}. The diffusion can be neglected below 200 km. The ion velocity orthogonal to $\boldsymbol{B}_0$ in the F region is given by (4. 1) which shows $\boldsymbol{V}_{\mathrm{i}\perp} \rightarrow [(\boldsymbol{E}\times\boldsymbol{B}_0)/B_0{}^2+(\nu_{\mathrm{i}}/\Omega_{\mathrm{i}})(\boldsymbol{W}\times\boldsymbol{B}_0)/B_0]$ for $\nu_{\mathrm{i}} \ll \Omega_{\mathrm{i}}$, suggesting that the electric field, almost orthogonal to $\boldsymbol{B}_0$, can be deduced from the measured $\boldsymbol{V}_\mathrm{i}$ as $\boldsymbol{E} \simeq -(\boldsymbol{V}_\mathrm{i} \times \boldsymbol{B}_0)$ with an error proportional to $(\nu_\mathrm{i}/\Omega_\mathrm{i})$, which is very small. Once $\boldsymbol{E}$ is found in the F region, we can obtain $\boldsymbol{W}_\perp$ from the measured $\boldsymbol{V}_\mathrm{i}$ in the lower region and the measured $\boldsymbol{E}$ in the F region, as is seen by (4. 1). Considering errors met in the measurement due to decreasing $(\nu_\mathrm{i}/\Omega_\mathrm{i})$ with height, the method is applicable up to only 130 km (HARPER *et al.*, 1976). Since $|W_z| \ll |W_\phi|$, the deduced velocity is almost equal to W_ϕ.

Further, it is now possible to know the three-dimensional thermospheric wind field by integrating the equation of motion for $W_{\theta,\phi}$, once given the data of incoherent scatter radar observations (e.g. ANTONIADIS, 1976). For horizontal wind, by (1. 1)

$$\frac{\partial \boldsymbol{W}_{\theta,\phi}}{\partial t} + W_\phi \frac{\partial \boldsymbol{W}_{\theta,\phi}}{\partial y} = -\frac{1}{\rho_0} \nabla_\mathrm{H} p - 2(\boldsymbol{\Omega} \times \boldsymbol{W}_{\theta,\phi}) + \frac{\mu_0}{\rho_0} \frac{\mathrm{d}^2 \boldsymbol{W}_{\theta,\phi}}{\mathrm{d}z^2} + \boldsymbol{g} + (\boldsymbol{J} \times \boldsymbol{B}_0/\rho_0) \qquad (3.66)$$

where the advection term, the nonlinear term on the left-hand side, contains advection only for W_ϕ because of its effectiveness; ∇_H denotes the horizontal gradient. The viscosity coefficient μ_0 is given in terms of T as $\mu_0 = 4.5 \times 10^{-5}$ $(T/1{,}000)^{0.71}$ kg m^{-1} sec^{-1}. T is given as a function of T_∞ and z, equivalently with $\mathscr{Z}$, as in (1.19) and (1. 20). Applying (1.10) to each species of the neutral gas component in diffusive equilibrium, we obtain the density of each species, N_j, and the pressure, p_j. Since $p = \sum_j p_j = T \sum_j R_j N_j$ where R_j is the gas constant, it turns out that p is a function of x and y only through T_∞. We thus find $\nabla_\mathrm{H} p = \partial p / \partial T \, \nabla_\mathrm{H} T_\infty$ in which $(\partial T_\infty / \partial y)$, the zonal exospheric temperature gradient, is obtained from observed temperature which is assumed to vary only with local time; T_∞ is approximately equal to T at $z = 300$ km. The latitudinal temperature gradient, $(\partial T_\infty / \partial x)$, however, is determined so that the resulting wind produces a vertical ion drift which is consistent with direct observation. The electric force $(\boldsymbol{J} \times \boldsymbol{B}_0)$ in (3. 66) is given by a model. Note that T_0 is taken to be a constant 355° K for $z_0 = 120$ km. The boundary conditions are that $W_{\theta,\phi}$ vanishes at $z = z_0$, and $(\mathrm{d}W_{\theta,\phi}/\mathrm{d}z)$ vanishes at $z = 600$ km. The overall approach is very interesting for obtaining $\boldsymbol{W}$ in the F region, but is improper for finding the E-region tide which propagates from below, because of lower boundary condition that $\boldsymbol{W} = 0$ at $z = 120$ km. The vertical velocity after obtaining $\mathbf{W}_{\theta,\phi}$ is calculated by div $\boldsymbol{W} = 0$ which is a good approximation for tides (ROBLE *et al.*, 1977).

It has been recently discovered that large incoherent scatter type radars can measure wind motion in the lower atmosphere as well. The method of scattering from the lower atmosphere is not yet well understood. Scattering in the lower mesosphere apparently occurs from electron density irregularities or sharp gradients which occur in very narrow layers (RASTOGI and BOWHILL, 1976; FUKAO *et al.*, 1979). In the troposphere and stratosphere strong scattering occurs apparently from both very stable layers (RÖTTGER and LUI, 1978; GAGE and GREEN, 1978) and from turbulently enhanced density and temperature irregularities (WOODMAN and GULLEN, 1974). In all cases, the irregularities are moving with the background wind.

The outstanding characteristic of this radar observation is its high accuracy in determining ***W*** due to the very long correlation times of the echoes. It is possible to detect the Doppler velocity as to accuracies of a few cm sec^{-1} in the stratosphere and a few tens of cm sec^{-1} in the lower mesosphere. The stratospheric tide in the vertical velocity can be detected (FUKAO *et al.*, 1978). Future radar measurements of this type will be of paramount importance in directly detecting steady meridional general circulation in the middle atmosphere. One drawback is the small number of facilities that now exist, which is inadequate to properly study the general circulation.

3. 3. 2 *Diurnal tide*

The diurnal tide is excited by radiative heating due to water vapor near the ground and also due to ozone in the stratosphere and mesosphere (Fig. 3. 18). Although qualitative agreement between theory and observation about the tide near the ground has been obtained, the details are not yet in agreement. The main reason could be local excitation, depending on topography, which is disregarded in the tidal theory to date. Above the surface, no continuous data of a global nature is available. But, using long records of 2- and 4- daily rawinsonde data of wind at various stations, WALLACE and TADD (1974) have determined some features of the tides from the surface up to 30 km. They calculated the difference of data which was obtained 12 hr apart to remove shorter period tides and longer period incoherent waves. Their conclusions are as follows:

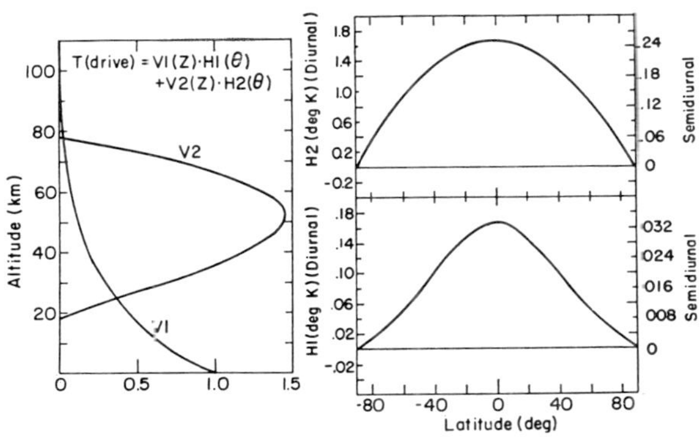

FIG. 3. 18. Vertical distribution of thermal excitation due to water vapor (V1) and ozone (V2), and latitudinal distribution of water vapor (H1) and ozone (H2). The heat input is given in terms of temperature variation with the amplitude shown in the diagrams for diurnal and semidiurnal tide (CHAPMAN and LINDZEN, 1970).

a) Topographical influences generate diurnal tides in the troposhere much larger than suspected previously.

b) The upper troposphere and lower stratosphere are considerably drier than previously assumed, and the thermal forcing for diurnal tides due to the water vapor near the ground (Fig. 3. 18) falls off rapidly. In latitudes lower than 30°, the tide excited by topographic irregularities of the planetary boundary layer heating can be as large as that by the radiative heating in amplitude and become larger in 20–30 km. The vertical wavelength is 10 km.

c) The diurnal tidal winds at 30-mb level (24 km) change very widely between summer and winter, whilst they exhibit little seasonal change at 900-mb level (1 km). It is likely that the prevailing winds change the transmission properties of the atmosphere. Note that the diurnal tides in this height region are very large in velocity, but can be different from those which have wave number 1 as $S(1, 1)$. They may not propagate horizontally but oscillate over the heat source with period of 24 hr.

Recent observation by VHF radar at Jicamarca has shown (Fukao *et al.*, 1978) that the stratospheric tide has a downward phase velocity with a wavelength consistent with that discussed by Wallace and Tadd. The result is

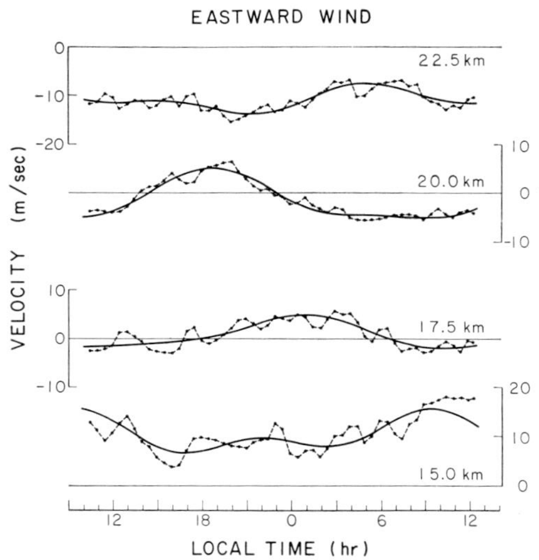

Fig. 3. 19. Observed tidal wind in the stratosphere over Jicamarca (Fukao *et al.*, 1978). Eastward wind velocity was obtained by the VHF Doppler radar at Jicamarca at every 1 min and then, the running average over 30 min was gotten for those velocities. The result was fit for a combination of diurnal and semidiurnal sinusoidal oscillations as shown. The vertical wavelength is about 10 km for the diurnal tide.

shown in Fig. 3. 19 which was obtained by a curve fitting to the observed eastward velocity in every one minute for a full day. This may suggest a tide excited locally below the stratosphere. In high latitudes where the (1, −2)-mode of the global tide is theoretically expected to dominate, the vertical profile is in agreement with LINDZEN's calculation (1967). This is illustrated in Fig. 3.20 (WALLACE and HARTRANFT, 1969) where only the amplitude (range) is shown, because no phase variation with height was observed. The calculated result is one-third or two-thirds of the observed one depending on heights. Note that the phase below the 500-mb level (5 km) was irregular. In the polar regions no topographical influences seem important.

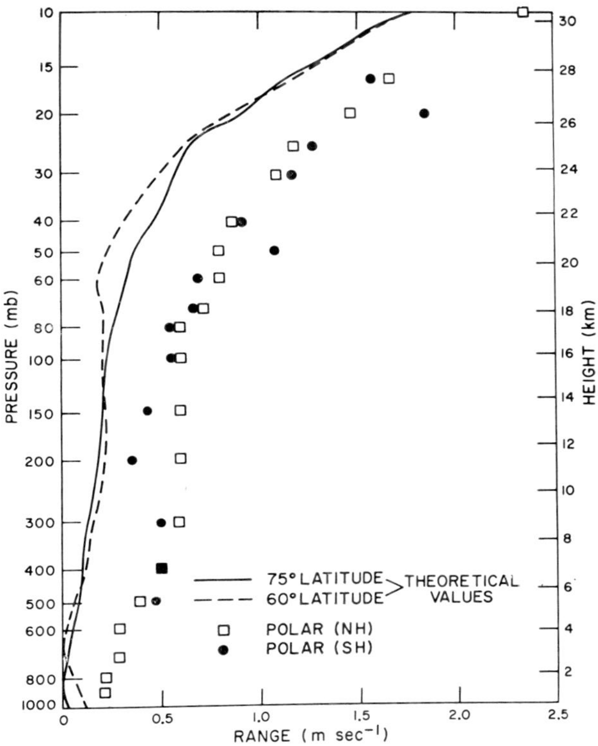

FIG. 3. 20. Comparison of theoretical and observed values of tidal range (2×amplitude) in polar region for z in 0–30 km. The data points were obtained by averaging data for all stations poleward of 60°. Theoretical values are from LINDZEN's tide tables (1967). The theoretical values are fairly smaller than the observed. Note that the observed value both at the Arctic and the Antarctic are in a good agreement (WALLACE and HARTRANFT, 1969).

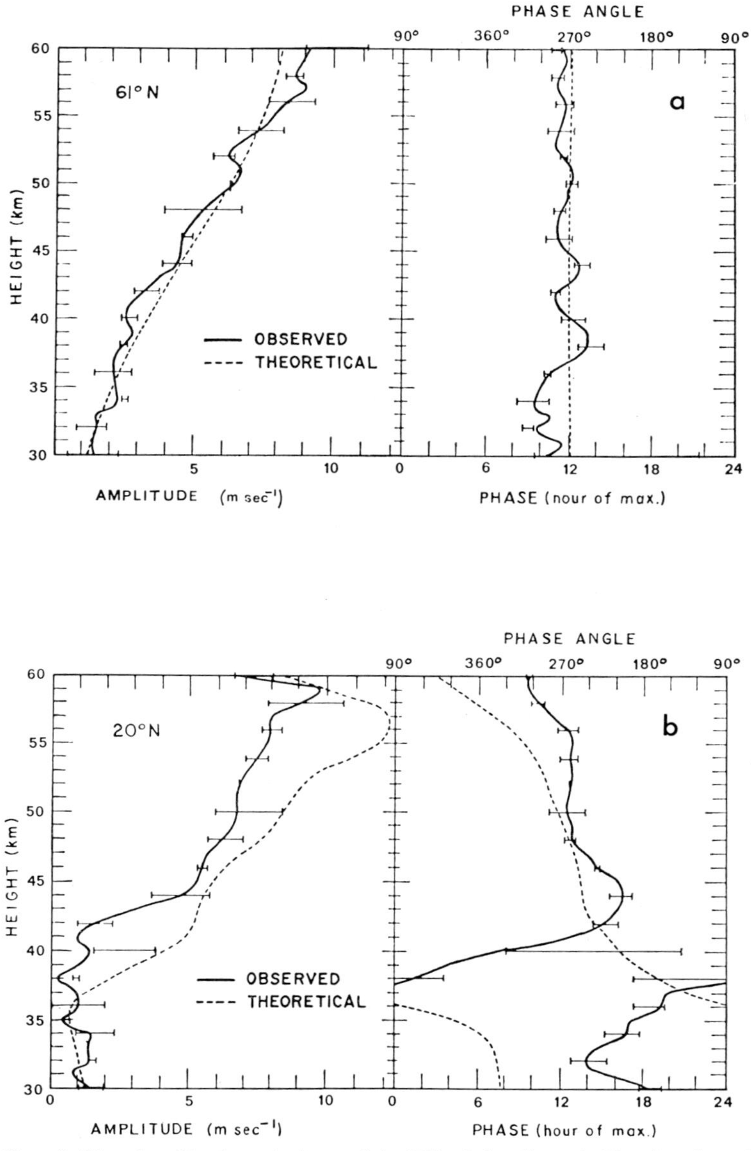

FIG. 3. 21. Amplitude and phase of $(-W_\theta)$ of the diurnal tide, the phase being the time at which the *northward* velocity is maximum. (a) 61 °N, (b) 20 °N. The season is equinox (CHAPMAN and LINDZEN, 1970).

Between 30–60 km, agreement with the model explaining surface pressure tides becomes better. This is because the non-migrating tides excited in the boundary layer tend to diminish at these heights (Fig. 3. 21). Note that there is a big difference of the meridional tidal velocity between latitudes 20°N and 60°N. This shows again that the (1, 1)-mode, a propagating mode which suffers a phase reversal around the source, predominates at low latitudes, whilst the (1, −2)-mode, a non-propagating mode, predominates in high latitudes. Model calculation can be improved by introducing more energy input at the source in the (1, −2)-mode, a realistic fearture which corresponds to the observed ozone content maximum at a latitude 60°N (e.g. DÜTSCH, 1971); the (1, −2)-mode excitation due to ozone increases with height.

Between 60–80 km, as mentioned above, no tidal observations are now available. Ample data from meteor radars are available between 80–110 km (e.g. FELLOUS *et al.*, 1975; CLARK, 1975). Figure 3. 22 shows the observation only in solstices. Assuming that the observed local time oscillation is of global westward migrating tide as exp $[i(\Omega t+\phi)]$, we can analyse the observed result in terms of the vertical wavelength which is characteristic to each $h_{\omega,-n,n}$ and T_0 as in (3.17′) (Fig. 3. 4). For the diurnal tide it is found that (1) the winter $W_{\theta,\phi}$ is mainly of $S\,(1, 1)$ with downward phase propagation; (2) the summer result is of $S\,(1, -2)$; (3) the spring result is complex, perhaps containing various modes; (4) the autumn result is similar to that of winter. The existence of negative modes has also been found by a meteor radar in Illinois (GELLER *et al.*, 1977). The amplitudes of these winds can be explained by calculation based on the classical theory (e.g. CHAPMAN and LINDZEN, 1970). However, the observed seasonal variation needs more refining of theoretical models. Anti-symmetric modes should be introduced. It is possible (Glass, private communication, 1973) that there are two evanescent modes $S\,(1, -2)$ plus $S\,(1, -3)$, both being in phase in summer and out of phase in winter. Steady zonal wind is necessary as discussed by LINDZEN and HONG (1974) (see Fig. 3. 16). No theory has been successful in explaining all of these features of the diurnal tide at meteor height. Of course, the theory must explain not only tides at the meteor height but those at higher altitudes.

The profile of the tidal wind $W_{\theta,\phi}$ at Arecibo has been obtained by HARPER (1977b) who measured the ion drift both orthogonal and parallel to $\boldsymbol{B}_0$. His result is illustrated in Fig. 3. 23 where some contour maps of W_θ between 100 and 300 km and that of W_ϕ between 100 and 130 km are given. Though the observation is limited to daylight hours, between 100 and 110 km a predominant diurnal wind can be inferred from Fig. 3. 23 (a)–(d); (c) of the figure shows the phase to vary almost by 180° in 10 hours. The vertical wavelength of about 25 km suggests that the wind is of $S\,(1, 1)$. Above 120 km the semi-diurnal wind clearly predominates (Fig. 3. 23(a)), but a diurnal wind appears to contribute near 150 km, because (1) the wind variation is asymmetric around the maximum northward velocity and (2) the contour of zero velocity,

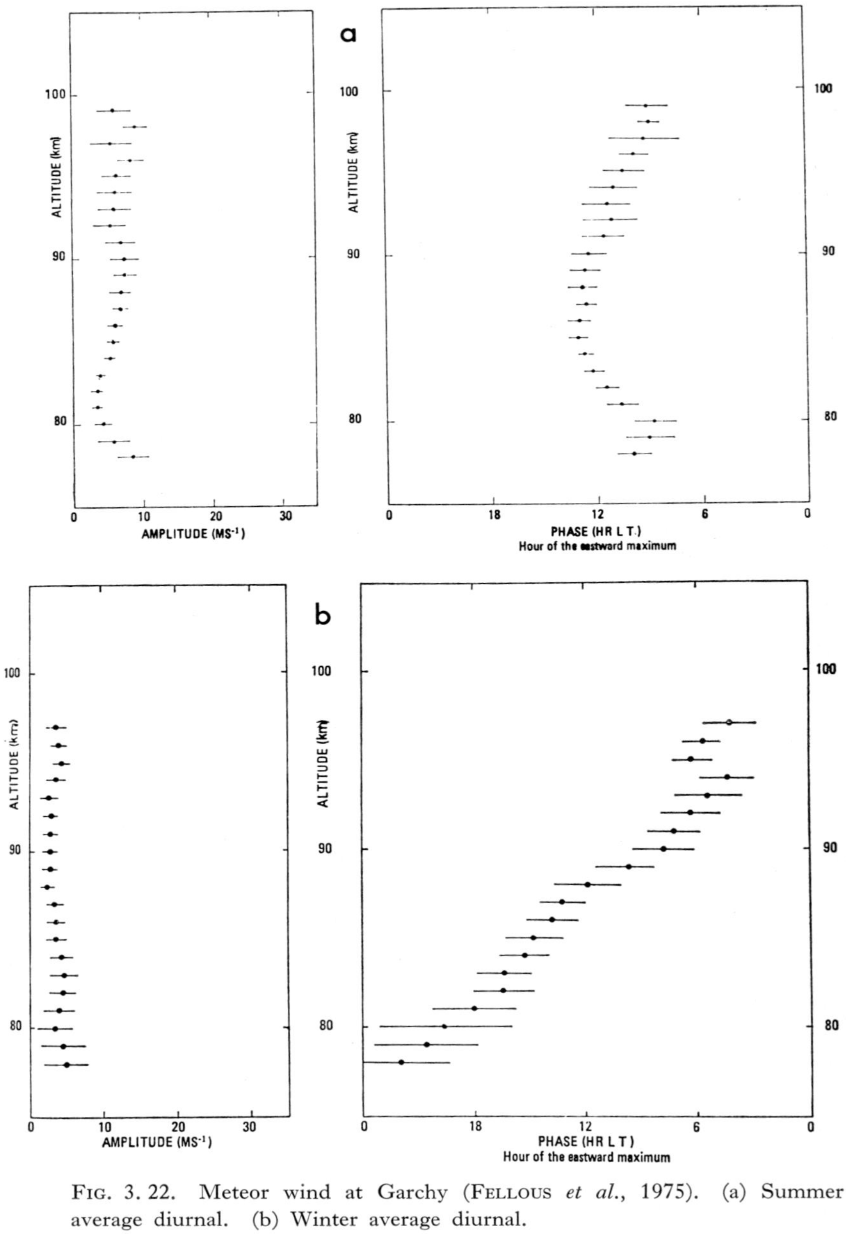

Fig. 3. 22. Meteor wind at Garchy (Fellous *et al.*, 1975). (a) Summer average diurnal. (b) Winter average diurnal.

the boundary between the shaded and unshaded regions, is not parallel with separation of 6 hours. Figure 3. 24 shows a distribution of the diurnal tide (W_θ) at St. Santin (AMAYENC, 1974). Because of the limited length of daily observation time as at Arecibo, the result would be uncertain. But it is argued (HARPER, 1977b) that the analysed phase is fairly reliable and is consistent with that at Arecibo around 150 km. We can say at least that the diurnal tide $S(1, 1)$ is much smaller than the semidiurnal between 130–180 km. This is due to dissipation by ion drag and viscosity which are more effective on $S(1, 1)$-mode of the short vertical wavelength than on $S(2, 2)$ and $S(2, 4)$, the main components of the semidiurnal tide at these heights. In the 100–130 km region it is not clear why $S(1, 1)$ predominates in summer, not in winter as in the meteor region. Arecibo is a low latitude station, and thus not suitable for detecting the $(1, -2)$-mode. Theoretically, the $(1, -2)$-mode at Arecibo's latitude (18°N) has a meridional velocity of only a quarter of that at St. Santin's (45°N). Note that the $(1, 1)$-mode reaches the maximum at Arecibo's latitude (Fig. 3. 2(B)).

Recently, REES (1979) has analysed approximately 400 wind-profiles of the lower thermosphere based on chemical releases from rockets since 1957 at $(30\pm5)°$ and $(40\pm5)°$ latitudes. Though the experiments are restricted mainly in twilights and nighttime, he attempted to obtain the best fitting of wind profiles consisting of constant, diurnal and semidiurnal components. The results were averaged over all seasons for two latitudes at heights between 90 and 140 km (Fig. 3.25). The existence of the $(1, -2)$-mode was clear. This $(1, -2)$-mode was found enhanced with magnetic activity, an interesting characteristic which should be elucidated in future. It is likely that this mode is sensitive to the energy input in high latitudes where the corresponding Hough function has a maximum. The excitation of this mode can maximize above 140 km, since the observed amplitude of W_θ increases with height at least at 140 km, which is consistent with the requirement from the dynamo theory (e.g. RICHMOND *et al.*, 1976) and FORBES and GARRETT's calculation (1976).

Above 180 km the in situ excitation of the diurnal tide becomes important and is dominant above 250 km, the phase remaining almost unchanged with height. Then, the winds can be determined by the method previously mentioned in calculating the pressure gradient from a model. This has been done for Millstone Hill by ROBLE *et al.* (1974b, 1977) and for St. Santin by ANTONIADIS (1976). The former used a simpler method, neglecting the advection term in (3. 66). Considering the lower boundary conditions that the wind vanishes at 120 km, the result is realistic only in the upper thermosphere where the in situ excitation is dominant. The inferred winds (Fig. 3. 26) blow southwards (equatorwards) at night and northwards (polewards) in day, although the nighttime characteristic lasts longer in summer than in winter. Eastward and westward winds are characteristic of the night and day, respectively. It has been found also that there is a steady southward

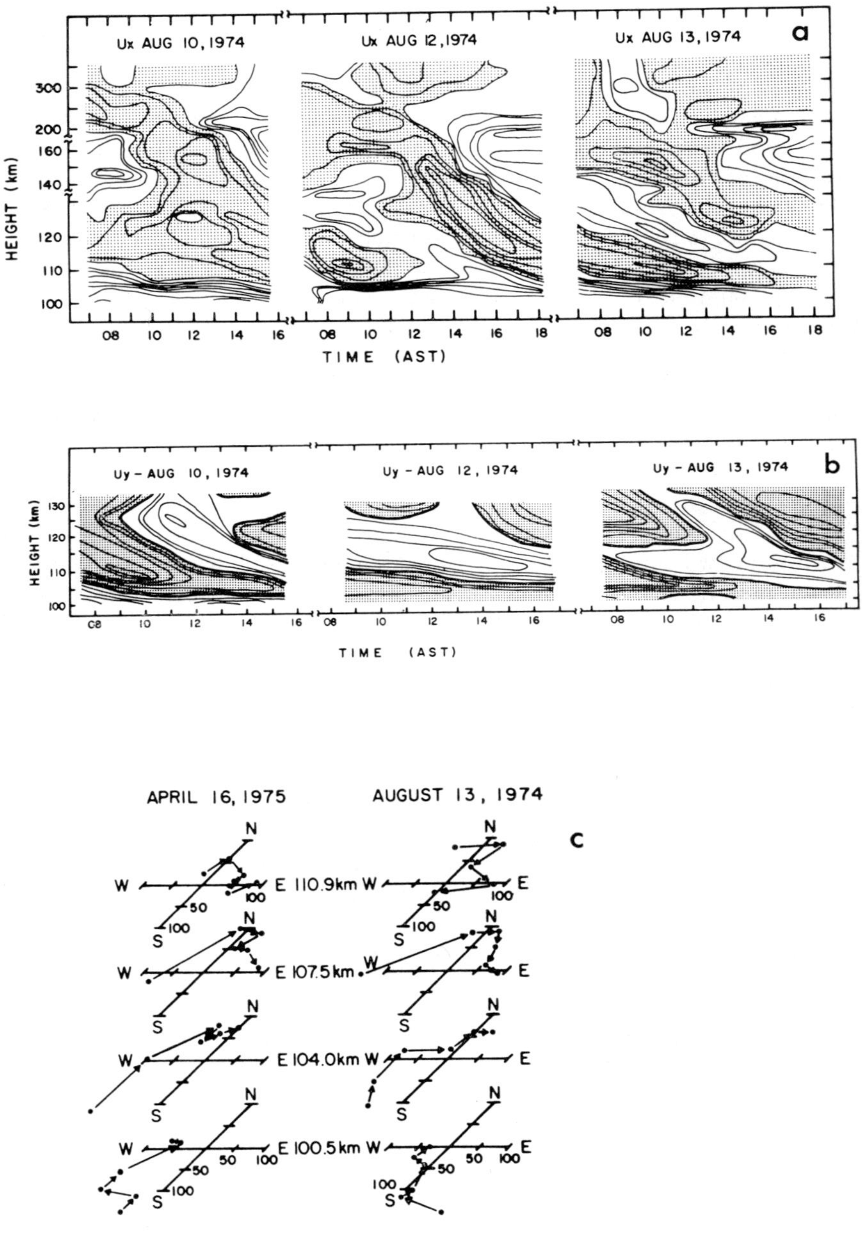
Ux AUG 10, 1974
Ux AUG 12, 1974
Ux AUG 13, 1974
a
HEIGHT (km)
TIME (AST)
Uy - AUG 10, 1974
Uy - AUG 12, 1974
Uy - AUG 13, 1974
b
HEIGHT (km)
TIME (AST)
APRIL 16, 1975
AUGUST 13, 1974
c
110.9 km
107.5 km
104.0 km
100.5 km

FIG. 3.23

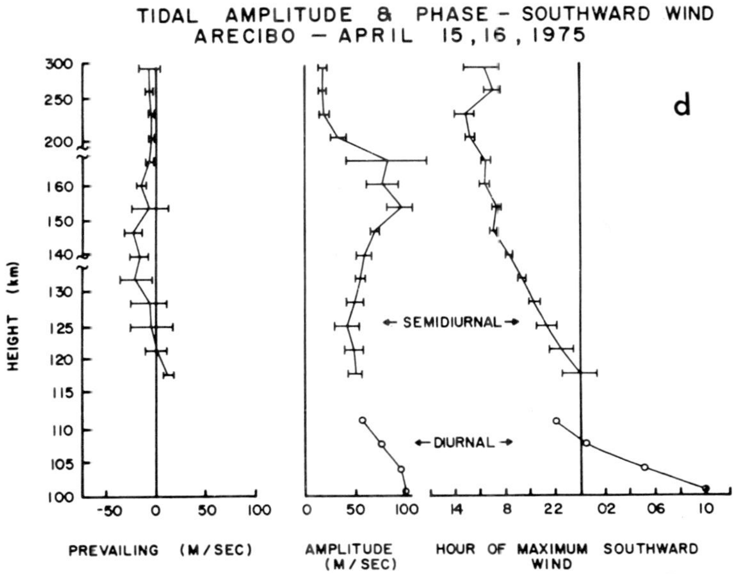

FIG. 3. 23. Tidal winds over Arecibo. (a) Meridional winds (positive southward). The contour spacing is 20 m sec^{-1}; the shaded parts denote positive, scale changes above 135 km, August 10, 12 and 13, 1974. (b) Similar to (a) for zonal velocity (positive eastward). (c) Wind vector at 2-hr intervals from 0800 to 1800 hr between 100 and 111 km for April 16, 1975 and August 13, 1974. (d) Amplitudes and phase for the August, 1975 data. The error bars are due to the averaging over the two days in the diagram (HARPER, 1977b).

(equatorward) wind in all seasons, except at high altitudes in winter; the zonal steady component is weak and westward, except for winter, when it is strong and eastward. Allowing for the steady component, the observed or, more exactly, observation-based wind is to be compared with the work by FORBES and GARRETT (1978) who discussed theoretically the in situ excited thermospheric diurnal tides. While the overall characteristics, equatorward winds at night and their seasonal variation, are consistent with each other, the precise comparison, for various latitudes and solar activities, would be a matter in future studies.

The basic problem here as well as with the semidiurnal tide is that no global scale observations are available, while no very realistic models are available to explain all details of observation at particular geographical points. Besides, in the lower atmosphere below 100 km, the outstanding problem on diurnal tides is to elucidate both in theory and experiment the locally-excited non-

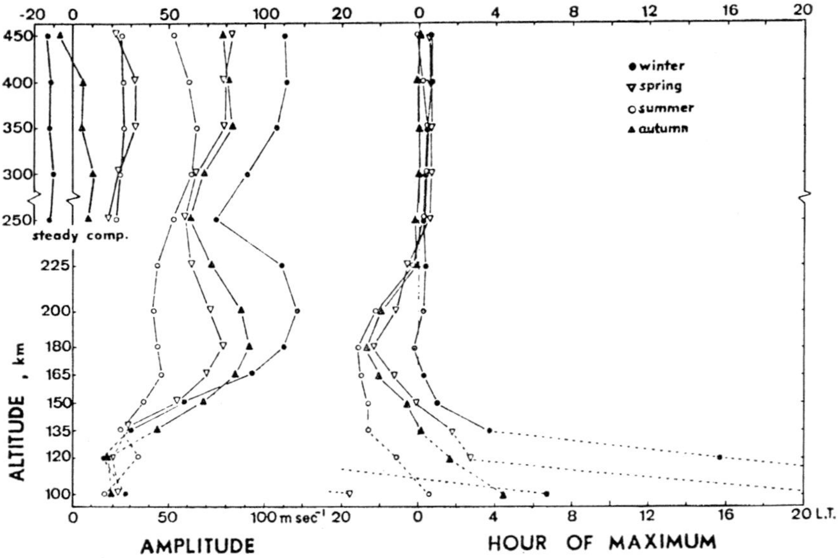

FIG. 3. 24. Mean seasonal height structures of the diurnal tide, amplitude and phase of W_θ observed at St. Santin between 1971–1972 (note the break in the height scale above 250 km). Left top: steady meridional southward velocity (AMAYENC, 1974).

migrating tide. Between 100 and 200 km the observation technique must be explored for obtaining good continuous data for a complete day both under quiet and disturbed conditions. Excitation of the tides during magnetic disturbances present an interesting problem. The model including the zonal wind as done by LINDZEN and HONG (1974) for the semidiurnal tide should be explored considering *the effects of tides on the zonal wind* (MIYAHARA, 1978).

3. 3. 3 *Semidiurnal tide*

The semidiurnal tide is, generally, much better understood than the diurnal, both theoretically and observationally. Large amplitude and regular phase distribution of the semidiurnal surface pressure variation stimulated people's curiosity as early as in the seventeenth century. However, it is rather recent that the basic agreement between theory and observation has been established. BUTLER and SMALL (1963) have shown that the semidiurnal tide is driven mainly by heating of ozone in the upper stratosphere and lower mesosphere. Now, the observed surface pressure tide is well explained mainly in terms of $S(2, 2)$-mode with a large vertical wavelength except for 40 min

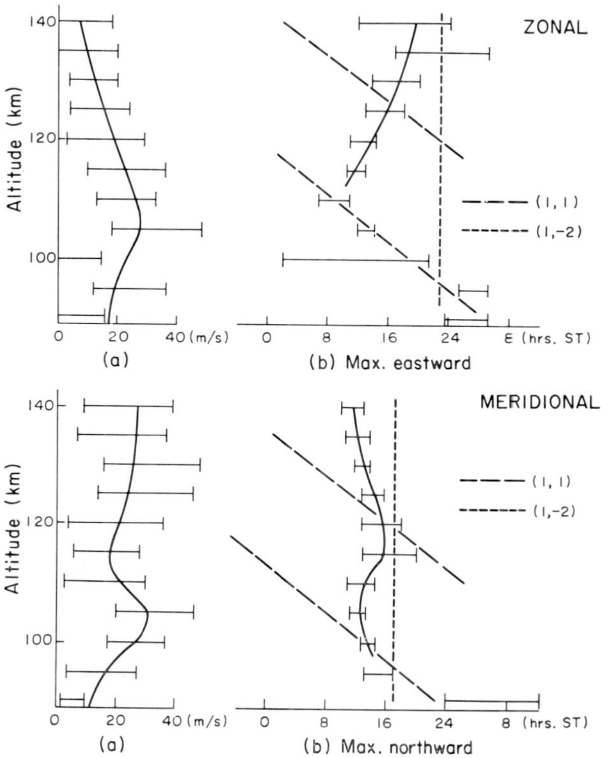

FIG. 3. 25. Tidal wind measured by chemical release from rockets at 30° latitude (REES, 1979). Top, diurnal, zonal (eastward, W_ϕ): Bottom, diurnal meridional (northward, $-W_\theta$).

of phase lag behind the theoretical prediction. The loss due to turbulence in the boundary layer was suggested to explain this disagreement (CHAPMAN and LINDZEN, 1970). Up to 25 km (30-mb) the agreement is similar to that at the surface; the non-migrating tide seems to be much weaker in the semi-diurnal component than in the diurnal (WALLACE and TADD, 1974). Above this level the observation shows (Fig. 3. 27) that the semidiurnal tidal velocity in midlatitudes remains almost unchanged both in amplitude and phase up to 50 km where a phase reversal occurs in summer. Contamination of large synoptic scale fluctuations makes it impossible to investigate the vertical structure of the tide in other seasons. The classical theory, however, predicted the phase reversal to occur at a much lower height at 28 km. LINDZEN and HONG (1974) appear to have in a way explained this difference (Section 3. 3). The height of the phase reversal may depend on the distribution

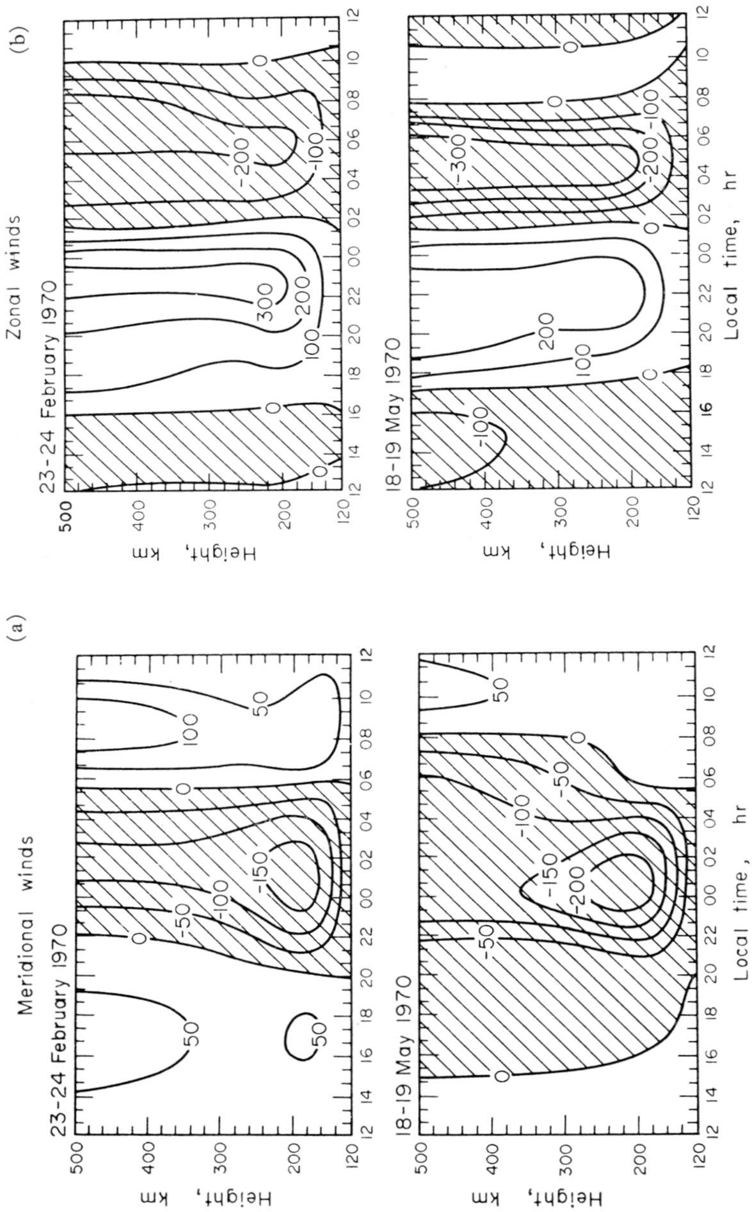

(a)
Meridional winds
23-24 February 1970
18-19 May 1970
Height, km
Local time, hr
(b)
Zonal winds
23-24 February 1970
18-19 May 1970
Height, km
Local time, hr

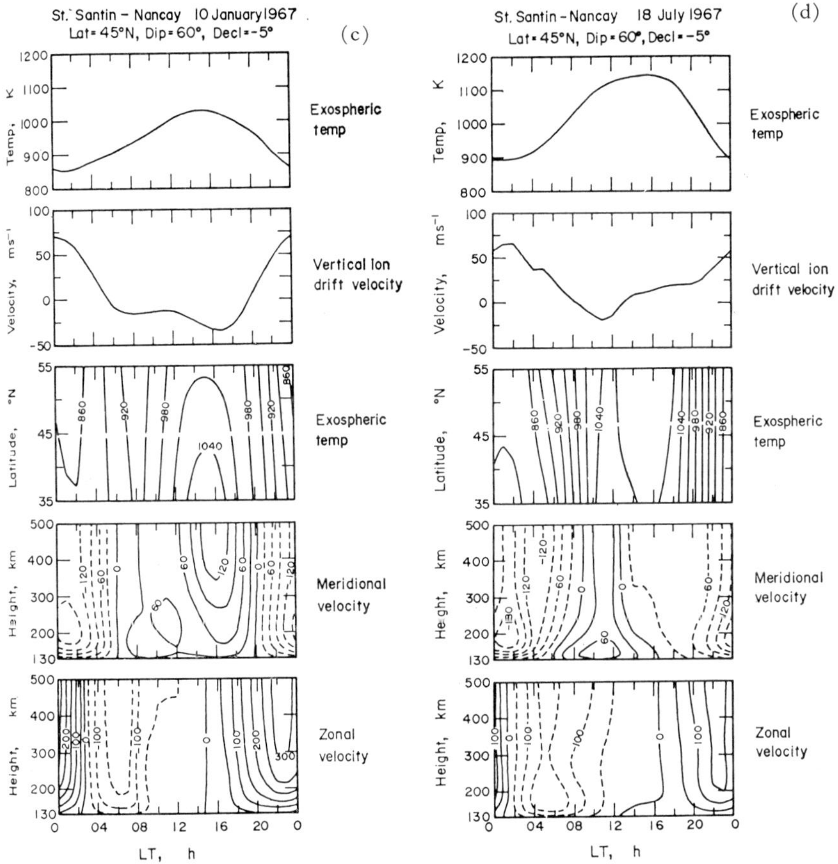

FIG. 3. 26. Diurnal wind based on incoherent scatter radar observation. (a) Meridional wind ($-W_\theta$) (m sec^{-1}) over Millstone Hill on 23–24 February, 1970 and 18–19 May, 1970 (ROBLE *et al.*, 1977). (b) Similar to (a) except for zonal wind, W_ϕ. (c) Day 10 January, 1967. St. Santin. From top to bottom: Observed exospheric temperature and wind induced vertical ion drift (positive upwards), calculated T_∞ geographic distribution, northward meridional and eastward zonal wind velocity contours; T in degrees, velocity in m sec^{-1}, the solid and broken curves correspond to positive and negative values respectively. (d) Similar to (c) except for day 18 June, 1967 (ANTONIADIS 1976).

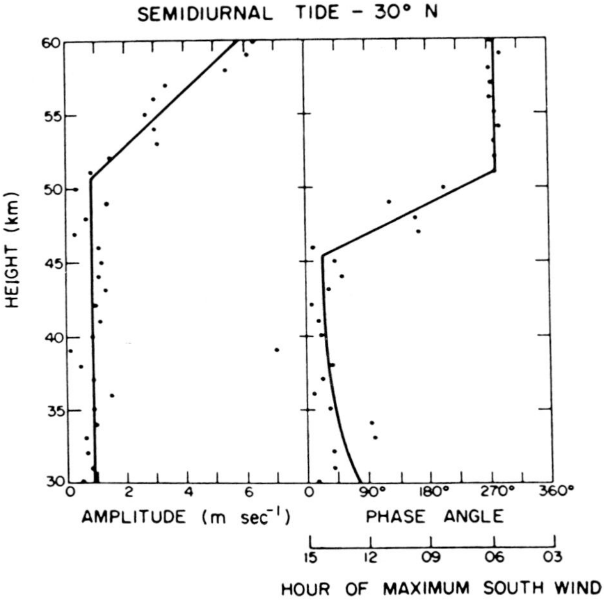

FIG. 3. 27. Semidiurnal tide in meridional wind (northward velocity) at 30°N based on data for White Sands (32.4°N) and Cape Kennedy (28.5°N) (REED, 1967).

of the zonal wind. Details regarding the phase reversal in the stratosphere seems still inconclusive. The discrepancy between observation and classical theory persists up to 60 km.

As in the diurnal tide almost no data are available between 60 and 80 km. In the region between 80 and 110 km the meteor radar is the main source of useful data. Also electron density in daytime makes it possible for incoherent scatter radars to observe above 70 km. Generally, the semidiurnal tide is stronger than the diurnal at midlatitudes. The seasonal variation of tidal winds is remarkable at Garchy (FELLOUS *et al.*, 1975) in that a short vertical wavelength tide is predominant in winter, the phase velocity being downward, whilst a long vertical wavelength tide is dominant in other seasons especially in summer (Fig. 3. 28). A similar result has been obtained in Illinois (GELLER *et al.*, 1977). It is still unknown why such variation takes place. It should be stressed that the horizontal structure cannot be determined from the vertical wavelength in a realistic atmosphere. Thus, the real question here is why the vertical wavelength varies with season as is observed. The relation between the vertical wavelength and the global distribution of tide may be a different question. The steady zonal wind in the middle atmosphere

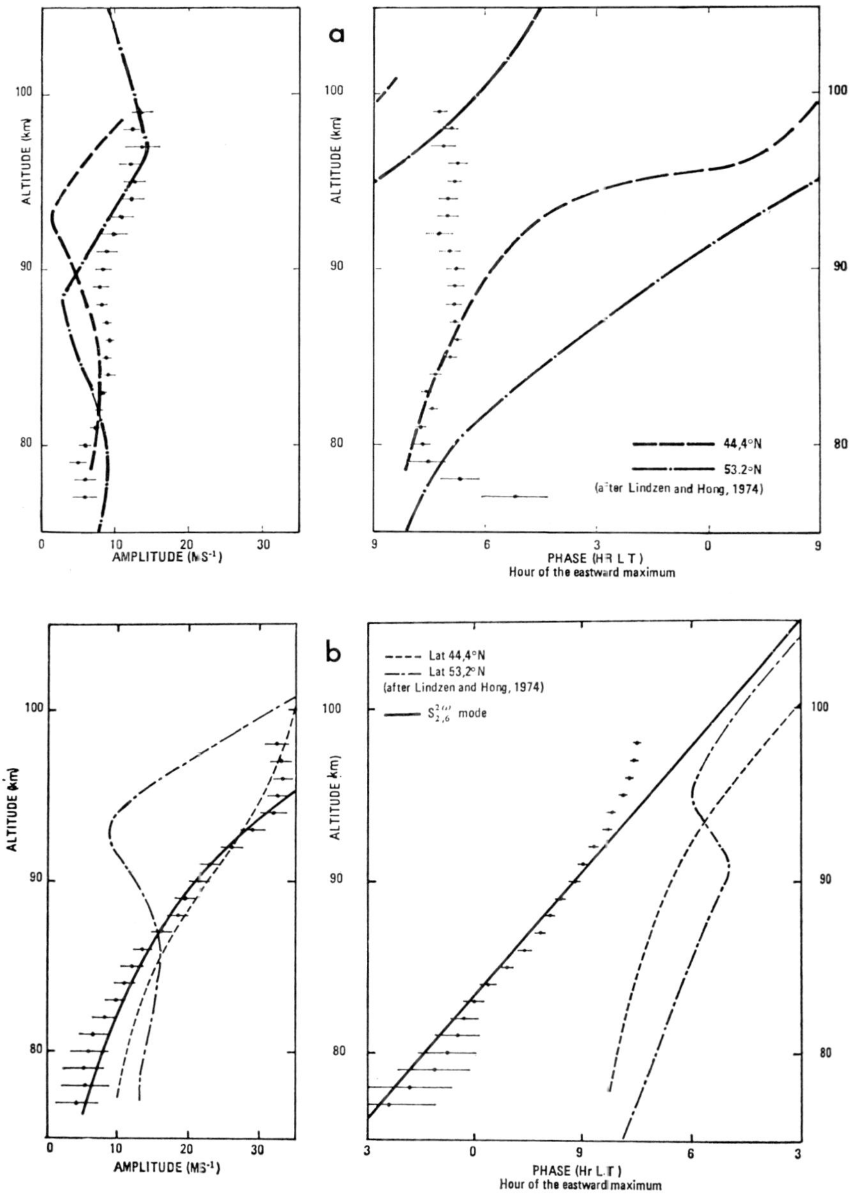

FIG. 3. 28. Meteor radar observation of semidiurnal tide at Garchy, W_ϕ (FELLOUS *et al.*, 1975). (a) Summer, (b) Winter. Theoretical results by LINDZEN and HONG (1974) and classic tidal mode (2, 6) are also shown.

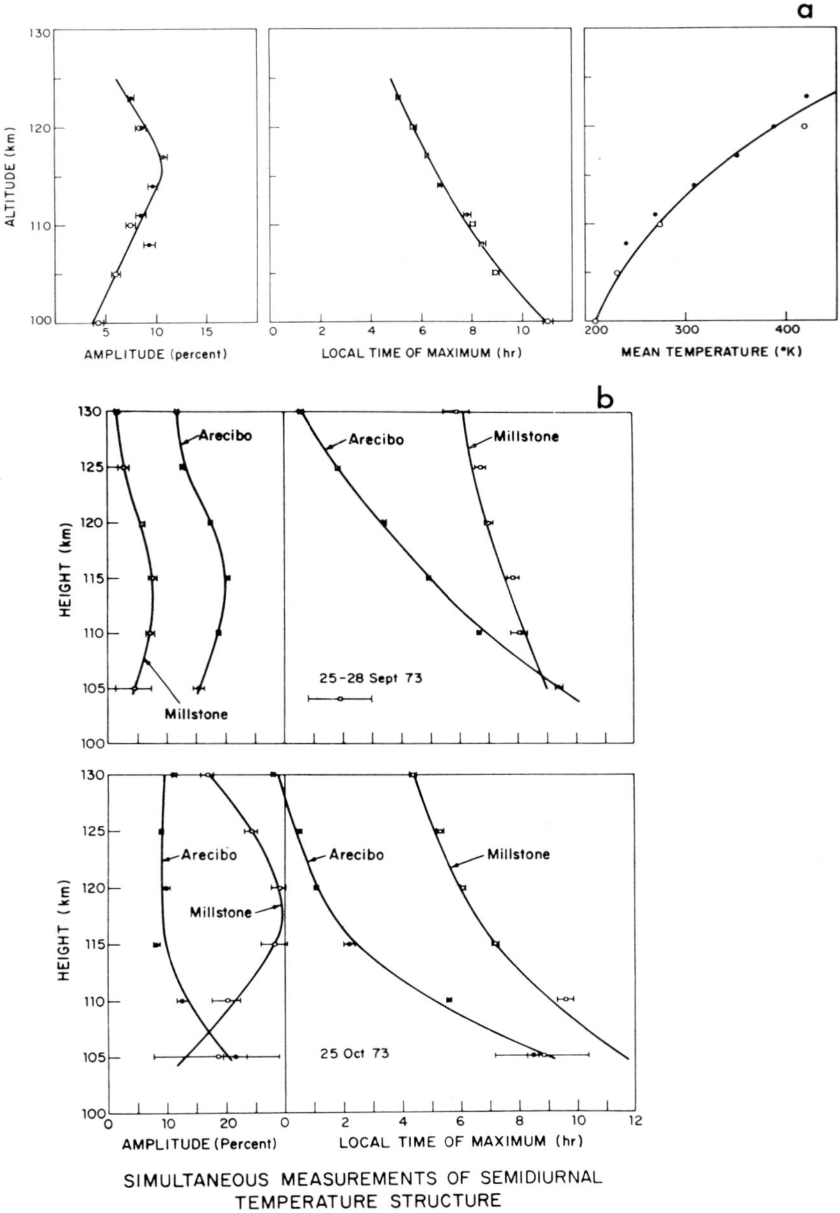

FIG. 3. 29. Semidiurnal temperature tide observed by incoherent scatter radars. (a) Millstone Hill by • and St. Santin by ∘ (SALAH and WAND, 1974). (b) Arecibo and Millstone Hill (SALAH *et al.*, 1975).

seems to be a good candidate to explain the variation, as the calculation of LINDZEN and HONG (1974) has suggested. Further, the global horizontal tidal structure should be clarified more precisely in observation to find the global structure. Day-to-day variation is fairly large, and a simple theory has difficulty to explain the observed features.

Between 100–130 km, the temperature tides have been measured during daylight hours by incoherent scatter radars at Millstone Hill, St. Santin and Arecibo. Since the ion temperature is very close to the neutral particle temperature below 200 km, one can regard the ion temperature to be T. The semidiurnal variation of T is remarkable in all observations in any season. On the average, the amplitude and phase variations with height are fairly similar between St. Santin and Millstone Hill, reaching a maximum around 115 km and the wavelength consistent with that of the (2, 4)-mode (Fig. 3. 29).

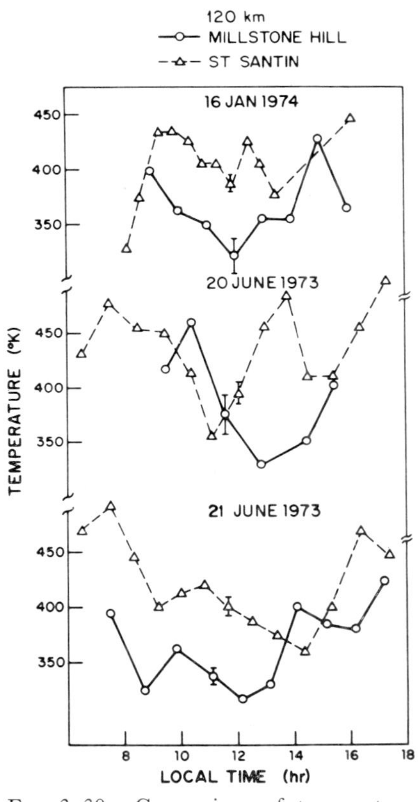

FIG. 3. 30. Comparison of temperature variation observed at 120 km over Millstone Hill and St. Santin (SALAH *et al.*, 1977).

However, on each day the temperature varies differently at each station (Fig. 3. 30). Further, similarity between Arecibo and Millstone Hill is less remarkable (Fig. 3.29(b)). The amplitude and phase differ occasionally by large amount on the same day of measurement. It is likely that there is certain contamination due to suspected diurnal components at Arecibo where $S(1, 1)$ is large in the wind tide. WAND (1976) has discussed seasonal variation of temperature, finding a large seasonal change which shows not only the existence of anti-symmetric components but also the existence of certain effects to produce a difference between the vernal and autumnal equinoxes (Fig. 3. 31). The temperature tides show that the phase velocity is downward, the energy flow being upward. The predominance of the (2, 4)-mode below 130 km has also been confirmed by the wind measurement based on the ion drift measurement (e.g. BERNARD, 1974). Using the data of rocket observation of wind, REES (1979) has obtained, as in the diurnal component, the semidiurnal components (Fig. 3. 32) which consists mainly of the (2, 4)-mode below 120 km in zonal winds.

There is an apparent difficulty (Harper, private communication) between the meteor radar observation and the incoherent scatter radar observation

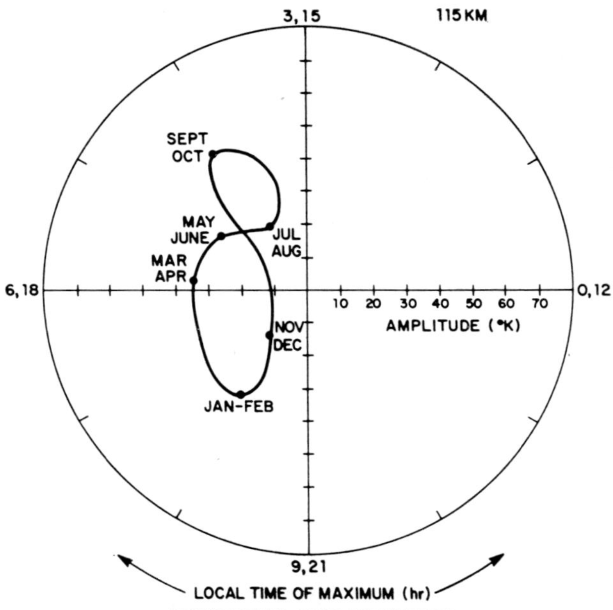

FIG. 3. 31. Amplitude and phase of semidiurnal tide at 115 km over Arecibo plotted on a 12-hr harmonic dial. Each point denotes a measurement averaged over the 2-month periods (WAND, 1977, see EVANS, 1978).

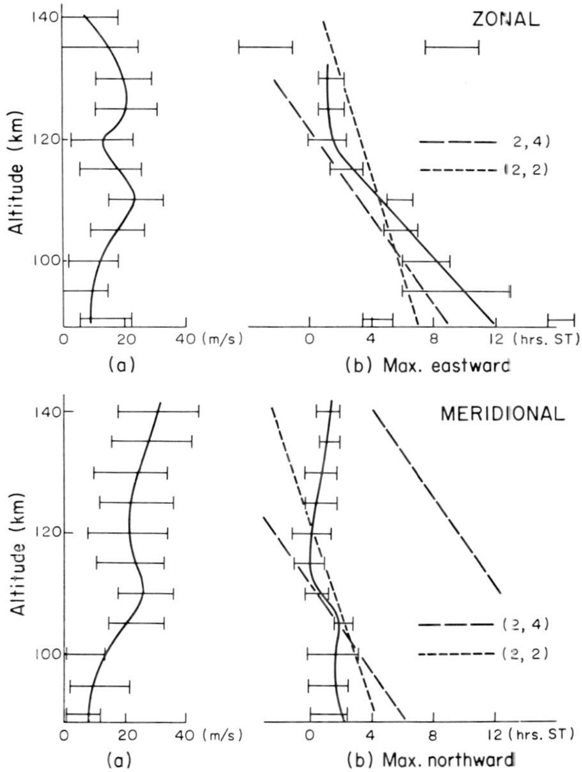

FIG. 3. 32. Same to Fig. 3. 25 except for semidiurnal tide (REES, 1978).

in the lower thermosphere. If the result of meteor radar observation at Garchy means variation of the tidal structure from summer to winter on global scale, as is expected in classical theory, the variation at Arecibo (18°N) and that at Garchy (46°N) should be different, because the (2, 4)-mode corresponding to the winter situation produces a velocity of opposite sign between these two stations, whilst the (2, 2)-mode corresponding to the summer situation produces a velocity of equal sign between the stations. However, the temperature at Arecibo behaves quite similarly as the wind velocity at Garchy. Here again the answer needs a future study of horizontal structure, vertical wavelength, and its seasonal variation of the actual tide in the thermosphere.

The semidiurnal meridional wind above 120 km has been observed by incoherent scatter radars by AMAYENC (1974) (Fig. 3. 33) and more recently by HARPER (1977b). During summer equinox the (2, 2)-mode appears to dominate between 130–175 km (with the amplitude of 70 m sec^{-1} at 150 km)

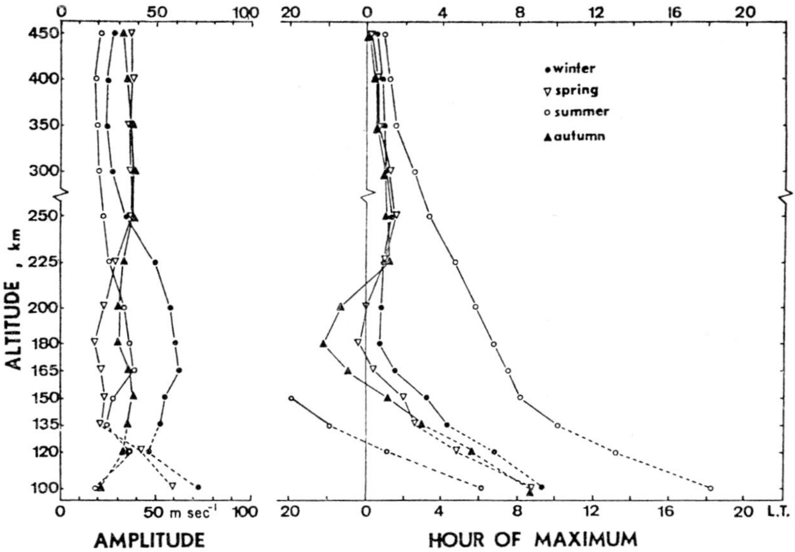

Fig. 3. 33. Semidiurnal southward wind tide (W_θ) over St. Santin (Amayenc, 1974).

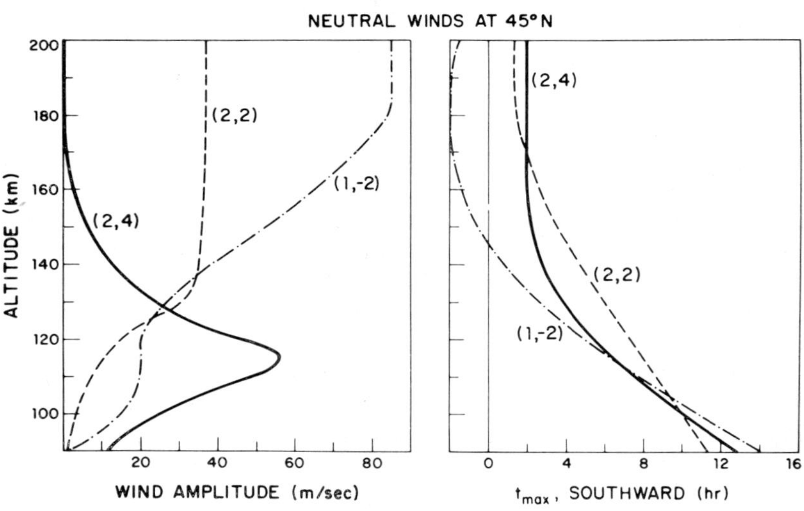

Fig. 3. 34. Tidal wind model in amplitude and phase for the main symmetric modes (1, −2), (2, 2), (2, 4) in W_θ (Salah and Evans, 1977).

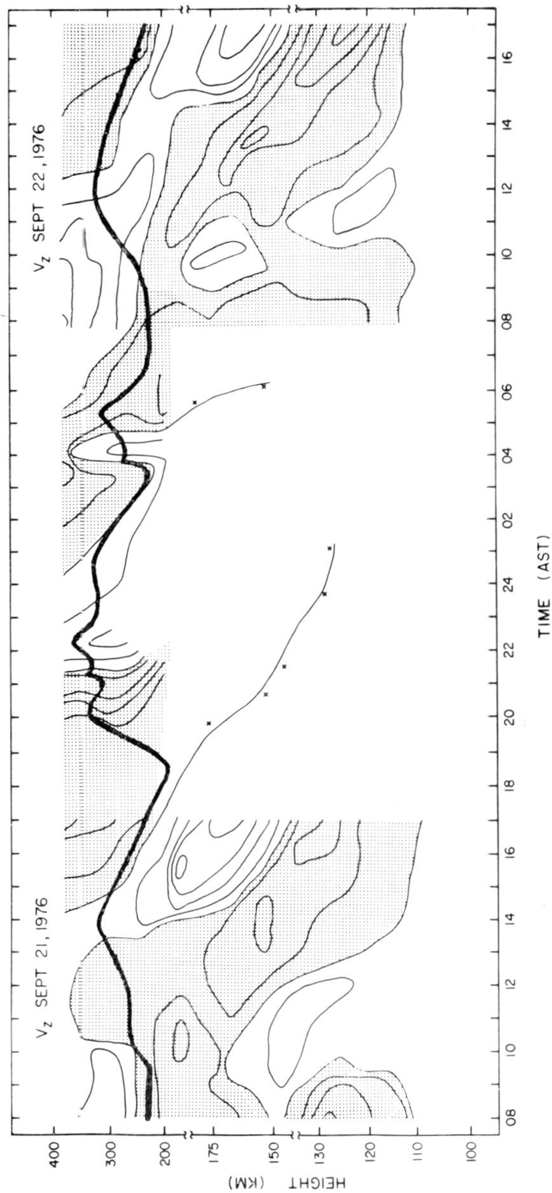

FIG. 3. 35. Ion drift velocity contour plot versus height and time for September 21–22, 1976. The contour spacing is 10 m sec^{-1}. Shaded parts show downward velocity. The heavy line is the F layer peak. Crosses show the height of observed intermediate layers (HARPER, 1979).

in sunspot minimum period (Fig. 3. 23).

Based on the observation of both diurnal and semidiurnal tides, and, further, on the tidal dynamo which requires the (1, −2)-mode wind for S_q currents, SALAH and EVANS (1977) have attempted to make a model for the thermospheric wind as shown in Fig. 3. 34. Another model has been used by RICHMOND *et al.* (1976) for a tidal dynamo simulation; the two models are similar in amplitude, but not in phase.

The semidiurnal tide has been found to produce an important effect on the F layer. According to the wind model based on satellite drag data (e.g. KOHL and KING, 1967), the wind velocity should be equatorward and large in post midnight sector. However, the F layer was observed to descend after midnights. This phenomenon is called " midnight collapse." This is caused by the semidiurnal tides which decrease or frequently reverse the equatorward meridional wind (HARPER, 1979, Fig. 3. 35). The wind shear associated with the semidiurnal tide produce a convergence of ions around " the convergent nulls " (Section 4. 2) which descend continuously from F regions to E regions. The semidiurnal tide appears to be as strong as or stronger than the diurnal tide for solar cycle minimum conditions. From the model in Fig. 3. 34, the meridional wind of $S(2, 2)$ becomes 30 m sec^{-1} at a latitude 18°N, whilst that of $S(1, -2)$ becomes 20 m sec^{-1} or less at a latitude 18°N. Actually, the observed meridional northward wind was 50 m sec^{-1}.

FUJITAKA and TOHMATSU (1973) have explained the formation of intermediate layers between E and F1 regions in midlatitudes by wind shear due to the (2, 4)-mode. An interesting finding (FUKUYAMA, 1976) was that the air glow of 5,577Å of atomic oxygen varies with seasons in a good agreement with the semidiurnal tidal wind at meteor heights in midlatitudes. Besides the intensity variation, the time, at which the maximum of the air glow intensity occurred, was found to remain in all seasons delayed just by 6 hr from the time at which the northward velocity of the semidiurnal wind at meteor heights occurred. PETITDIDIER and TEITELBAUM (1977) discussed in detail the problem. These may suggest that the semidiurnal tide is effective on the variation of atmospheric composition in the upper atmosphere.

CHAPTER 4

ELECTRODYNAMICS OF THE UPPER ATMOSPHERE

The subject of this chapter will be the electromagnetics of the thermosphere. Although the dynamic energy resides mainly in the neutral gas component, the behavior of the ionized gas is important in producing the electric current and electric field.

Since neutral gas particles are much more numerous than charged particles, the behavior of charged particles is controlled by the presence of neutral particles especially in the lower ionsphere. Of course, the charged particle motion is influenced by the electric field $\boldsymbol{E}$ and magnetic field $\boldsymbol{B}$ which are both quasi-static. While the electric field is produced in association with plasma motion in the ionosphere and the magnetosphere far above the ionosphere, the magnetic field is mainly the geomagnetic field $\boldsymbol{B}_0$. The magnetic field which is produced by the electric current in the ionosphere amounts only to 1%, at most, of the geomagnetic field. The gravity $\boldsymbol{g}$, which controls the neutral gas behavior, can usually be neglected in the plasma motion because of the large ratio between e and $m_{e,i}$, the electromagnetic force being proportional to e and gravitational force proportional to $m_{e,i}$. This is correcter to electrons than to ions because $m_e \ll m_i$. The electron inertia can also be neglected for the same reason. In a quasi-stationary state, as we are mainly concerned herewith, the ion inertia can usually be neglected also. Under these circumstances we can define mobilities for ions and electrons.

4.1 Motion of Charged Particles, Mobilities, Electric Current and Field

The velocity of charged particles of the ionosphere under the influence of quasi-static electric field and stationarily moving neutral particles, which are more numerous than charged particles, is given by mobilities $\mu_{e,i}{}^{(1),(2),(0)}$ ((A. 4. 1) and (A. 4. 1′)) as

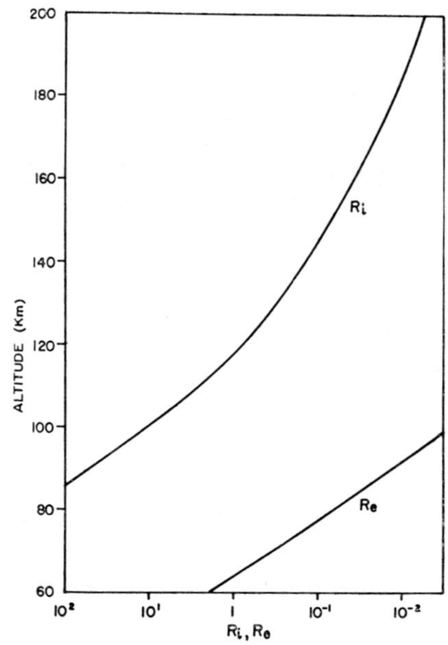

FIG. 4. 1. Collision frequency $\nu_{e,\,i}$ versus height. $R_{e,\,i}=\nu_{e,\,i}/\Omega_{e,\,i}$ (AXFORD, 1963).

$$\boldsymbol{V}_{e,i} = \mu^{(1)}{}_{e,i}\hat{\boldsymbol{E}}_{\perp} - \mu^{(2)}{}_{e,i}\frac{\boldsymbol{B}_0\times\hat{\boldsymbol{E}}}{B_0} + \mu^{(0)}{}_{e,i}\boldsymbol{E}_{\parallel} + \boldsymbol{W} \tag{4.1}$$

where $\hat{\boldsymbol{E}}=\boldsymbol{E}+\boldsymbol{W}\times\boldsymbol{B}_0$; subscripts e, i denote electrons andions ; $\mu_{e,i}{}^{(1)}$, $\mu_{e,i}{}^{(2)}$, and $\mu_{e,i}{}^{(0)}$ are given in terms of collision- and gyro-frequencies, $\nu_{e,i}$ and $\Omega_{e,i}$, respectively, in (A. 4. 2). The distribution of $\nu_{e,i}$ is shown in Fig. 4. 1.

Firstly, consider ions and electrons in static electromagnetic field in the absence of neutral particles. Then, it is well known that charged particles, in a plane orthogonal to $\boldsymbol{B}_0$, gyrate with the angular frequency $\Omega_{e,i}$ around $\boldsymbol{B}_0$ and also drift with the velocities $(\boldsymbol{E}\times\boldsymbol{B}_0)B_0{}^{-2}$ where $\Omega_{e,i}$ is as large as megahertz for electrons and a few hundred hertz for ions. Along $\boldsymbol{B}_0$, charged particles move as if there were no magnetic field.

The presence of stationary neutral particles will now be taken into consideration. If neutral particles are moving with a velocity $\boldsymbol{W}$, we introduce a moving coordinate system on which neutral particles are stationary and $\boldsymbol{E}$ is Lorentz-transformed to $\hat{\boldsymbol{E}}$. We assume that the neutral particle motion is not disturbed by charged particles, because $n_e \ll n$. The stationary neutral particles impede, by collision, the motion of charged particles under the influence of $\hat{\boldsymbol{E}}$ and $\boldsymbol{B}_0$. However, the control of $\boldsymbol{B}_0$ on electrons is still large as known from $\Omega_e \gg \nu_e$ i.e. there are many gyrations between collision, implying that electrons move as if in the absence of collision. The situation is given by (4. 1) as $\mu_e^{(1)} \to 0$ and $\mu_e^{(2)} \to 1/B_0$ implying that electrons cannot move along $\hat{\boldsymbol{E}}$, but can move orthogonally both to $\hat{\boldsymbol{E}}$ and $\boldsymbol{B}_0$, with the velocity $(\hat{\boldsymbol{E}} \times \boldsymbol{B}_0/B_0^2)$. Here, the gyration is not found, because mobilities are obtained only by neglecting inertia which produces the gyration. The same result is obtained also by averaging the instantaneous motion of electrons in the absence of collision as above. The average is taken for more than one rotation of the gyration which takes only 10^{-6} sec in the ionosphere. One can represent the electron motion by that of the gyration center for such *slow* phenomena as we are concerned herewith.

Ions, in a plane orthogonal to $\boldsymbol{B}_0$, move differently in the lower E and D region where $\nu_i \gg \Omega_i$ i.e. there are no gyration between collisions. These ions move almost exactly with the neutral particles. The situation is given $\mu_i^{(1),(2)} = 0$. In the F region where $\nu_i \ll \Omega_i$, ions move like electrons. At intermediate heights where $\nu_i \sim \Omega_i$ ions move peculiarly depending on $\mu_i^{(1),(2)}$ partly along $\hat{\boldsymbol{E}}$ and partly along $(\hat{\boldsymbol{E}} \times \boldsymbol{B}_0)$. Along $\boldsymbol{B}_0$ the motion of charged particles is just impeded by collision. It is significant here to know that the mobilities are defined for $\hat{\boldsymbol{E}}$ instead of $\boldsymbol{E}$.

The electric current is defined as

$$\boldsymbol{J} = (n_i \boldsymbol{V}_i - n_e \boldsymbol{V}_e)\, e. \tag{4.2}$$

An important condition relative to (4. 2) is that ions and electrons are exactly equal in number, i.e. $n_e = n_i$, for those phenomena which have larger scales than the Debye shielding distance d_D defined as $\sqrt{\pi \sigma T_e / e^2 n_e}$; d_D is of the order of cm at ionospheric heights. In large scale phenomena, a very slight violation of the neutrality condition leads to the occurrence of prohibitively large electric fields. By Maxwell's equation $\operatorname{div} \boldsymbol{E} = (e/\mu_0)(n_i - n_e)$, $\mu_0 = 8.8 \cdot 10^{-12}$ F m^{-1}, even for a small inequality, $|n_i - n_e| = 1$ cm^{-3}, the resultant electric field $\boldsymbol{E}$ would be as strong as 10^2 V m^{-1} for a scale of 10 km. Note that the observed ionospheric electric field is usually of the order of 10^{-3} V m^{-1}. This electric field, which plays an important role in the ionosphere dynamics, is produced by an accumulation, on the boundary, of very small quantities of charge.

By (4. 1) and (4. 2), Ohm's law is given as

$$\boldsymbol{J} = \sigma_1 \hat{\boldsymbol{E}}_\perp + \sigma_2 \frac{\boldsymbol{B}_0 \times \hat{\boldsymbol{E}}}{B_0} + \sigma_0 \boldsymbol{E}_{/\!/} \tag{4.3}$$

$$\begin{aligned} \sigma_1 &= n_e e(\mu_i^{(1)} - \mu_e^{(1)}) \\ \sigma_2 &= n_e e(\mu_e^{(2)} - \mu_i^{(2)}) \\ \sigma_0 &= n_e e(\mu_i^{(0)} - \mu_e^{(0)}) \end{aligned} \tag{4.4}$$

where $\sigma_{1,2,0}$ are called Pedersen, Hall and parallel conductivity and given in terms of collision- and gyro-frequencies in (A. 4. 3). In (4. 2), only one species of ions is considered. For more than one species of ions the conductivity

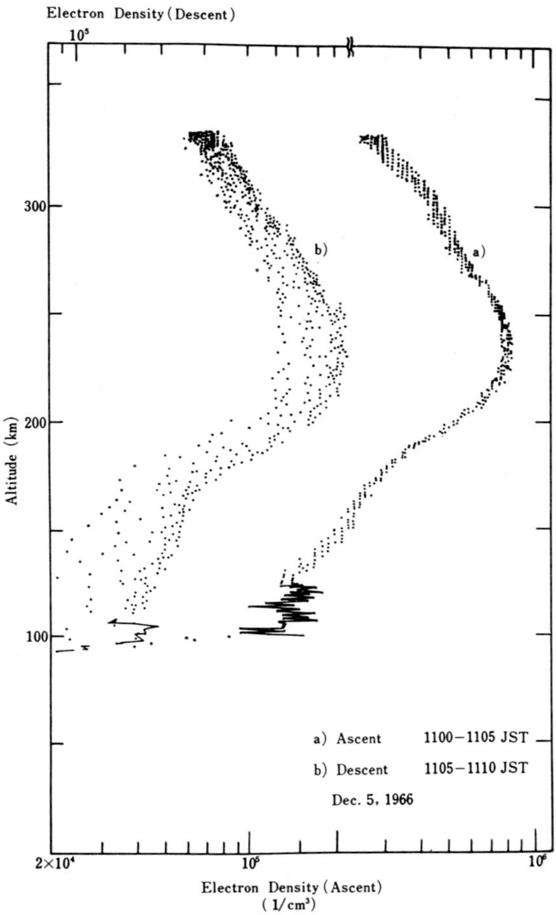

FIG. 4. 2. Electron density distribution. Rocket observation by OYA and ASO (1972). Sporadic E occurs around 100 km.

of each species of ions must be added, giving $\sigma_1 = e(\Sigma\, n_i\mu_i^{(1)} - n_e\mu_e^{(2)})$ etc. (see (A. 4. 4)). The distribution of n_e and $\sigma_{0,1,2}$ is given in Figs. 4. 2 and 4. 3, respectively.

For low frequency phenomena as we are concerned with, the displacement current can be neglected in Maxwell's equation, giving

$$\mathrm{curl}\,\boldsymbol{B} = \vartheta_0 \boldsymbol{J} \tag{4. 5}$$

where ϑ_0 is the permeability of free space; m.k.s. is used. Taking the divergence of (4. 5) gives

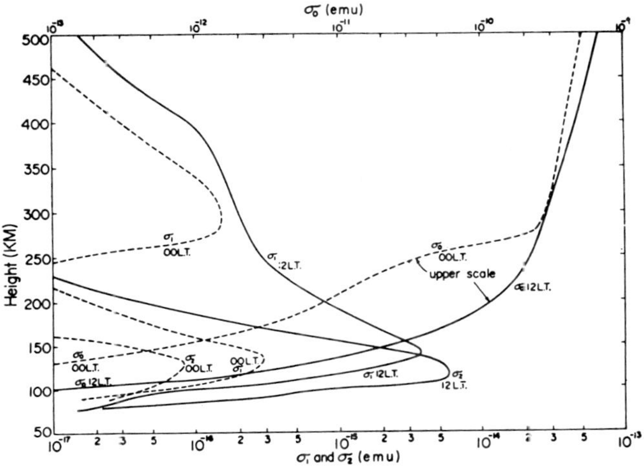

FIG. 4. 3. Conductivity $\sigma_{0,1,2}$ versus height (MAEDA and KATO, 1966). Multiply by 10^{11} to obtain, in m.k.s. units, mho m^{-1}. LT denotes local time.

$$\operatorname{div} \boldsymbol{J} = 0. \tag{4. 6}$$

Biot-Savart's law, which is the basis for producing an equivalent current system in geomagnetism, is obtained by integrating (4. 5) (e.g. STRATTON, 1941). Another Maxwell's equation is

$$\frac{\partial \boldsymbol{B}}{\partial t} = -\operatorname{curl} \boldsymbol{E} \tag{4. 7}$$

which is simplified usually by

$$\operatorname{curl} \boldsymbol{E} = 0. \tag{4. 7'}$$

But in the F region $\partial \boldsymbol{B}/\partial t$ cannot be neglected (e.g. VOLLAND and GRELLMANN, 1978). A different expression about $\boldsymbol{V}_{e,i}$ is by (A. 4. 5)

$$\begin{aligned}
\boldsymbol{V}_e &= \frac{\boldsymbol{J}\times\boldsymbol{B}_0}{n_e(m_e\nu_e+m_i\nu_i)} - \frac{m_i\nu_i\boldsymbol{J}}{n_e e(m_e\nu_e+m_i\nu_i)} + \boldsymbol{W} \\
\boldsymbol{V}_i &= \frac{\boldsymbol{J}\times\boldsymbol{B}_0}{n_e(m_e\nu_e+m_i\nu_i)} + \frac{m_e\nu_e\boldsymbol{J}}{n_e e(m_e\nu_e+m_i\nu_i)} + \boldsymbol{W} \\
\boldsymbol{C}_0 &= \frac{\boldsymbol{J}\times\boldsymbol{B}_0}{n_e(m_e\nu_e+m_i\nu_i)}.
\end{aligned} \tag{4. 8}$$

Mathematically, this expression is just a rearrangement of (4. 1) or a decomposition of $\boldsymbol{V}_{e,i}$ into different directions. Physically, however, one can now find direction, along which both ions and electrons move together. This gives the velocity denoted as $\boldsymbol{C}_0$ parallel to $(\boldsymbol{J}\times\boldsymbol{B}_0)$. The differential motion

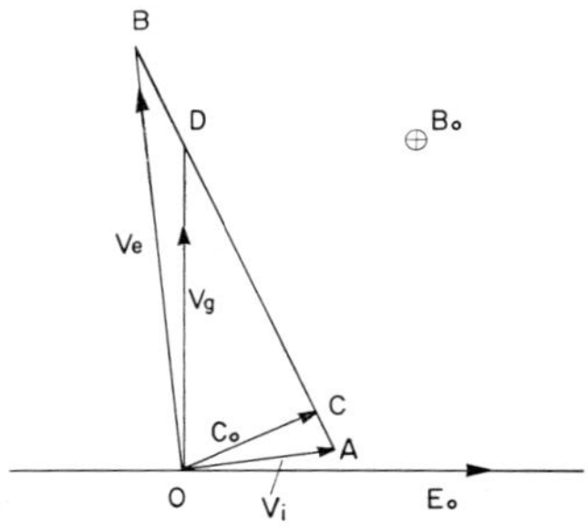

FIG. 4. 4. Transverse drift velocity of charged particles OA=ion velocity $\boldsymbol{V}_i$; OB=electron velocity $\boldsymbol{V}_e$; OC=neutral ionization drift $\boldsymbol{C}_0$; BA=electric current $\boldsymbol{J}$; CD/OC=(σ_2/σ_1). $\boldsymbol{B}_0$ is orthogonal and inward to the diagram (MARTYN, 1953). Motion parallel to $\boldsymbol{B}_0$ is proportional to $\mp\Omega_{e,i}/\nu_{e,i}$, negative for electrons and positive for ions. $\boldsymbol{V}_g$ is the phase velocity of irregularities (KATO, 1963a). The diagram is based on the E region parameters.

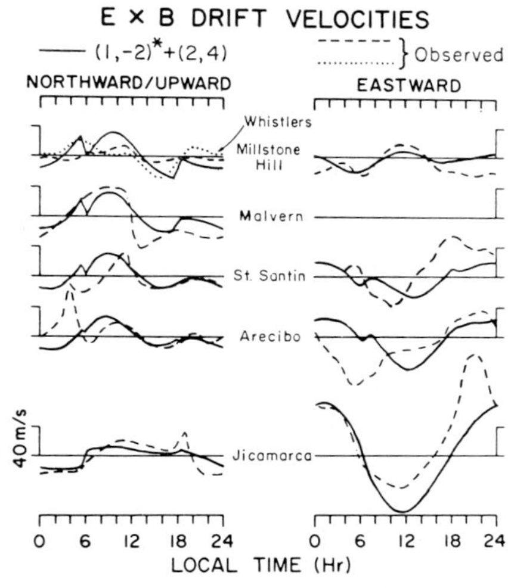

FIG. 4. 5. $\boldsymbol{E}$ in terms of the $(\boldsymbol{E}\times\boldsymbol{B})$ drift velocity at various latitudes as a function of local time. Computed results from the combined tidal wind as in Figs. 5. 11, 5. 12 (solid line), and observed results at various stations by incoherent scatter radars as shown (broken and dotted lines). 40 m sec^{-1} corresponds to about one mV·m^{-1} (RICHMOND *et al.*, 1976).

between ions and electrons, producing $\boldsymbol{J}$, is therefore orthogonal to $\boldsymbol{C}_0$. $\boldsymbol{C}_0$ tends to be $(\hat{\boldsymbol{E}}\times\boldsymbol{B}_0)/B_0^2$ for $\nu_{i,e}\ll\Omega_{i,e}$. $\boldsymbol{C}_0$ was called the neutral ionization drift by MARTYN (1953, Fig. 4. 4).

The momentum transfer from charged particles to neutral particles per unit volume per unit time is $n_e[m_e\nu_e(\boldsymbol{V}_e-\boldsymbol{W})+m_i\nu_i(\boldsymbol{V}_i-\boldsymbol{W})]$ which gives the force to be exerted by charged particles upon neutral particles. By (4. 8) this force, which is called the Ampere force, becomes $(\boldsymbol{J}\times\boldsymbol{B}_0)$ and plays a role in the hydromagnetic tide.

The ionospheric electric field (Fig. 4. 5) has two sources of production. One originates in the ionosphere itself, and is due to the small accumlation of charge in regions where the conductivity changes rapidly. The region where substantial currents flow is called the dynamo region. It extends from about 90 to 200 km. The lower boundary of the dynamo region is fairly well defined because of the rapidly decreasing conductivities with decreasing z below 80 km due to the decreasing electron densities (Fig. 4. 3). But the upper boundary might be difficult to define. In $\sigma_{0,1,2}$, σ_0 increases rapidly with z; σ_2 decreases rapidly after a maximum around 110 km; and σ_1 also decreases after a maximum, which is at about 125 km and is broader than that of σ_2. Since the loss of electrons due to recombination and attachment decreases with height, the electron density decreases at night by larger amount in the lower region than in the higher region. This is also the case for conductivities which are proportional to electron density by (4. 4). Thus, the ionospheric electric current vertical distribution changes accordingly between day and night, although the nighttime total current integrated with height is only a fraction of the daytime one.

The other region thought to produce an important electric field is the magnetosphere (Fig. 4. 6). This electric field is produced through a dynamic interaction between the solar wind and the geomagnetic field in the magnetopause. The large scale quasi-electrostatic field penetrates the ionosphere along field lines almost without attenuation due to high electric conductivity σ_0. Any large potential drop along $\boldsymbol{B}_0$ should produce an extremely intense current, producing a destructive heating or impact ionization. That $\boldsymbol{E}$ is almost orthogonal to field lines, as is proven later in Section 5. 4, is due to high σ_0. Thus, the quasi-electrostatic field of magnetospheric origin comes down along field lines to the high latitude ionosphere where an ionospheric electric current is produced and a secondary field is also established down to low latitudes. However, anomalous resistivity along $\boldsymbol{B}_0$ might produce a different situation on some occasions under disturbed conditions (FÄLTHAMMER, 1977). The electric field of ionospheric origin also affects the magnetosphere. These two fields cannot be separated in observations. KAMIIDE and HORWITZ (1978) have worked in various cases on this problem, but the discussion of their works is outside the scope of the present concern. At the present time it appears that the magnetospheric contribution is small, as far as the electric field relevant to the S_q is concerned.

In special configuration of magnetic field and electric field, interesting phenomena occur. The electrojet is one of them, flowing both at the magnetic

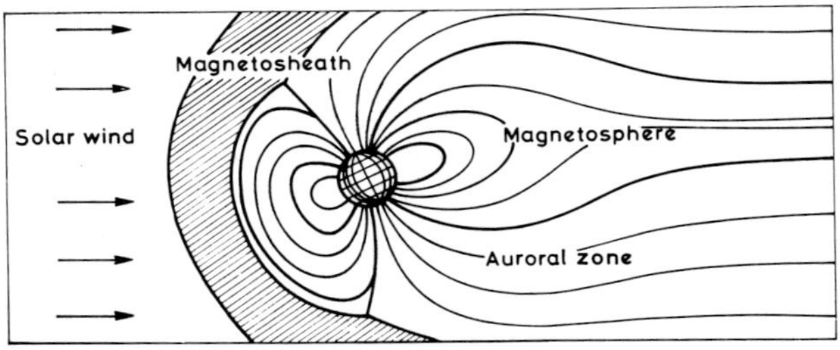

FIG. 4. 6. Solar wind, magnetosheath and magnetosphere.

equator and the auroral zone where the electric current must be almost strictly orthogonal to $\boldsymbol{B}_0$. Consider the equatorial condition as $J_z=0$ (x, y, and z are positive southward, eastward, and upward, respectively) because of rapidly decreasing $\sigma_{1,2}$ above the lower E region (Fig. 4. 3). Then, by (4. 3) and $W_z \simeq 0$

$$E_z = \frac{\sigma_2}{\sigma_1} E_y \tag{4. 8'}$$

which gives

$$J_y = \sigma_3 E_y \tag{4. 8''}$$

where σ_3 is called the Cowling conductivity as

$$\sigma_3 = \sigma_1 \left(1 + \frac{\sigma_2{}^2}{\sigma_1{}^2}\right). \tag{4. 8'''}$$

σ_3 is given in terms of $\nu_{e,i}$ and $\Omega_{e,i}$ in (A. 4. 3). Figure 4. 7 shows the distribution of (σ_2/σ_1) which reaches 10 or more at its maximum near 100 km (MAEDA and KATO, 1966). Primarily, E_y can be produced by the global tidal dynamo or applied partly from the magnetosphere as mentioned above. This implies that an intense electric current (jet) is driven by E_y. The intense current is mainly due to a Hall current by the secondarily produced field E_z in (4. 8'). Since $\mu_e{}^{(2)} \gg \mu_i{}^{(2)}$ around 100 km by (A. 4. 2), the jet current consists mainly of electrons. It is found that for $E_y = 10^{-3}\,\mathrm{Vm}^{-1}$, $(\sigma_2/\sigma_1)=10$, and $B_0 = 5\times 10^{-5}$ Wbm^{-2}, we have $V_{ey} \simeq -2\times 10^2\,\mathrm{m\,sec}^{-1}$. The real situation may not be as simple as $J_z=0$ (e.g. SUGIURA and POROS, 1969), but the mechanism is still in operation in essence. The equatorial electrojet requires a horizontal $\boldsymbol{B}_0$. Even at a dip angle I as small as $5°$, E_z decreases drastically, and no jet current flows. This equatorial electrojet flows usually eastwards, but sometimes westwards. This westward electrojet is called " counter electrojet ". The counter electrojet can be caused by some semidiurnal tidal winds in the dynamo region (e.g. SCHIELDGE *et al.*, 1973; STENING, 1977).

A similar situation happens at the auroral zone. The conductivity (σ_1)

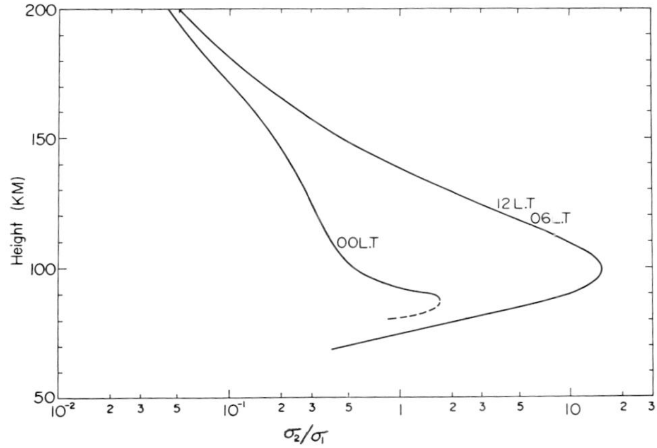

FIG. 4. 7. Ratio (σ_2/σ_1) versus height (MAEDA and KATO, 1966).

is enhanced by increased electron density due to disturbances along the auroral oval. The electric current tends to be enhanced along the oval, producing a Hall current orthogonal to the oval. Since the conductivity decreases rapidly outside the disturbed oval, as above the dynamo region at the magnetic equator, the Hall current tends to be inhibited, producing a large electric field across the oval as E_z in (4. 8′). This polarization field due to the Hall current drives further a Hall current along the disturbed oval. Then, the resultant current is σ_3 times the electric field along the oval.

These electrojets are responsible for various dynamic disturbances in the upper atmosphere. It is now well-known that the equatorial electrojet produces small scale field-aligned irregularities of types I and II. The type I irregularity is produced by two stream plasma instabilities due to the interpenetrating ion- and electron-stream of the jet (SATO, 1973). The type II irregularity is produced by the cross field plasma instability due to $\boldsymbol{E}_y$ and $\boldsymbol{B}_0$, which is antiparallel to x-axis at the equator, in the presence of the ambient electron density gradient along the z-axis i.e. along ($\boldsymbol{E}_y \times \boldsymbol{B}_0$) (KATO, 1972). Similar situation also occurs at the auroral zone. Besides the plasma instability, there is an intense joule heating which heats the ionosphere (e.g. COLE, 1962). KATO (1962) pointed out a possible excitation of some airglow due to elevated electron temperature at the auroral zone. He (1963b) also suggested possible formation of certain electron irregularities along the equatorial jet by decreased recombination due to the elevated electron temperature.

It is now established that the auroral electrojet excites GW through the joule heating which can amount to 10^{-6} Wm^{-3} and also by the Ampere force (e.g. RICHMOND and MATSUSHITA, 1975). As in the example in Section 2. 3, supersonic motion of the auroral electrojet can also excite the auroral in-

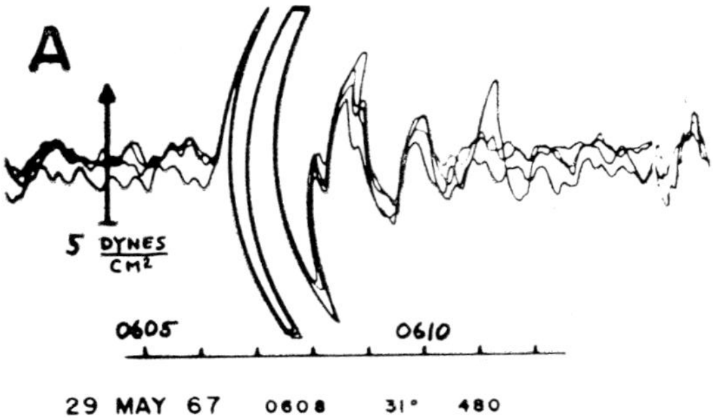

FIG. 4. 8. AIW Alaska (WILSON, 1969). The pressure fluctuation was recorded by micro-barograph.

frasonic wave (AIW, Fig. 4. 8) as discussed by WILSON (e.g. 1969) and SWIFT (1973).

4.2 Wind Shear Effects

If $\boldsymbol{W}$ in (4. 1) has a shear, the motion of charged particles can produce a divergence and convergence of ionization. Consider $\boldsymbol{V}_{\mathrm{i}}$ in the absence of $\boldsymbol{E}$ in (4. 1), which gives

$$\boldsymbol{V}_{\mathrm{i}} = \mu_{\mathrm{i}}^{(1)}(\boldsymbol{W}\times\boldsymbol{B}_0) - \mu_{\mathrm{i}}^{(2)} \frac{\boldsymbol{B}_0\times(\boldsymbol{W}\times\boldsymbol{B}_0)}{B_0^2} + \boldsymbol{W} \tag{4.9}$$

or

$$\boldsymbol{V}_{\mathrm{i}} = \frac{\nu_{\mathrm{i}}\Omega_{\mathrm{i}}}{\nu_{\mathrm{i}}^2+\Omega_{\mathrm{i}}^2}\frac{(\boldsymbol{W}\times\boldsymbol{B}_0)}{B_0} + \frac{\nu_{\mathrm{i}}^2}{\nu_{\mathrm{i}}^2+\Omega_{\mathrm{i}}^2}\boldsymbol{W}_{\perp} + \boldsymbol{W}_{/\!/}. \tag{4.9'}$$

The first term in the right-hand side of (4.9′) shows that the ion moves partly *orthogonally to* the applied wind i.e. the colliding neutral particles velocity; the second term shows that the ion moves partly *with* the wind orthogonally to $\boldsymbol{B}_0$; the last term shows that the ion moves *exactly with* the wind along $\boldsymbol{B}_0$. The situation is schematically depicted in Fig. 4. 9.

Large scale neutral winds are mostly horizontal, and div $\boldsymbol{W}\approx 0$. Thus, if $\boldsymbol{W}$ has a shear, the first and last terms in (4. 9′) contribute to the accumulation of ions along the convergent nulls of $\boldsymbol{W}$. In the F region where $\nu_{\mathrm{i}} \ll \Omega_{\mathrm{i}}$ (Fig. 4. 1), only the last term is important. The basic idea of this wind shear theory has been put forward by DUNGEY (1958) and WHITEHEAD (1961).

The electron motion is not discussed above, but it should be such that no divergence of current exists i.e. div $\boldsymbol{J}=0$. Let us consider a simple situation where $\boldsymbol{W}$ is horizontal, varying only vertically as e^{ikz} (x, y, and z are positive

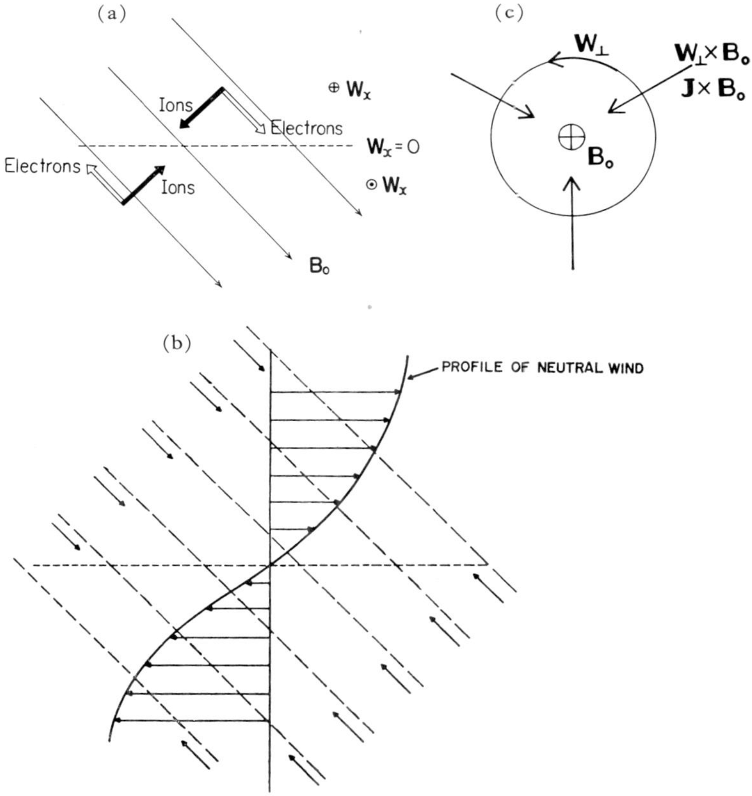

FIG. 4. 9. Wind shear. (a) Wind shear effect orthogonal to $\boldsymbol{B}_0$, important in the E region below 150 km; $\boldsymbol{W}_x$ is horizontal and zonal. (b) Wind shear effect parallel to $\boldsymbol{B}_0$, important above 150 km; $\boldsymbol{W}$ is horizontal and meridional (AXFORD, 1963). (c) Wind shear effect orthogonal to $\boldsymbol{B}_0$; $\boldsymbol{W}$ is toroidal.

southward, eastward, and upward, respectively). Then, div $\boldsymbol{J}=0$ means now $J_z=0$, because only sinusoidal part concerns us now. By (4. 3) and $J_z=0$

$$E_z = \frac{-B_0 \cos I}{\sigma_0 \sin I}(\sigma_1 W_y - \sigma_2 \sin I \, W_x) \qquad \text{for } \tan I \gg \frac{\sigma_1}{\sigma_0} \tag{4. 10}$$

because the polarization takes place only along the z direction; we have an infinite extension horizontally. E_z in (4. 10) is so small as $(\sigma_1+\sigma_2)/\sigma_0$ $|WB_0|$ by $(\sigma_{1,2}/\sigma_0) \ll 1$ that the ion motion is affected only very slightly for $\Omega_i \sim \nu_i$. However, E_z is very important to drive electrons (mainly along $\boldsymbol{B}_0$) as

$$V_{e,\parallel} = \mu_e^{(0)} E_z \sin I. \tag{4. 11}$$

For $\Omega_e \gg \nu_e$ we have

$$\sigma_0 \simeq -n_e e \mu_e^{(0)}, \ \sigma_1 \simeq \left(\frac{n_e e}{B_0}\right) \frac{\nu_i \Omega_i}{\nu_i^2+\Omega_i^2} \text{ and } \sigma_2 \simeq \frac{n_e e}{B_0}\left(1-\frac{\Omega_i^2}{\nu_i^2+\Omega_i^2}\right)$$

(see (A.4.2) and (A.4.3)). Then, we obtain by (4. 10)

$$V_{e,\parallel,z} = \cos I\left[\frac{\Omega_i \nu_i}{\nu_i^2+\Omega_i^2} W_y + \left(\frac{\Omega_i^2}{\nu_i^2+\Omega_i^2}-1\right)\sin I\, W_x\right]. \tag{4. 12}$$

Thus, by $[\boldsymbol{B}_0 \times (\boldsymbol{W} \times \boldsymbol{B}_0)]_z = -\sin I \cos I B_0^2 W_x$ and $\mu_e^{(2)} \simeq 1/B_0$, and $\mu_e^{(1)} \simeq 0$

$$V_{e,z} = -\frac{[\boldsymbol{B}_0 \times (\boldsymbol{W} \times \boldsymbol{B}_0)]_z}{B_0^2} + V_{e,\parallel,z}$$

$$= \frac{\Omega_i \nu_i}{\nu_i^2+\Omega_i^2} \cos I\, W_y + \frac{\Omega_i^2}{\nu_i^2+\Omega_i^2} \cos I \sin I\, W_x \tag{4. 13}$$

and by (4. 9′)

$$V_{ez} = V_{iz}\ (J_z = 0). \tag{4. 14}$$

It is found by (4. 14) that only the mass (of equal number of ions and electrons) flows vertically. Since $\boldsymbol{W}$ varies vertically, the mass flux also varies, converging ionization (1) at the height where W_y in the first term in (4. 13) changes from positive to negative with z in the lower ionosphere with $\nu_i \geq \Omega_i$ and (2) that where W_x in the last term changes from positive to negative with z in the upper ionosphere. This implies that in the northern hemisphere the convergence occurs (1) for westward winds overlying eastward winds and (2) for northward winds overlying southward winds. In the southern hemisphere the wind direction for (2) i.e. the north-south winds must be reversed, because the magnetic dip becomes negative (upward). One can use (4. 8) instead of (4. 1) and get the identical result. In this case the convergence or divergence of ionization flux, $(n_e V_{e,i})$, is produced by $(n_e \boldsymbol{C}_0) \propto (\boldsymbol{J} \times \boldsymbol{B}_0)$ because of div $\boldsymbol{J}=0$ and div $\boldsymbol{W}=0$ as is assumed above.

We shall illustrate by a simple case how much ionization piles up at the convergent nulls. The equation of continuity (1. 2) of ionization includes production Q_p due to solar radiation and recombination $\alpha_{eff} n_e^2$ where α_{eff} is the effective recombination coefficient for molecular ions, as

$$\frac{\partial n_e}{\partial t} = Q_p - \alpha_{eff} n_e^2 - \operatorname{div}(n_e \boldsymbol{V}_i) \tag{4. 15}$$

$\boldsymbol{V}_i$ in (4.15) is the velocity given by (4. 9) or (4. 13). In stationary state, $\partial n_e/\partial t = 0$ and

$$\alpha_{eff} n_e^2 + n_e \frac{dV_{iz}}{dz} + V_{iz} \frac{dn_e}{dz} - Q_p = 0. \tag{4. 16}$$

The maximum of n_e, n_m, can be easily obtained by putting $(dn_e/dz)=0$ in (4.16) as

$$n_m = \frac{-(dV_{iz}/dz)+\sqrt{4\alpha^2_{eff}n_0^2+(dV_{iz}/dz)^2}}{2\alpha_{eff}} \tag{4.17}$$

where n_0 is n_e in the absence of the wind as $Q_p=\alpha_{eff}n_0^2$. An extreme case is

$$\left|\frac{dV_{iz}}{dz}\right|^2 \gg (\alpha_{eff}n_0)^2 \tag{4.18}$$

for which

$$n_m/n_0 = -\frac{dV_{iz}/dz}{\alpha_{eff}n_0} \tag{4.19}$$

where $(dV_{iz}/dz)<0$ for the convergent nulls of V_{iz}. This case can occur for normal ions at night in the E region for $|V_{iz}|\simeq 3\times 10$ m sec^{-1} and the vertical wavelength $\lambda_z\simeq 15$ km which give $|dV_{iz}/dz|\simeq 2\times 10^{-3}$ sec^{-1} and $\alpha_{eff}=10^{-13}$ m^3 sec^{-1} and $n_0\simeq 10^9$ m^{-3} which give $\alpha_{eff}n_0\simeq 10^{-4}$ sec^{-1} (HARPER, 1979). For atomic metallic ions this is the case even in day time in the E region for $\alpha_{eff}\simeq 10^{-14}$ m^3 sec^{-1} and $n_0=10^{10}$ m^{-3} sec^{-1}. It is thought that intense sporadic E at midlatitudes is formed by such a mechanism. The metallic ions have been detected in sporadic E by direct mass spectroscopic experiment on board on rockets (YOUNG *et al.*, 1967). It is likely that the metallic ions are of extraterrestrial origin, entering the earth's upper atmosphere as meteors and meteorites. The above simple example is instructive for understanding the wind shear effect, though it is a gross oversimplification.

KATO *et al.* (1972) have attempted to compare the result of n_e distribution measured by rocket-borne probe with that calculated from simultaneously measured wind profiles (Fig. 4. 10). The time-dependent solution and the observed n_e are shown in Fig. 4. 9(b), (c), (d), and (e) where

$$n_0 = n_{00}\exp[(z-150)/10] \tag{4.20}$$

in which z is in km; $n_{00}=1.4\times 10^{11}$ m^{-3}. In the integration of the equation of continuity (4. 15) for t and z, the method of characteristics was adopted. The assumed condition at $t=0$ is a sudden application of the wind shear on an initial distribution given by (4. 20). If $W_{x,y}$ is due to a tide or GW traveling vertically, such sudden disturbance over the entire range of heights does not really happen. Upon meeting the tide and GW, the ionization begins to be disturbed gradually. Thus, only the result at large t, which gives almost a stationary state, is significant. Even so, the present approach is useful, because the boundary condition is generally unknown. In comparison between the observed and calculated result, we see that the calculation for $\alpha_{eff}=10^{-14}$ m^3 sec^{-1} and $t=90$ min explains the observation well; 90 min gives an almost stationary state. The largest peak at 101 km has almost the same magnitude

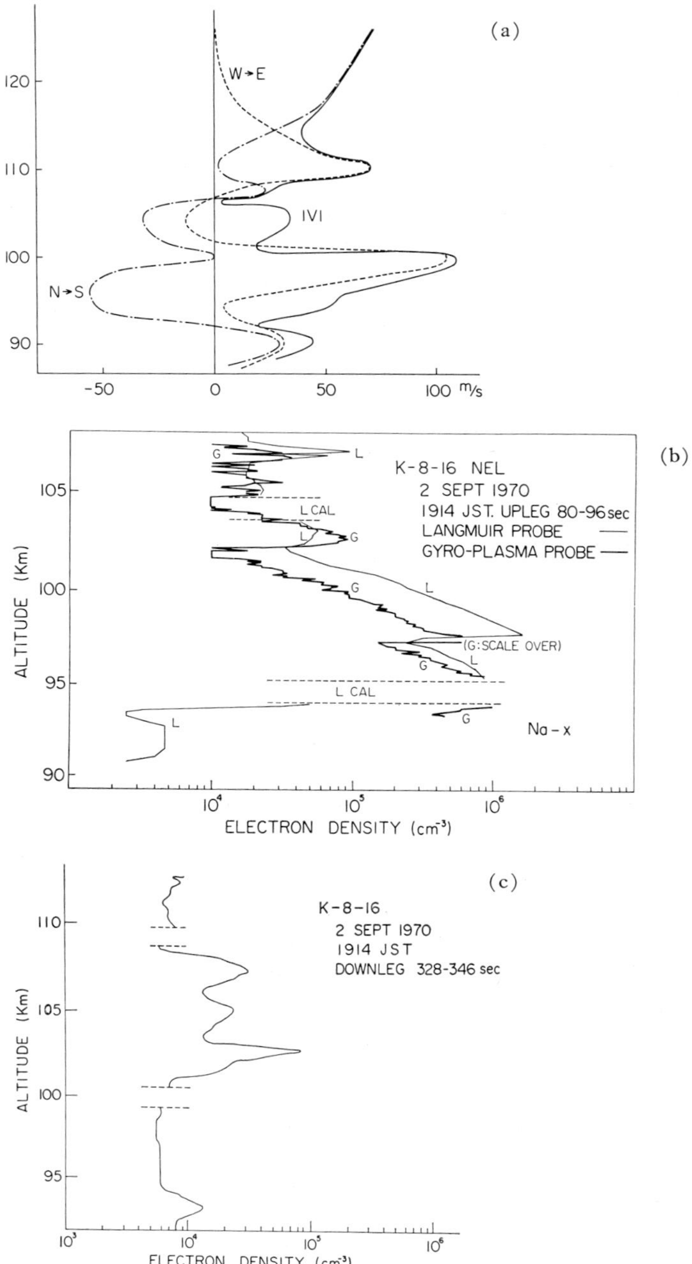
(a)
W→E
IVI
N→S
120
110
100
90
-50
0
50
100 m/s
(b)
K-8-16 NEL
2 SEPT 1970
1914 JST. UPLEG 80-96 sec
LANGMUIR PROBE
GYRO-PLASMA PROBE
L CAL
(G:SCALE OVER)
Na - x
ALTITUDE (Km)
ELECTRON DENSITY (cm⁻³)
(c)
K-8-16
2 SEPT 1970
1914 JST
DOWNLEG 328-346 sec
ALTITUDE (Km)
ELECTRON DENSITY (cm⁻³)

FIG. 4. 10

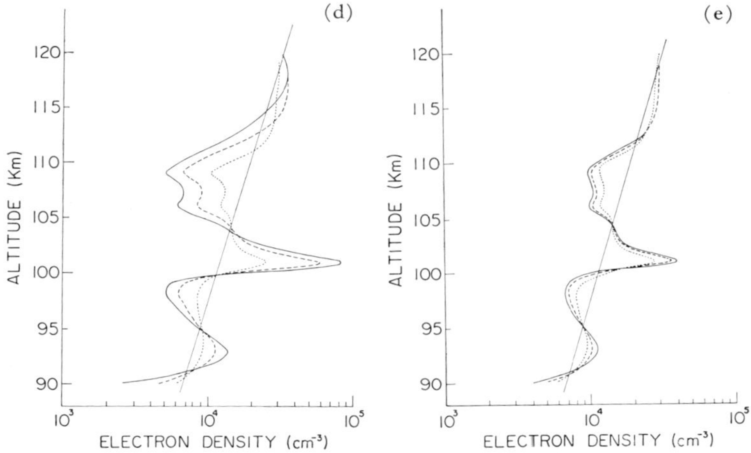

FIG. 4. 10. Sporadic E, observation and theory. (a) Observed wind profile; southward velocity (chain curve), eastward wind (broken curve) and the total wind velocity (solid curve). The result is based on a sodium vapor release from rocket K-8-16 which was launched from Kagoshima Space Center (31°15′N 131°04′E) at 1914 JST on 2nd Sept., 1970. (b) Observed electron density by probe on board. Note that upon chemical release a large increase of electron density occurred around 98 km; G is the result by the gyroplasma probe and L that by the Langmuir probe. The result was obtained during ascent. (c) Similar to (a) except that the result was obtained during descent by L. The missing part corresponds to the calibration period. (d) Calculated electron density distribution $\alpha_{eff}=10^{-14}$ m^3 sec^{-1}, at 15 min (dotted), 60 min (broken), 90 min (solid). The thin straight line is the initial distribution. In the calculation we used the relation as $R_i=(\nu_i/\Omega_i)=80 \exp[-(z-90)/10]$ where z is in km. (e) Similar to (d) except that $\alpha_{eff}=3\times10^{-14}$ m^3 sec^{-1} and the dotted curve in n_e at 10 min instead of 15 min in (d), (KATO *et al.*, 1972).

as that observed at the same height. A good coincidence also occurs for the second largest peak at 93 km. It is remarkable that a small peak at 107 km in the calculation corresponds to that at 103 km in the observations. The smaller magnitude in the calculation results from the constant recombination coefficients with z. If α_{eff} decreases with height, the observed magnitude can be better explained. The peak at 116 km in the calculation, does not correspond well with the peak at 106 km in the observation. Considering, however, that the horizontal distance between the location of the sodium cloud and the location where the density is measured is several tens of kilometers, the coincidence is surprisingly good.

Another important factor to be taken into account is ambipolar diffusion

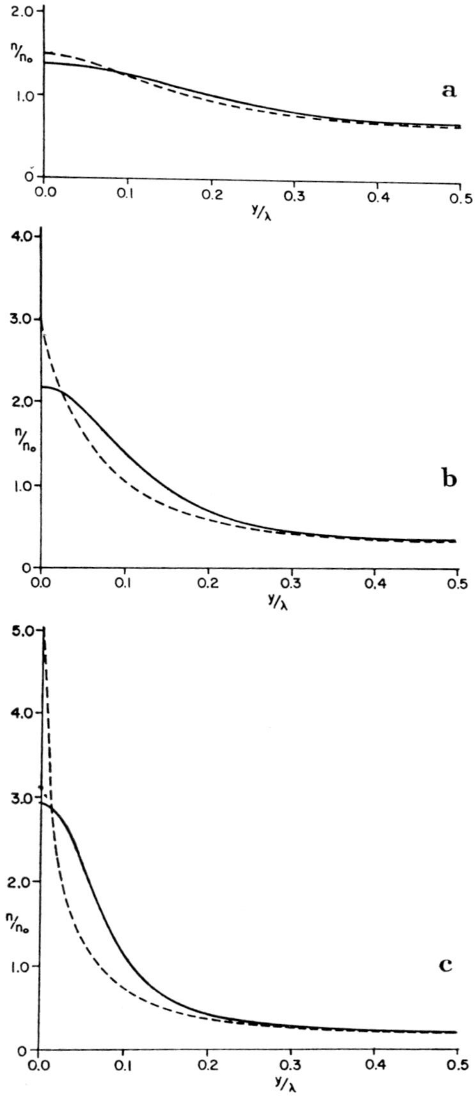

FIG. 4. 11. Calculated electron density distribution for stationary layers (AXFORD, 1963; the original figure is redrawn). The non-disturbed density is $Q_p = 4 \times 10^7\,\mathrm{m}^{-3}\,\mathrm{sec}^{-1}$, $\alpha_{\mathrm{eff}} = 10^{-14}\,\mathrm{m}^3\,\mathrm{sec}^{-1}$, $T_\mathrm{i} + T_\mathrm{e} = 500\,°\text{—}2000\,°\mathrm{K}$, $m_\mathrm{i} = 2.67 \times 10^{-26}\mathrm{Kg}$. The three cases are distinguished by $(2\pi \sin I \cos I)\ (\lambda^{-1} W / \alpha_{\mathrm{eff}} n_0)$ which amounts to (a) 0.835, (b) 2.662 and (c) 4.804 ; W is the eastward velocity and λ is the vertical wavelength. The ratio between maximum and minimum is (a) 1.5 (b) 3.0 (c) 5.0. The broken curves correspond to the case without diffusion and the solid curves with diffusion. For diffusion, $2\pi\sigma(T_\mathrm{i} + T_\mathrm{e})/(m_\mathrm{i}\nu_\mathrm{i}\alpha_{\mathrm{eff}}\, n_0)\ \lambda^{-1} = 0.01$. In this figure the y axis is upward.

which limits the peak density. The diffusion velocity is an additional term in (4. 1), when one considers " hot plasma" for which the pressure force $\nabla p_{e,i}$ is important. The additional velocity $\Delta \boldsymbol{V}$ is given by (A. 4. 8) as

$$\Delta \boldsymbol{V} = \frac{\Omega_e \Omega_i \nu_e \nu_i}{(\nu_e \nu_i + \Omega_e \Omega_i)(\Omega_e \nu_i + \Omega_i \nu_e)} \frac{\nabla_\perp (p_e + p_i)}{e n_e B_0} - \frac{\Omega_e \Omega_i}{\Omega_e \nu_i + \Omega_i \nu_e} \frac{\nabla_{/\!/} (p_e + p_i)}{e n_e B_0} \tag{4.21}$$

where $p_{e,i} = n_e \sigma T_{e,i}$; the diffusion parallel to $\boldsymbol{B}_0$ grows with height, while the orthogonal diffusion to $\boldsymbol{B}_0$ can be neglected above 150 km. The introduction of the diffusion velocity makes the continuity equation be of second order with respect to the space coordinate. AXFORD (1963) has discussed such a case (Fig. 4. 11).

The wind shear effect in midlatitudes has been discussed above; I must be larger than a few degrees for (4. 10) to be valid. Near the magnetic equator where I is less than a few degrees, the vertical polarization field tends to balance the applied $(\boldsymbol{W} \times \boldsymbol{B}_0)$-field, diminishing the effective field $[\boldsymbol{E} + (\boldsymbol{W} \times \boldsymbol{B}_0)]_z$. According to the simple geometry with straight field lines as in midlatitudes, $E_z = -W_y B_0$ at $I = 0$. However, actually, the curved field line constitutes a closed circuit, along the field line down in the lower region and across the field line dissipating the polarization to certain extent (Fig. 4. 12). Then, the ion and electron are able to move together, producing the convergence as in midlatitudes. KATO (1973) has discussed the problem showing

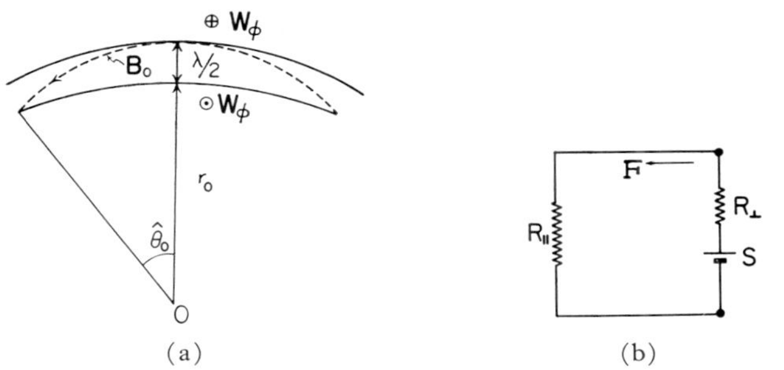

FIG. 4. 12. Dissipation of polarization along $\boldsymbol{B}_0$ (KATO, 1973). (a) Vertical potential drop at the equator dissipates along $\boldsymbol{B}_0$ down to the lower level and across $\boldsymbol{B}_0$. Here λ is the vertical wavelength; r_{C} is the earth's radius plus the ionosphere height; $\hat{\theta}_0$ is the latitude at which the field line meets the reverse wind. (b) Equivalent circuit; $S = \int_{z_1}^{z_2} (\boldsymbol{W} \times \boldsymbol{B}_0)\, dz$ where $z_{1,2}$ corresponds to the upper and lower level. $R_{/\!/}$ = total resistance along field line between the lower and upper levels. $R_\perp$ = total resistance across $\boldsymbol{B}_0$ in the lower level. $\boldsymbol{F}$ is the electrical current.

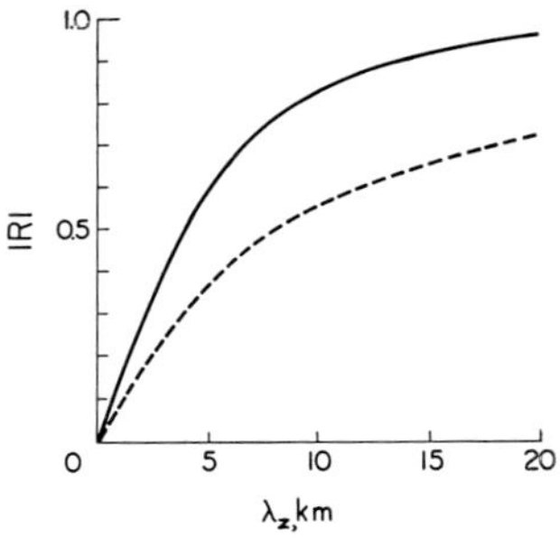

FIG. 4. 13. The effective electric field at the equator for the wind shear (KATO, 1973); here, $|R|=|\hat{E}_z/WB_0|$ versus $\lambda_z=2\pi/k$. The solid curve corresponds to the case for uniform electron density n_0 in the absence of the perturbation and the broken curve to the case with an exponential increase of n_0 proportional to e^{kz} where $k=2\pi/\lambda_z$.

a possible application of the wind shear mechanism even near the magnetic equator. The expected total field effective for the ionization accumulation is dependent on the vertical wavelength of $\boldsymbol{W}$, λ_z; the longer λ_z is, the larger the effective field (Fig. 4. 13). Later, the result was confirmed by Reddy and Devasia (private communication, 1977). It has been found that an intense peaky structure of n_e occurs in reality (PRAKASH *et al.*, 1970) (Fig. 4. 14).

If $\boldsymbol{W}$ is of tidal or GW origin, the wind profile propagates, usually downwards, with a phase velocity $\boldsymbol{V}_p$. Then, the ionization perturbation by the wind shear is expected to propagate with $\boldsymbol{V}_p$. However, the maximum and minimum are displaced from the original convergent nulls of $\boldsymbol{W}$ in the absence of propagation. This is readily understood by a transformed coordinate system moving with $\boldsymbol{V}_p$ with the waves. On this coordinate, $\boldsymbol{W}$ remains stationary as assumed above. If the background of n_0 and $\nu_{e,i}$ is uniform, the maximum occurs at the convergent nulls on the *moving coordinates* i.e. $V_{iz}=V_p$. Since $V_p<0$, the new convergent nulls are located above the original ones where $V_{iz}<0$. However, the situation is more complicated, because, on the moving system, $n_0(z)$ and $\nu_{i,e}(z)$ are all time-dependent. In the case of tides, $|V_p|\ll|V_{iz}|$ and the displacement is fairly small. FUJITAKA and TOHMATSU (1973) have attempted to interprete the formation and the descending motion of the intermediate E layer or $1(^1/_2)$ E layer in terms of the moving wind shear effect of the semidiurnal tide.

The motion of ionization caused by the propagating wind shear has been examined in greater detail by CHIMONAS and AXFORD (1968). They have found that after the convergence of ionization at the new convergent nulls the ionization is transported down to the lower E region around 90 km, where (ν_i/Ω_i) becomes so large that the magnetic field is unable to deflect the ioniza-

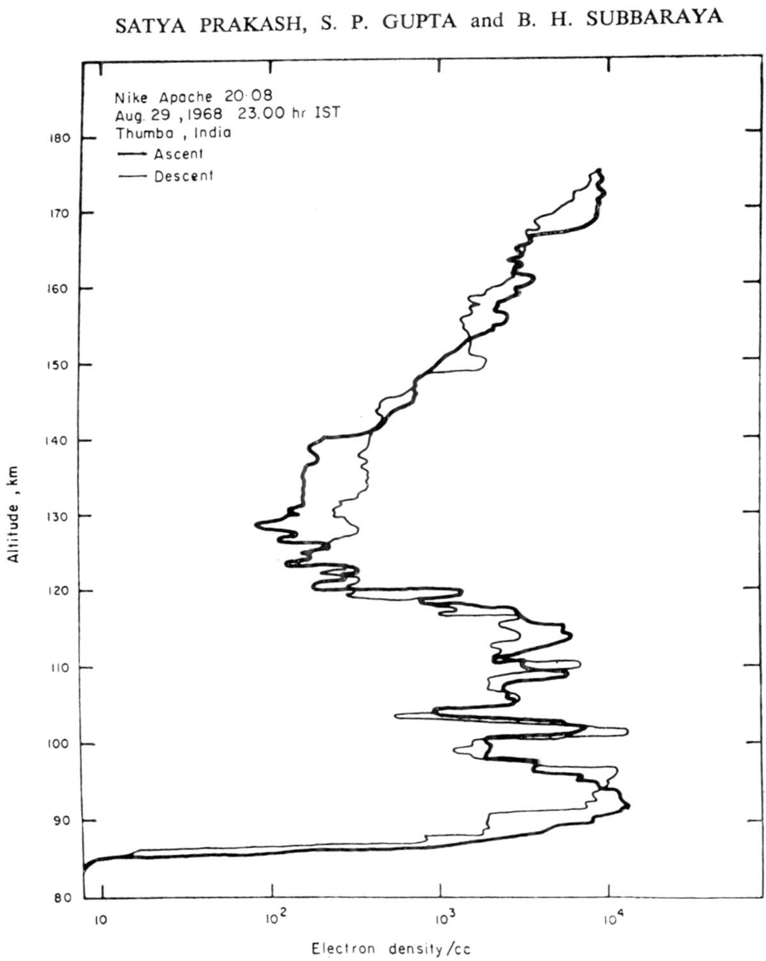

Fig. 4. 14. Observed electron density profile: Rocket-borne probe detected various peaks above Thumba, India, 2300 IST, August 29, 1968. The heavy and thin lines show the observation on the ascent and descent path, respectively (Prakash *et al.*, 1970).

tion orthogonally to ***W*** and the ions move almost with neutral particles. The transported metallic ions are thus dumped, giving rise to an intense layer; V_p is so small that only metallic ions with long life time can survive such a long trip down to the D region from above 120 km; W_y can contribute to such transportations from the region where $\nu_i \geq \Omega_i$, whilst W_x can contribute from the

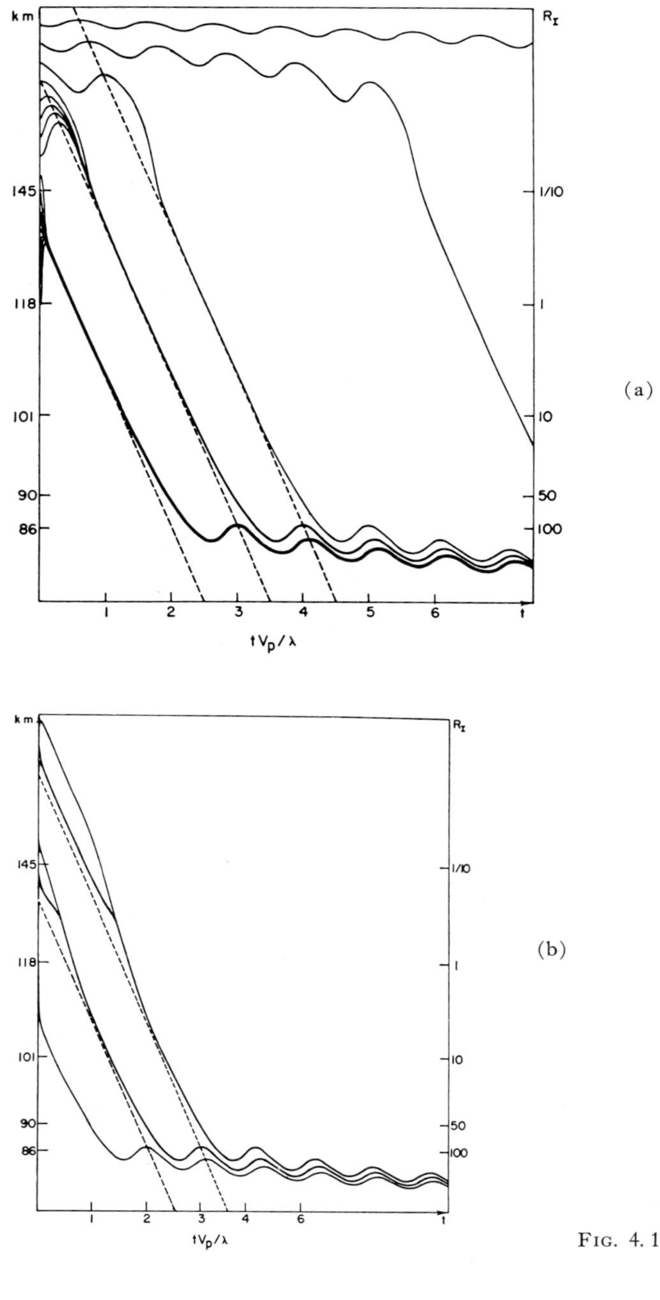
km
R_I
145
118
101
90
86
1/10
1
10
50
100
1
2
3
4
5
6
t
tVp/λ
(a)
km
R_I
145
118
101
90
86
1/10
1
10
50
100
1
2
3
4
6
t
tVp/λ
(b)

Fig. 4.15

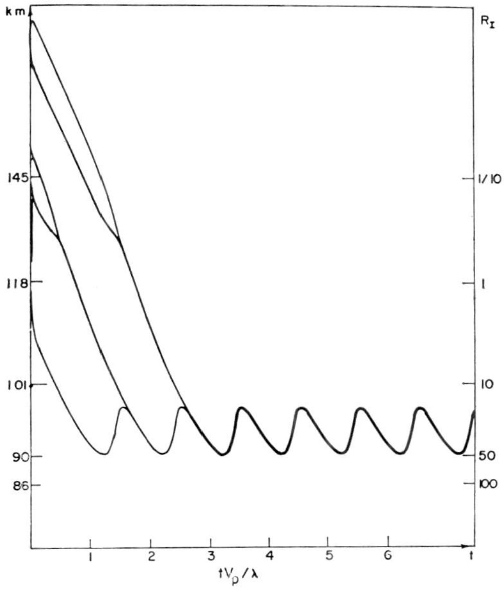

FIG. 4. 15. (a) Trajectories of ionization transported by W_y. Thirteen representative points equally spaced between $z=0$ and $z=2\lambda_z$ at $t=0$ are followed where $z=0$ correspond to about 118 km with $\Omega_i=\nu_i$; $\lambda_z=3$ scale heights. $W_y=A \sin [(k(z-V_p t)]$ where $A=100\ V_p$. The wind dumps its ionization in the 85–90 km region. The broken straight lines indicate the phase velocity. The peak is located above the lines. (b) Similar to (a) except for W_x as well as W_y. $W_x=A \sin [k(z-V_p t)+67°]$, $I=27°$. Note that the ionization is collected much faster at high altitudes by W_x and, with parameters and phase angle used here, forms into double layers that merge as the profile descends. (c) Similar to (a) and (b) except for the existence of C_{oz} which amounts to $(A/2)$. The dumping takes place around 90–95 km higher than in (a) and (b). The oscillation at low altitude is remarkable (CHIMONAS and AXFORD, 1968); λ_z is given as λ on the diagram abscissa.

higher region as with $\Omega_i \gg \nu_i$. Figure 4. 15 illustrates the trajectories of ionizations in the presence of tidal wind. The trajectories were obtained by numerical integration of $V_{iz}=dz/dt$ from initial height z_0 and $\boldsymbol{W} \propto e^{i(\omega t+kz)}$.

The presence of large scale quasi-electrostatic field produces a corresponding large scale uniform motion $\boldsymbol{C}_0$ of ionization, which tends to be $(\boldsymbol{E}\times\boldsymbol{B}_0)/B_0^2$ in the F region for $\nu_{e,i} \ll \Omega_{e,i}$. It was pointed out by KATO (1966c) that if C_{0z} exceeds the vertical phase velocity, which is true always for tide and occasionally for GW, the peak becomes smaller and the location is shifted from the original convergent nulls (Fig. 4. 16). Fig. 4. 15 (c) shows that a constant vertical drift velocity is also effective in displacing the dumped height of the transported ions with the velocity V_p as mentioned above (CHIMONAS and AXFORD, 1968).

Kato also suggested a possible resonance, if the phase velocity and the drift velocity are close. If this occurs, sporadic E formation is expected to be correlated to the ionospheric current, as was found by MATSUSHITA (1962). WHITEHEAD (1971) discussed more precisely the possible resonance both in the E and F regions. His work revealed the following complications in the F region:

1) The wave vector $\boldsymbol{k}$ of $\boldsymbol{W}$ (with tidal or GW origin), especially the vertical component, varies with height due to varying atmospheric structure. So, if a resonance exists, it would be restricted to some height range.

2) Generally, the background wind is present and the wind varies with height, destructing the resonance condition. Moreover, the ionization motion in turn reacts on the wind, slowing it down.

3) The wave and the background wind interact with each other tending to destruct the original resonance condition.

Whitehead suggested that the resonance could possibly increase the densities at a peak by 20%.

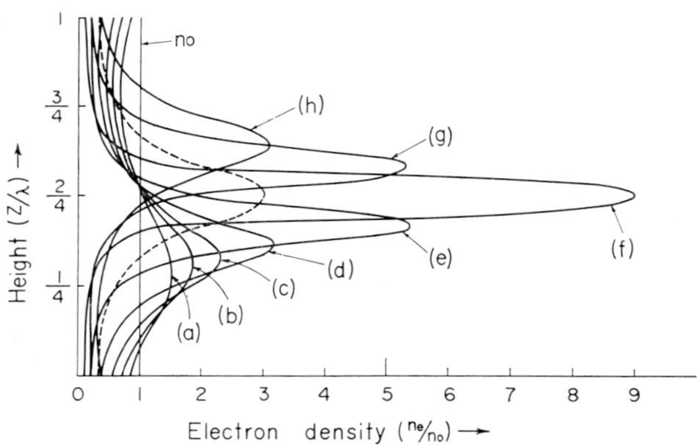

FIG. 4. 16. Density profile for time dependent layer, when a uniform electrostatic field is suddenly applied in the presence of horizontal wind which varies sinusoidally along z. Cases (a)–(h) are distinguished by the time, normalized by the wind period, after application of the electric field:

a	b	c	d	e	f	g	h
$\frac{\pi}{6}$	$\frac{\pi}{4}$	$\frac{\pi}{3}$	$\frac{\pi}{2}$	$\frac{5\pi}{6}$	$\frac{5\pi}{3}$	$\frac{17\pi}{6}$	$\frac{19\pi}{6}$

The ratio between the ion velocity and C_0 is 0.8 (KATO, 1966c). The broken curve shows the case with no electric field.

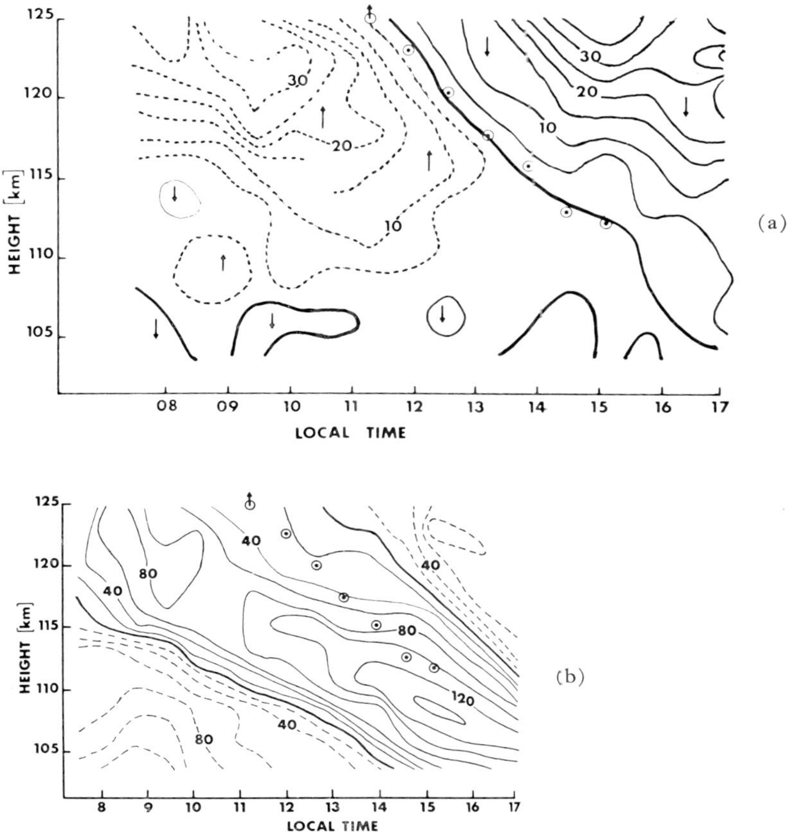

FIG. 4. 17. Measurement of ion drift and the height of sporadic E over Arecibo (HARPER *et al.*, 1975). (a) The vertical *downward* ion drift velocity (m sec^{-1}). The contour spacing is 5 m sec^{-1}. (b) The sourthward ion drift velocity (m sec^{-1}). The contour spacing is 20 m sec^{-1}. The solid and broken curves are the contour of the positive and negative velocities, respectively. ⊙ denotes the height of sporadic E.

As to the E region, $\boldsymbol{C}_0$ is generally much smaller than the phase velocity of GW's V_p. However, the background horizontal wind $\boldsymbol{W}_0$ should be responsible for the space resonance, if

$$\boldsymbol{k} \cdot \boldsymbol{W}_0 = V_p. \tag{4.22}$$

The condition, however, is also that of the critical level. And an intense interaction takes place between the wave $\boldsymbol{W}$ and the general wind $\boldsymbol{W}_0$, complicating the analysis. It is an interesting open question to date.

RÖTTGER (1978) has recently found that the space resonance theory can explain his observed equatorial spread F. He noticed that the spread F irregularity appears, when the electric field becomes westward, the resultant vertical drift being downward like the GW phase velocity. Scales in time and space seem to be consistent with the GW produced irregularities.

HARPER *et al.* (1975) directly confirmed that the sporadic E layer occurs at the convergent null in V_z. He measured simultaneously $\boldsymbol{V}_i$ in the magnetic meridian plane and the height of sporadic E by incoherent scatter radar. It was found that the sporadic E occurred at the convergent nulls of $\boldsymbol{V}_i$ (Fig. 4. 17) and near the maximum northward winds. HARPER *et al.* later showed (1976) that the neutral wind probably had the required null in the east-west wind associated with maximum northward winds, provided that the wind is tidal. This is consistent with the wind shear theory. It is fairly common to see between the E and F regions at night a downward moving layer of ionization which is caused by the wind shear of tidal wind with the (2, 2)-mode (HARPER, 1979).

CHAPTER 5

TIDAL DYNAMO

The magnetic field at the surface of the earth $\boldsymbol{B}_0$, which has an amplitude of about $10^5\gamma$, (10^{-4} Wbm^{-2}) shows variations of the order of 10^0 to $10^3\gamma$ on time scales of seconds to days. Some of these variations have been long recognized to be due to magnetic activity in the high magnetic latitudes. However, even on days when there is no significant magnetic activity, a systematic variation of the earth's magnetic field is observed. This solar daily geomagnetic variation, which is usually denoted as S_q, is derived as the average of the daily variation with solar time of only geomagnetically quiet days on worldwide scales. Technically, five internationally quiet days per month are selected for low and midlatitude stations. For high latitude station, further restrictions are considered (CHAPMAN and BARTELS, 1940). The concept of S_q represents a limiting state which can be attained, when disturbances would be all removed. The S_q variation is mainly diurnal, the phase varying with latitudes as in Fig. 5. 1. It is straightforward to express the variation in terms of the horizontally flowing electric current by Biot-Savart's law in electromagnetics. The overall chracteristic of this equivalent current system is fairly simple (Fig. 5. 2). The system consists of current vortices on each hemisphere. Between the northern and southern hemispheres, the direction of the current around the vortices is opposite (counterclockwise in the northern hemisphere in daytime). The same is true between the day and night hemispheres. In summer the vortices become stronger than that in winter, invading the winter hemisphere. The weak nighttime vortices depend strongly on the definition of the base line of daily geomagnetic variation, one problem which cannot be solved only from geomagnetic data at sea-level. It is not the subject in this book to discuss any further details of S_q except to note that S_q can vary from day-to-day (Fig. 5. 3).

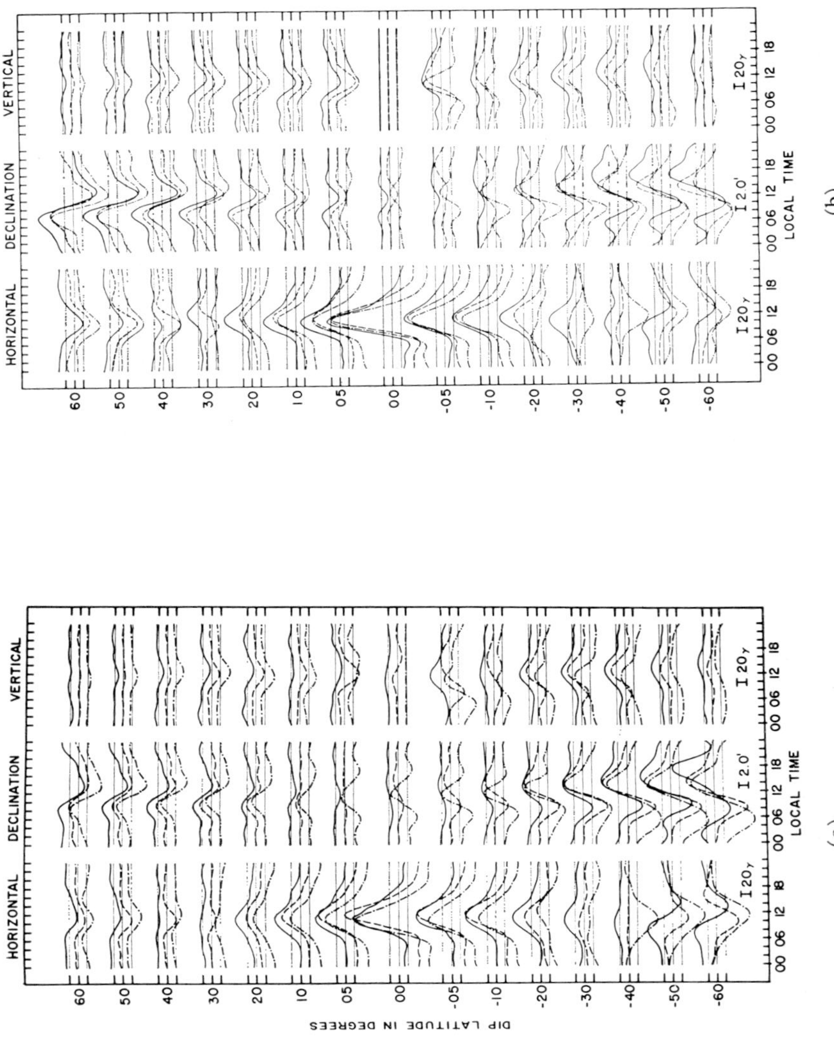
HORIZONTAL
DECLINATION
VERTICAL
60
50
40
30
20
10
05
00
-05
-10
-20
-30
-40
-50
-60
I 20γ
I 2.0'
I 20γ
00 06 12 18
LOCAL TIME
(b)
DIP LATITUDE IN DEGREES
(a)

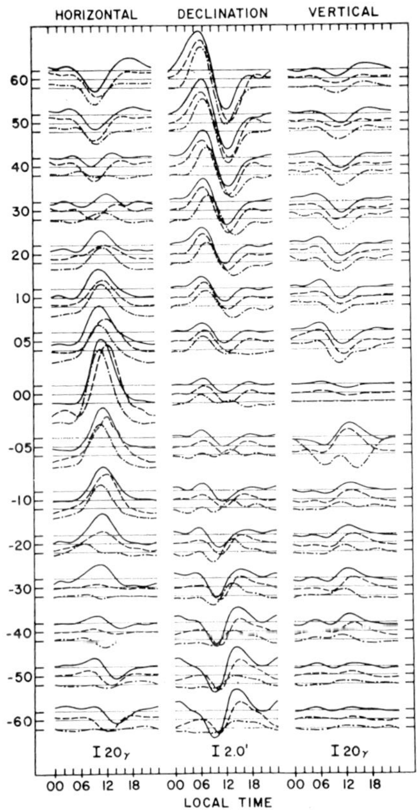

(c)

FIG. 5. 1. Solar quiet daily variation of geomagnetic three components (S_q) at several dip latitudes from 60° to −60° in zones 1 (solid curves), 2 (broken curves) and 3 (chain curves) for D, E and J months (MATSUSHITA and MAEDA, 1965). Zone 1: The Europe-Africa zone, 45°–165° E geomagnetic longitude. Zone 2: the Asia-Australia zone, 165°–285° E geomag. long. Zone 3: the North America-South America zone, 285°–45° E geomag. long. (a) D month: January, February, November and December. (b) E month: March, April, September and October. (c) J month: May, June, July and August.

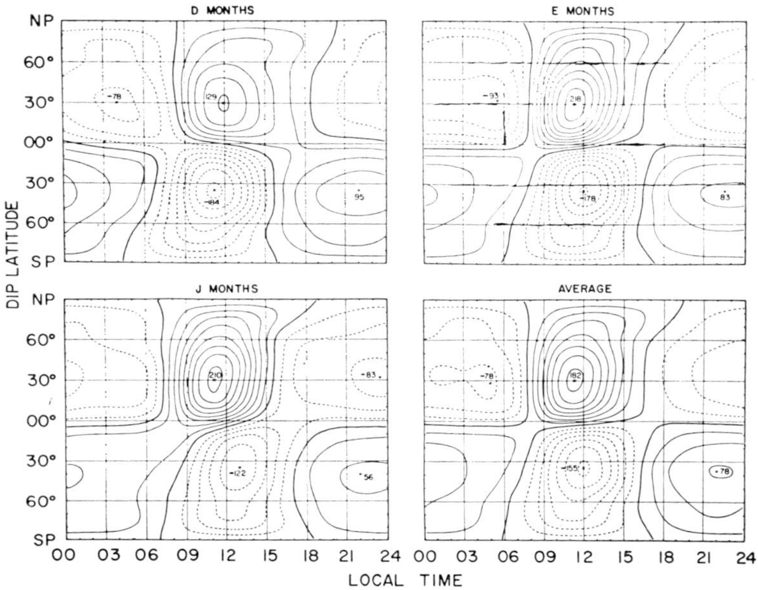

FIG. 5. 2. S_q current system (MATSUSHITA and MAEDA, 1965). External part is shown in D, E, and J months and in the yearly average. The current intensity between two consecutive lines is 25×10^3 A. The thick solid curves indicate the zero line, the thin solid curves counterclockwise current, the broken curves clockwise current and the numbers near the small cross the total current of each vortices in units of 10^3 A.

5.1 Separation of S_q between Internal and External Parts

It was STEWART (1883) who first suggested the existence in the earth's upper atmosphere of the above equivalent current system. He further went on saying that the electromotive force necessary to drive the current would be a motion of possibly electrically conductive upper atmosphere across the geomagnetic field. This idea of the atmospheric dynamo was conceived a few decades before the discovery of the ionosphere by Appleton in 1924. SCHUSTER (1889) proved that the S_q is mainly caused by some source above the earth's surface. This was an important step forward on the study of S_q. Schuster attempted to separate S_q into two parts, corresponding to the two independent solutions of the Laplace equation,

$$\nabla^2 \mathscr{S} = 0 \tag{5.1}$$

where $\mathscr{S}$ is a potential function of S_q; the S_q magnetic field variation $\Delta \boldsymbol{H}$ ob-

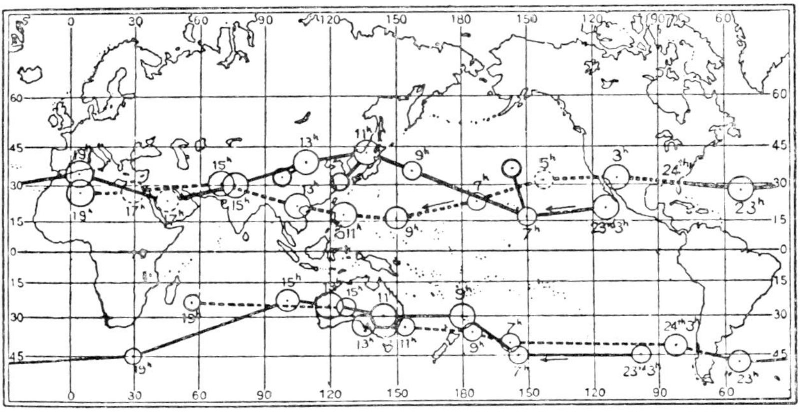

FIG. 5. 3. Day to day variability of S_q current system (HASEGAWA, 1936). Map showing the tracks of the centers of the equipotential curves of the field of the daily magnetic variation on September 23 (full line) and 24 (broken line) at various Japanese Mean Times (beginning at 3-hr JMT on September 23, i.e. 18-hr GMT on September 22).

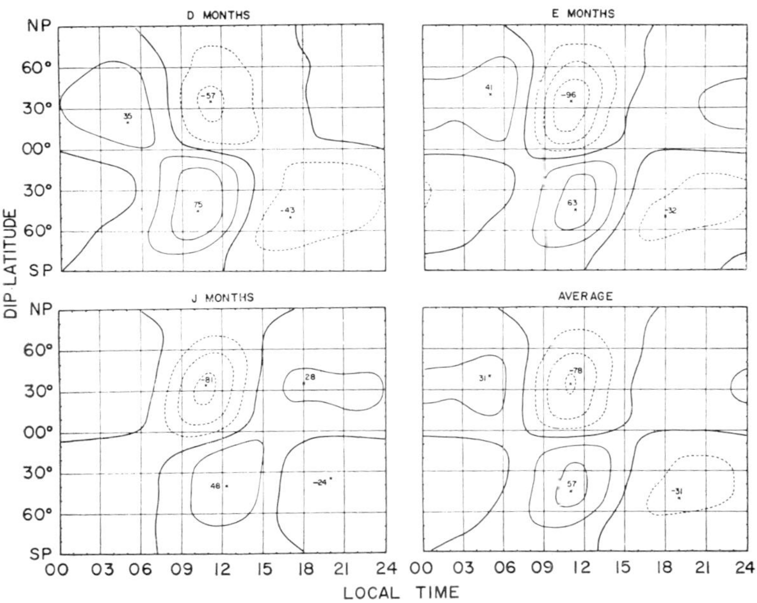

FIG. 5. 4. Internal part of S_q current system relative to that in Fig. 5. 2 (MATSUSHITA and MAEDA, 1965).

served at the ground should be derived from this scalar potential, because no electric current causing S_q seems to flow at the ground and by (4. 5) curl $(\Delta \boldsymbol{H})=0$. The solution of (5.1) consists of $r^n P_{n,m}(\theta)\,e^{im\phi}$ and $r^{-(n+1)} P_{n,m}(\theta)\,e^{im\phi}$ where $n=1,2,\ldots$ ((A 5.1)–(A 5.4)). The former ($\propto r^n$) becomes large with increasing r, whilst the latter ($\propto r^{-n}$) becomes large with decreasing r. Thus, the former gives the field due to sources external to the earth's surface with $r=a$, and the latter is due to internal sources. Generally, $\mathscr{S}$ is a combination of the two parts. Here, the time derivative of $\Delta \boldsymbol{H}$ is neglected, as the variations are so slow that the displacement current is negligible. If we assume S_q to vary purely with local time, the time dependency of the solution is $e^{im(\Omega t+\phi)}$ and we have

$$\mathscr{S} = \sum_{m,n} [a_{m,n} r^n + b_{m,n} r^{-(n+1)}]\, P_{n,m}\, e^{im(\Omega t+\phi)}. \tag{5.2}$$

In (5. 2) $a_{m,n}$ and $b_{m,n}$ are arbitrary constants; $m=1,2$, etc. correspond to diurnal, semidiurnal, etc. The observed $\Delta \boldsymbol{H}$ at many stations and at various local times is expressed by Fourier series about $(\Omega t+\phi)$ as $\sum_m \Delta \boldsymbol{H}_m(\theta)\, e^{im(\Omega t+\phi)}$ where $\Delta \boldsymbol{H}_m(\theta)$ is the latitude dependent vector amplitude. Each $\Delta \boldsymbol{H}_m$, being independent, corresponds to each term of $\mathscr{S}$ with equal m as

$$\Delta \boldsymbol{H}_m = -\text{grad}\, \mathscr{S}_m \qquad \text{at} \qquad r = a \tag{5.3}$$

where $\mathscr{S}_m = \sum_n [a_{m,n} r^n + b_{m,n} r^{-(n+1)}]\, P_{n,m}(\theta)$ and $\mathscr{S} = \sum_m \mathscr{S}_m\, e^{im(\Omega t+\phi)}$. It is

TABLE 5. 1

COMPARISON BETWEEN THE EXTERNAL AND INTERNAL PARTS OF $\Delta H_{m,n}$

$E_n{}^m$ and $I_n{}^m$ are the amplitude of the external and internal part for each m and n, respectively. $\epsilon_n{}^m$ and $i_n{}^m$ are the corresponding phase, respectively. The result of various workers is shown (MATSUSHITA and MAEDA, 1965).

		Chapman [1919]				*McNish* [1937]		*Hasegawa and Ota* [1950]		Matsushita and Maeda	
Year		1902		1905		1923		1932–1933		1958	
Sunspot no. and activity ( )		5 (min.)		64 (max.)		6 (min.)		8 (min.)		185 (max.)	
No. of stations		21		21		5		46		69	
m	n	$E_n{}^m/I_n{}^m$	$\epsilon_n{}^m-\iota_n{}^m$	$E_n{}^m/I_n{}^m$	$\epsilon_n{}^m-\iota_n{}^m$	$E_n{}^m/I_n{}^m$	$\epsilon_n{}^m-\iota_n{}^m$	$E_n{}^m/I_n{}^m$	$\epsilon_n{}^m-\iota_n{}^m$	$E_n{}^m/I_n{}^m$	$\epsilon_n{}^m-\iota_n{}^m$
					Mean of Solstices						
1	2	3.0	−20°	2.8	−03°	...	...	2.3	−09°	2.7	−13°
2	3	2.2	−18	2.3	−19	...	...	2.4	−10	2.3	−15
3	4	2.4	−21	2.7	−20	...	...	2.2	−14	2.2	−09
Mean*		2.5	−20	2.6	−14	...	...	2.3	−11	2.4	−12
Weighted mean†		2.6	−20	2.6	−10	...	...	2.3	−10	2.5	−13
					Equinoxes						
1	2	2.7	−23	2.9	−05	2.4	−21°	...	...	2.8	−13
2	3	2.0	−17	2.4	−18	2.1	−13	...	...	2.3	−11
3	4	2.5	−21	2.4	−21	1.9	−02	...	...	2.2	−14
Mean*		2.4	−20	2.6	−15	2.1	−12	...	...	2.4	−13
Weighted mean†		2.5	−21	2.7	−11	2.2	−16	...	...	2.6	−13

* Simple average of the above three values.
† Weights of 4, 2, and 1 are given to the above three values, respectively.

possible to determine $a_{m,n}$ and $b_{m,n}$ as the solution of the first degree linear simultaneous equation with $\Delta \boldsymbol{H}_m$ as the knowns. Of $\Delta \boldsymbol{H}_m{}^{x,y,z}$ dependent on θ and z, only two components are independent because of (5. 3). However, we have to use either $(\Delta \boldsymbol{H}_m{}^x, \Delta \boldsymbol{H}_m{}^z)$ or $(\Delta \boldsymbol{H}_m{}^y, \Delta \boldsymbol{H}_m{}^z)$ for various θ's, because otherwise the determinant of the coefficient of the simultaneous equations becomes singular (see the expression inside the bracket in (5. 2) being independent of θ and ϕ) and $a_{m,n}$ and $b_{m,n}$ are inseparable. In actual analysis we follow more sophisticated computation scheme decomposing $a_{m,n}$, $b_{m,n}$, and introducing additional parameters. Figures 5.2 and 5.4 illustrate a result of the analysis (MATSUSHITA and MAEDA, 1965), where $\Delta \boldsymbol{H}$ at 69 stations for three longitudinal zones during each of three seasons was used. Table 5. 1 shows the amplitude ratio between external and internal S_q field. We can conclude that a substantial part of S_q is due to external source, a result which is important for the atmospheric dynamo theory.

5.2 Basic Structure of the Atmospheric Dynamo Theory

The atmospheric dynamo theory is based on the following equations (x, y, and z are positive southward, eastward and upward, respectively):

$$\operatorname{div} \boldsymbol{J} = 0 \tag{5. 4}$$

$$\boldsymbol{J} = [\sigma](\boldsymbol{E} + \boldsymbol{W} \times \boldsymbol{B}_0) \tag{5. 5}$$

$$\boldsymbol{E} = -\operatorname{grad} S \tag{5. 6}$$

$$[\sigma] = \begin{bmatrix} \sigma_{xx} & \sigma_{xy} & \sigma_{xz} \\ \sigma_{yx} & \sigma_{yy} & \sigma_{yz} \\ \sigma_{zx} & \sigma_{zy} & \sigma_{zz} \end{bmatrix} \tag{5. 5'}$$

where each element of $[\sigma]$ is a function of $\sigma_{0,1,2}$. Using $\sigma_{0,1,2}$ one can also express (5. 5) generally as (4. 3). The basic structure of the dynamo theory using (5. 4)–(5. 6) is justified in the discussion in Chapter 4, especially in the introductory remarks and Section 4. 1 ((4. 1)–(4. 7′)). In the centered dipole approximation $\boldsymbol{B}_0$ is approximated by a dipole at the earth's center along the rotational axis. Then, each element of (5. 5′) is given by

$$\begin{aligned} \sigma_{xx} &= \sigma_0 \cos^2 I + \sigma_1 \sin^2 I \\ \sigma_{xy} &= -\sigma_{yx} = \sigma_2 \sin I \\ \sigma_{xz} &= \sigma_{zx} = (\sigma_0 - \sigma_1) \cos I \sin I \\ \sigma_{yy} &= \sigma_1 \\ \sigma_{yz} &= -\sigma_{zy} = \sigma_2 \cos I \\ \sigma_{zz} &= \sigma_0 \sin^2 I + \sigma_1 \cos^2 I. \end{aligned} \tag{5. 5''}$$

Figure 4. 3 illustrates $\sigma_{0,1,2}$ which varies with t, θ, ϕ. (5. 4)–(5. 6) consist of seven relations with ten dependent (scalar) variables as $\boldsymbol{J}$, $\boldsymbol{E}$, $\boldsymbol{W}$ and S, of which seven can be expressed in terms of the remaining three variables. Thus, $\boldsymbol{J}$, $\boldsymbol{E}$ and S can be obtained for a given $\boldsymbol{W}$. The procedure is to substitute (5. 6) into (5. 5) which is further substituted into (5. 4). The resulting equation is

$$\operatorname{div}\,[[\sigma]\operatorname{grad} S] = \operatorname{div}\,[[\sigma](\boldsymbol{W}\times\boldsymbol{B}_0)]. \tag{5.7}$$

The equation is a linear second order differential equation in S, and, in principle, the proper boundary conditions should give a unique solution; (5. 7) becomes a Poisson equation for uniform isotropic conductivity. The complete solution of the three-dimensional boundary value problem of (5. 7) has not yet been achieved due to the mathematical complexity. For instance, $[\sigma]$ varies with t, θ, ϕ and z; σ_0 changes by a factor of 10^6 between $z=100$ and 300 km.

Actually, the height-integrated current is considered as

$$I_{\theta,\phi} = \int_{z_1}^{z_2} J_{\theta,\phi}\,\mathrm{d}z \tag{5.8}$$

where $z_{1,2}$ are the lower and upper boundaries of the ionosphere. If we impose the boundary condition that the electric current is confined between the boundaries as $J_z=0$ at $z=z_{1,2}$, (5. 4) gives

$$\frac{1}{a\sin\theta}\frac{\partial}{\partial\theta}(\sin\theta I_\theta)+\frac{1}{a\sin\theta}\frac{\partial I_\phi}{\partial\phi} = 0\,. \tag{5.9}$$

This equation is not really independent of z because of anisotropic conductivity $[\sigma]$, which produces a term containing E_z in $I_{\theta,\phi}$. MARTYN and BAKER (1953) have attempted to avoid the difficulty by assuming

$$J_z = 0 \qquad z_1<z<z_2 \tag{5.10}$$

which by (5. 5) leads to

$$\hat{E}_z = -\frac{\sigma_{zx}\hat{E}_\theta+\sigma_{zy}\hat{E}_\phi}{\sigma_{zz}}. \tag{5.11}$$

Then, $[\sigma]$ reduces to $[\bar{\sigma}]$ which is two-dimensional in the horizontal plane as

$$\boldsymbol{J} = [\sigma]\hat{\boldsymbol{E}} \tag{5.12}$$

where

$$[\bar{\sigma}] = \begin{bmatrix}\bar{\sigma}_{xx}, & \bar{\sigma}_{xy}\\ \bar{\sigma}_{yx}, & \bar{\sigma}_{yy}\end{bmatrix}$$

$$\left.\begin{aligned}\bar{\sigma}_{xx} &= \sigma_{xx}-\frac{\sigma_{xz}\sigma_{zx}}{\sigma_{zz}}\\ \bar{\sigma}_{xy} &= \sigma_{xy}-\frac{\sigma_{xz}\sigma_{zy}}{\sigma_{zz}}\\ \bar{\sigma}_{yx} &= \sigma_{yx}-\frac{\sigma_{yz}\sigma_{zx}}{\sigma_{zz}}\end{aligned}\right\} \tag{5.12'}$$

$$\bar{\sigma}_{yy} = \sigma_{yy} - \frac{\sigma_{yz}\sigma_{zy}}{\sigma_{zz}}.$$

In the centered dipole approximation, (5. 11) and (5. 12′) are given as

$$\hat{E}_z = \frac{-(\sigma_0 - \sigma_1)\cos I \sin I \hat{E}_\theta + \sigma_2 \cos I \hat{E}_\phi}{\sigma_0 \sin^2 I + \sigma_1 \cos^2 I} \tag{5.11′}$$

$$\bar{\sigma}_{xx} = \frac{\sigma_0 \sigma_1}{\sigma_0 \sin^2 I + \sigma_1 \cos^2 I} \simeq \frac{\sigma_1}{\sin^2 I}$$

$$\bar{\sigma}_{xy} = -\bar{\sigma}_{yx} = \frac{\sigma_0 \sigma_2 \sin I}{\sigma_0 \sin^2 I + \sigma_1 \cos^2 I} \simeq \frac{\sigma_2}{\sin I} \tag{5.12″}$$

$$\bar{\sigma}_{yy} = \frac{\sigma_0 \sigma_1 \sin^2 I + (\sigma_1{}^2 + \sigma_2{}^2)\cos^2 I}{\sigma_0 \sin^2 I + \sigma_1 \cos^2 I} \simeq \sigma_1$$

where the approximation holds for $I \gg \tan^{-1}\sqrt{\sigma_1/\sigma_0} \approx 3°$. Furthermore $\partial \boldsymbol{E}/\partial z = 0$ is assumed, so that $\boldsymbol{E} = -\mathrm{grad}_{\mathrm{H}} S$ where $\mathrm{grad}_{\mathrm{H}}$ is two-dimensional in the horizontal plane. Thus, (5. 9) becomes of second order in S for a given $\boldsymbol{W}$, the global wind distribution. We call it the dynamo equation.

Ohm's law for the height integrated current is

$$\boldsymbol{I} = [\boldsymbol{\Sigma}]\boldsymbol{E} + \int [\sigma](\boldsymbol{W} \times \boldsymbol{B}_0)\,\mathrm{d}z \tag{5.13}$$

where

$$[\boldsymbol{\Sigma}] = \begin{bmatrix} \boldsymbol{\Sigma}_{xx}, & \boldsymbol{\Sigma}_{xy} \\ -\boldsymbol{\Sigma}_{xy}, & \boldsymbol{\Sigma}_{yy} \end{bmatrix} \tag{5.14}$$

$$\boldsymbol{\Sigma}_{xx} = \int \bar{\sigma}_{xx}\,\mathrm{d}z \qquad \text{etc.}$$

As to $\boldsymbol{W}$, the tidal wind, which is periodic in local time, and the prevailing wind, which is constant with longitude, have been discussed. KATO (1957b) proved that a prevailing (steady) zonal wind can drive no current system. There were two conditions in his proof, Condition (a) being that $\boldsymbol{B}_0$ is centered-dipolar along the earth's axis and Condition (b) being that $W = W_y$ is constant with ϕ and z as was usually assumed at the time. His proof is as follows: By (5. 13), in which $(\boldsymbol{W} \times \boldsymbol{B}_0)$ is now taken out of integration,

$$\hat{\boldsymbol{E}} = \boldsymbol{E} + \boldsymbol{W} \times \boldsymbol{B}_0 = [R]\boldsymbol{I} \tag{5.15}$$

$$\boldsymbol{W} \times \boldsymbol{B}_0 = (W_y B_z, 0) \tag{5.15′}$$

where

$$[R] = \begin{bmatrix} R_{xx} & R_{xy} \\ -R_{xy} & R_{yy} \end{bmatrix} = [\boldsymbol{\Sigma}]^* = \text{total resistivity tensor}$$

$$R_{xx} = \frac{\boldsymbol{\Sigma}_{yy}}{\boldsymbol{\Sigma}_{xx}\boldsymbol{\Sigma}_{yy} + (\boldsymbol{\Sigma}_{xy})^2}$$

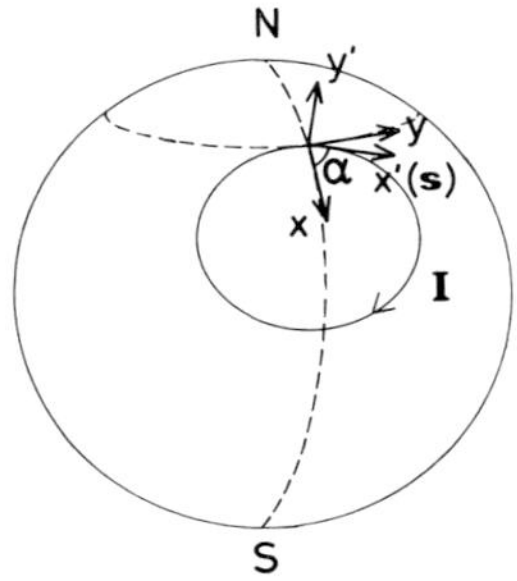

FIG. 5. 5. Coordinates (x, y, z) and (x', y', z'). The broken lines are a latitudinal and longitudinal circle. Here, $\boldsymbol{\alpha}$ is the angle between the x- and x'-axis.

$$R_{xy} = -\frac{\boldsymbol{\Sigma}_{xy}}{\boldsymbol{\Sigma}_{yy}} R_{xx} = -R_{yx}$$

$$R_{yy} = \frac{\boldsymbol{\Sigma}_{xx}}{\boldsymbol{\Sigma}_{yy}} R_{xx}. \tag{5. 16}$$

The asterisk which appears as a superscript of $[\boldsymbol{\Sigma}]$ denotes inverse matrix. We shall transform (x, y, z) to (x', y', z') where the x' axis is along $\boldsymbol{I}$. Then, $[R] \rightarrow [R']$ where

$$[R'] = \begin{bmatrix} \cos\boldsymbol{\alpha}, & \sin\boldsymbol{\alpha} \\ -\sin\boldsymbol{\alpha}, & \cos\boldsymbol{\alpha} \end{bmatrix} [R] \begin{bmatrix} \cos\boldsymbol{\alpha}, & -\sin\boldsymbol{\alpha} \\ \sin\boldsymbol{\alpha}, & \cos\boldsymbol{\alpha} \end{bmatrix} \tag{5. 17}$$

$$R'_{x'x'} = (R_{xx}\cos^2\boldsymbol{\alpha} + R_{xy}\sin^2\boldsymbol{\alpha})$$
$$R'_{x'y'} = [-(R_{xx}-R_{yy})\sin\boldsymbol{\alpha}\cos\boldsymbol{\alpha} + R_{xy}] \tag{5. 18}$$

in which $\boldsymbol{\alpha}=\boldsymbol{\alpha}(\theta, \phi)$=angle between the x- and x'-axes (Fig. 5. 5). Since div $\boldsymbol{I}=0$ in the horizontal plane, we can take a contour integral along $\boldsymbol{I}$ as

$$\oint_{\text{current}} \hat{\boldsymbol{E}}\cdot \mathrm{d}\boldsymbol{s} = \oint_{\text{current}} [R']\boldsymbol{I}'\cdot \mathrm{d}\boldsymbol{s} \tag{5. 19}$$

in which d$\boldsymbol{s}$ is a current line element ($\mathrm{d}s = \mathrm{d}x'$) and

$$\oint \hat{\boldsymbol{E}}\cdot \mathrm{d}\boldsymbol{s} = \oint (\boldsymbol{W}\times\boldsymbol{B}_0)\cdot \mathrm{d}\boldsymbol{s} = \oiint \operatorname{curl}(\boldsymbol{W}\times\boldsymbol{B}_0)\cdot \mathrm{d}\boldsymbol{S} \tag{5. 20}$$

by the Stokes theorem and curl $\boldsymbol{E}=0$; d$\boldsymbol{S}$ is the surface element. The double integral in (5. 20) is over any regular surface bounded by $\boldsymbol{s}$. By (5. 15′) and $B_z = -B_{00}\cos\theta$ where B_{00} is a constant.

$$[\operatorname{curl}(\boldsymbol{W}\times\boldsymbol{B}_0)]_z = -\frac{1}{a\sin\theta}\frac{\partial}{\partial\phi}(W_y B_z) = 0 \tag{5. 21}$$

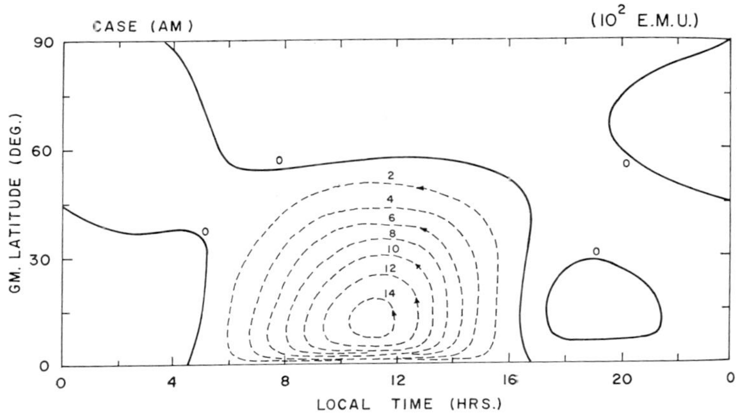

FIG. 5. 6. Current vortex produced by non-periodic winds (MAEDA and MURATA, 1968). Here, CASE (AM) assumes a meridional northward velocity as V_0 sin 2θ where V_0 is 10 m sec^{-1}. The resulting current is about (1/10) of the observed, though the shape is very much like that observed. The contour spacing is 10^2 emu (10^3 A).

which leads us to

$$\oint [R']\boldsymbol{I}'\cdot \mathrm{d}\boldsymbol{s} = \oint R'_{x'x'} I_{x'} \mathrm{d}x' = 0. \tag{5.22}$$

where by (5. 18) $R'_{x'x'} = R_{xx} \cos^2 \boldsymbol{\alpha} + R_{xy} \sin^2 \boldsymbol{\alpha} > 0$. And $I_{x'} \geq 0$, so $I_{x'} = \boldsymbol{I} = 0$.

If one of the above two Conditions (a) or (b) is violated, even the prevailing wind drives current. VAN SABBEN (1962) and, later, MAEDA and MURATA (1968) have considered a dipole field oblique to the earth's rotation axis, violating Condition (a). They have obtained current vortices which resemble, in shape, the S_q vortices (Fig. 5. 6). However, large amplitude winds are needed, velocities as large as 100 m/s for the zonal wind and 50 m/s for the meridional wind i.e. much larger than usually observed. Also, it is open to question whether the wind system of this type is dynamically stable. For instance the assumed meridional wind has a large divergence producing a large vertical motion which is difficult to drive from thermodynamic considerations.

The other case where the zonal prevailing wind drives current is that $\boldsymbol{W}$ varies with z violating Condition (b). In this case, the second term in the right-hand side of (5. 13) is expressed as

$$(B_z \int \bar{\sigma}_{xx} W_y \, \mathrm{d}z) \text{ for } I_\theta, \text{ and } (-B_z \int \bar{\sigma}_{xy} W_y \, \mathrm{d}z) \text{ for } I_\phi.$$

If we assume

$$\frac{\int \bar{\sigma}_{xy} W_y \mathrm{d}z}{\int \bar{\sigma}_{xx} W_y \mathrm{d}z} = \frac{\boldsymbol{\Sigma}_{xy}}{\boldsymbol{\Sigma}_{xx}} \tag{5.23}$$

which approximates

$$\frac{\int \sigma_2 W_y \mathrm{d}z}{\int \sigma_1 W_y \mathrm{d}z} = \frac{\boldsymbol{\Sigma}_2}{\boldsymbol{\Sigma}_1}, \tag{5.23'}$$

we can define the total *two-dimensional electric field* for the two-dimensional

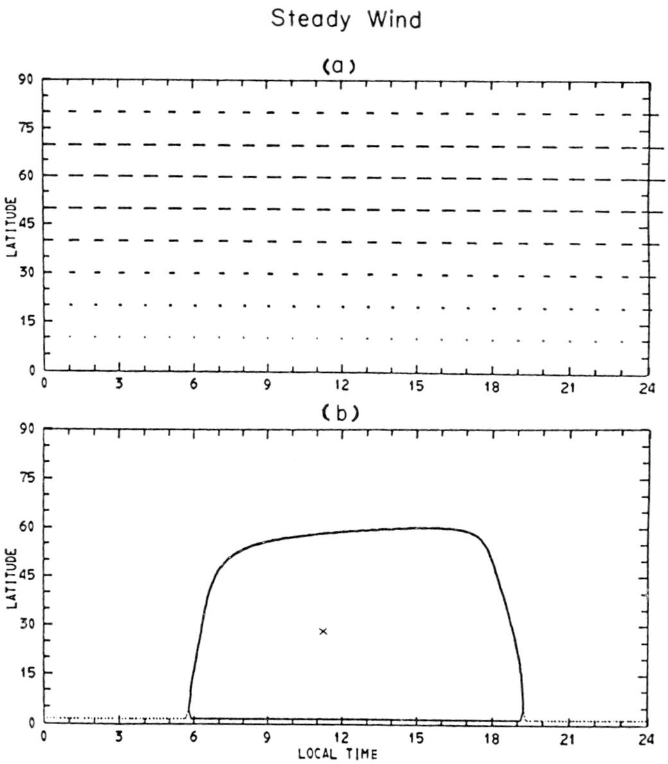

FIG. 5. 7. (a) Latitude-local time (or longitude) pattern of a steady wind at 120-km altitude. A vector length equal to 5° of latitude represents a wind speed of 50 m/sec. (b) Current function pattern for the steady wind. The heavy solid line is the zero contour. The current function is positive inside this contour, with a maximum value of 4.5×10^3 A at ×. Current flows counterclockwise about × : Eastward wind (RICHMOND *et al.*, 1976).

current $\boldsymbol{I}$ as

$$\hat{\boldsymbol{E}} = \begin{bmatrix} E_\theta + \dfrac{B_{0z}\int \sigma_{xx} W_y \mathrm{d}z}{\boldsymbol{\Sigma}_{xx}} \\ E_\phi \end{bmatrix} \tag{5.24}$$

for which Ohm's law is given as

$$\hat{\boldsymbol{E}} = [R]\boldsymbol{I}. \tag{5.24'}$$

It is straightforward to follow a similar argument as in the preceding case after replacing $(W_y B_z)$ by $(B_z \int \sigma_{xx} W_y \, \mathrm{d}z / \boldsymbol{\Sigma}_{xx})$ in (5. 15′). However, upon reaching the point corresponding to (5. 21), one must impose the following condition to conclude $\boldsymbol{I}=0$:

$$\left(\int \frac{\partial \bar{\sigma}_{xx}}{\partial \phi} W_y \mathrm{d}z \Big/ \int \bar{\sigma}_{xx} W_y \mathrm{d}z \right) = \left(\frac{\partial \boldsymbol{\Sigma}_{xx}}{\partial \phi} \Big/ \boldsymbol{\Sigma}_{xx} \right) \tag{5.25}$$

which is necessary for $(\mathrm{curl}\ \hat{\boldsymbol{E}})_z=0$; curl $\hat{\boldsymbol{E}}$=curl $(\boldsymbol{W}\times\boldsymbol{B}_0)$ in (5. 21) by curl $\boldsymbol{E}=0$. (5. 25) is satisfied, if $\sigma_{xx}=\mathscr{F}(\phi)\mathscr{G}(z)$ where $\mathscr{F}$ and $\mathscr{G}$ are some functions of only ϕ and z, respectively. This implies that the conductivity varies with longitude (equivalently local time) similarly at all heights, a situation which is unrealistic, as the conductivity decreases more in the E region than in the F region at night. It is, therefore, expected that some current flows for realistic conductivity distribution, if $\boldsymbol{W}_y$ is z-dependent. Recently, RICHMOND *et al.* (1976) have obtained a current system for the altitude dependent zonal wind system of DICKINSON *et al.* (1975). The result is shown in Fig. 5. 7. It is found that the current intensity is very small. This is due to (5. 23) and (5.25) which might be satisfied to some extent.

Note that the appearance of the main vortices in daytime is fairly common for almost *any* large scale wind assumed, because the nighttime conductivity is much smaller than the daytime one (about 1/30), so that the current, driven by $(\boldsymbol{W}\times\boldsymbol{B}_0)$ in daytime hemisphere, should close itself mainly in the daytime hemisphere. Thus, the question as to whether or not a particular wind can produce a current system agreeable to the geomagnetic observation at the earth's surface has no clear-cut answer. However, each wind system produces an electric field to drive stationary current satisfying div $\boldsymbol{J}=0$. This electric field can now be measured by incoherent scatter radars (e.g. BLANC *et al.*, 1977). The electric field seems to present a good checking of the theory (RICHMOND *et al.*, 1976).

The *periodic* stable wind is a good candidate to explain S_q by the atmospheric dynamo. The atmospheric tides belong to this category. SCHUSTER (1889) already pointed out the possibility. However, he also noticed that the observed pressure tide near the ground is mainly semidiurnal instead of diurnal as is found in the geomagnetic S_q. Thus, he inferred that the semidiurnal tide should predominate also in the upper atmosphere. Then, this meant that the conductivity varies with both diurnal and semidiurnal periods and can produce an initial current $[\sigma](\boldsymbol{W}\times\boldsymbol{B}_0)$ with both diurnal and semi-diurnal periods, but, eventually, we have a stable current system which is mainly diurnal. This had to be accomplished by the conductivity decreasing the semidiurnal variation in $\boldsymbol{J}$. This idea was accepted until the 1950's.

For $I_{\theta,\phi}$, the givens, we should be able to deduce $\boldsymbol{W}$ from (5.9), if $\boldsymbol{W}$ does not vary with height z. By (5. 13), in which $\boldsymbol{W}$ is tidal wind, and (5. 6)

$$\mathrm{curl}_H(\boldsymbol{W}\times\boldsymbol{B}_0) = \mathrm{curl}_H([\boldsymbol{\Sigma}]^*\boldsymbol{I}) \tag{5.26}$$

where the left-hand side is unknown and the right-hand side is known. More explicitly, (5. 26) is given as

$$\frac{\partial}{\partial\phi}(W_\phi B_{0z})+\frac{\partial}{\partial\theta}(\sin\theta\, W_\theta B_{0z}) = \frac{\partial\hat{E}_\theta}{\partial\phi}-\frac{\partial(\hat{E}_\phi\sin\theta)}{\partial\theta}. \tag{5.27}$$

In order to determine $\boldsymbol{W}$ we need a relation between $\boldsymbol{W}_\theta$ and $\boldsymbol{W}_\phi$. Note that in this approach (5. 5) (actually, its integrated form, (5. 13)) and (5. 6) are used, but (5. 4) is not. However, the deduction of $\boldsymbol{I}$ from the magnetic variation follows Biot-Savart's law which is based on Maxwell's equation as

$$\mathrm{curl}(\varDelta\boldsymbol{H}) = \boldsymbol{J}. \tag{5.28}$$

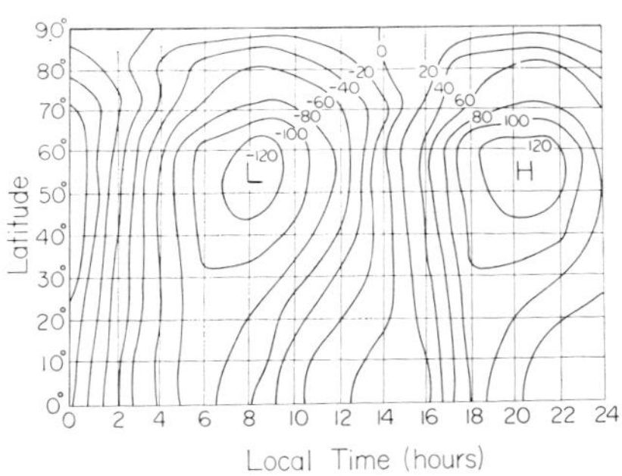

FIG. 5. 8

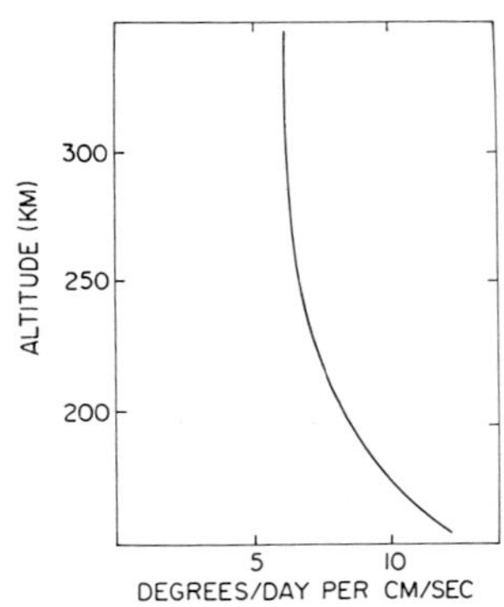

FIG. 5. 9

FIG. 5. 8. Contour of (p_1/ρ_0) deduced from geomagnetic S_q variation at equinox (KATO, 1956). Unit: 10^2 m^2 sec^{-2}.

FIG. 5. 9. Temperature variation by adiabatic warming due to vertical motion (DICKINSON and GEISLER, 1968). The variation is given by $-W_z(\partial T_0/\partial z+g/C_p)$ per unit time where C_p is the heat capacity of constant pressure. The diagram shows degree (day)$^{-1}$ for $W_z=1$ cm sec^{-1}.

It turns out that (5. 4) is obtained by operating div on (5. 28) i.e. the given $\boldsymbol{I}$ satisfies (5. 4). Thus, $\boldsymbol{W}$ can be determined from $\boldsymbol{I}(\theta, \phi)$ which is deduced from the geomagnetic S_q variation. The approach was attempted in the 1950's by MAEDA (1955, 1957), HIRONO and KITAMURA (1956), and KATO (1956, 1957a). Their approaches are different from each other in the assumed relation between $\boldsymbol{W}_\theta$ and $\boldsymbol{W}_\phi$. A simple relation is either curl $\boldsymbol{W}=0$ or div $\boldsymbol{W}=0$. A more proper relation is the dynamical equation of motion which should be satisfied by $\boldsymbol{W}$. All approaches indicated that the diurnal component of $\boldsymbol{W}$ is larger than the semidiurnal. The finding was in contrast to the conventional belief that the semidiurnal wind was dominant in the dynamo region (Fig. 5. 8).

In relation to the solution of (5. 26), a question (MÖHLMAN, 1974) may arise as to whether the solution is unique, because we can add any $\boldsymbol{W}'$ to the solution, if curl $(\boldsymbol{W}'\times\boldsymbol{B}_0)=0$. However, this implies (KATO, 1974)

$$\boldsymbol{W}'\times\boldsymbol{B}_0 = \text{grad}\Psi \tag{5. 29}$$

where Ψ is a scalar function. It turns out, then, $\boldsymbol{W}'$ can be either parallel to $\boldsymbol{B}_0$ as $\boldsymbol{W}'\times\boldsymbol{B}_0=0$, or

$$\boldsymbol{W}' = \frac{\boldsymbol{B}_0\times\text{grad}\Psi}{B_0{}^2} \tag{5. 30}$$

which gives

$$W'_z = \frac{1}{B_0{}^2}\left(B_{0\theta}\frac{1}{a\sin\theta}\frac{\partial\Psi}{\partial\phi} - B_{0\phi}\frac{\partial\Psi}{a\,\partial\theta}\right). \tag{5. 31}$$

For the centered dipole,

$$W'_z = \frac{-2}{B_{00}(1+3\cos^2\theta)}\frac{\partial\Psi}{a\,\partial\phi} \tag{5. 32}$$

which shows that $\boldsymbol{W}'_z$ is as large as horizontal components $\boldsymbol{W}'_{\theta,\phi}$, because

$$\begin{aligned} W'_\theta &= \frac{4\cot\theta}{B_{00}(1+3\cos^2\theta)}\frac{\partial\Psi}{a\,\partial\phi} \\ W'_\phi &= \frac{-4}{B_{00}(1+3\cos^2\theta)}\left(\cos\theta\frac{\partial\Psi}{a\partial\theta} - \frac{1}{2}\sin\theta\frac{\partial\Psi}{\partial z}\right) \end{aligned} \tag{5. 33}$$

and for diurnal and semidiurnal tide $\Psi\propto e^{i\phi}, e^{2i\phi}$, respectively. We thus find

$$|W'_z| \sim |W'_{\theta,\phi}| \tag{5. 34}$$

which is unacceptable in global motion. Consider $|\boldsymbol{W}'_z| \simeq |\boldsymbol{W}'_{\theta,\phi}| = 10$ m/s, which is still smaller than usually observed *horizontal* tidal velocity in the F region. Such a $\boldsymbol{W}'_z$ should produce an enormous daily heating and cooling of the order of a few thousands °K (Fig. 5. 9). It is always true that $|W'_z| \ll |W'_{\theta,\phi}|$ for large scale motion which is responsible for the atmospheric dynamo.

If $[\sigma]$ or $(\boldsymbol{W}\times\boldsymbol{B}_0)$ is not symmetric with respect to the equator, $\boldsymbol{J}_z$ cannot vanish at $z=z_2$ and (5. 9) contains the term $J_{z=z_2}$ in the left-hand side; $J_{z=z_1}=0$ is always relevant, because $[\sigma]$ is very small for $z<z_1$ where z_1 will be chosen, say, 70 km above the ground. Thus, the conventional approach of the height-integrated current would become difficult. However, one can avoid the difficulty by dividing $\boldsymbol{I}$ into symmetric and anti-symmetric parts, of which only the symmetric part satisfies (5. 9). The electrostatic field must be mainly symmetric, because otherwise the anti-symmetric part of the electric field would be quickly short-circuited along high conductive field lines connecting the hemispheres. One need not solve the differential equation for the anti-symmetric electric current. SCHIELDGE *et al.* (1973) and MAEDA (1974) obtained $J_{z=z_2}=-\mathrm{div}\boldsymbol{I}_a$ by

$$\boldsymbol{I}_{\mathrm{a}} = [\boldsymbol{\Sigma}]_{\mathrm{a}}[(\boldsymbol{W}\times\boldsymbol{B}_0)_{\mathrm{s}}+\boldsymbol{E}_{\mathrm{s}}]+[\boldsymbol{\Sigma}]_{\mathrm{s}}(\boldsymbol{W}\times\boldsymbol{B}_0)_{\mathrm{a}} \tag{5. 35}$$

where subscripts a and s denote anti-symmetric and symmetric parts, respectively. While $\boldsymbol{I}=\boldsymbol{I}_{\mathrm{a}}+\boldsymbol{I}_{\mathrm{s}}$ flows horizontally in the ionosphere, $J_{z=z_2}(\theta, \phi)$ was assumed to flow along field lines between the conjugate points. This treatment suggests a possible solution as is discussed in detail in Section 5. 4.

New developments in tidal theory occurred rapidly around the middle of the 1960's, when the powerful computers and the in situ experiment by rocket were available. We were then able to do computations which had previously been impossible, and we were also able to measure directly various quantities of the upper atmosphere, in particular, temperature which had never been correctly known. The structure of the diurnal tide could, for the first time, be studied. The negative modes, which were then discovered, have opened a new era for the tidal dynamo theory as will be explained in the next section.

5.3 Recent Development of the Dynamo Theory and Comparison with Observation

KATO (1966a) has discussed the contribution of the (1, −2)-mode of the solar tide, excited in the thermosphere, to the S_{q} current system. He has found that the latitudinal pressure distribution which was deduced from the observed S_{q} variation (KATO, 1956) seemed consistent with that by this mode (Fig. 3. 8). The finding has been also confirmed by STENING (1969), TARPLEY (1970) and VOLLAND (1971). All of them have attempted to derive the current vortices for a given tidal wind. Their common conclusion was that the (1, −2)-mode is necessary to explain the observations.

As is noted in Section 5. 2, agreement in the shape and location of current vortices is not sufficient, because similar vortices are generated by many different wind systems. In order to decide whether a particular wind system is driving currents, one must find that other derived quantities such as $\boldsymbol{E}$ also agree with observation. One can also use direct observations to determine the wind systems used in the models. Sophistication of the models of the

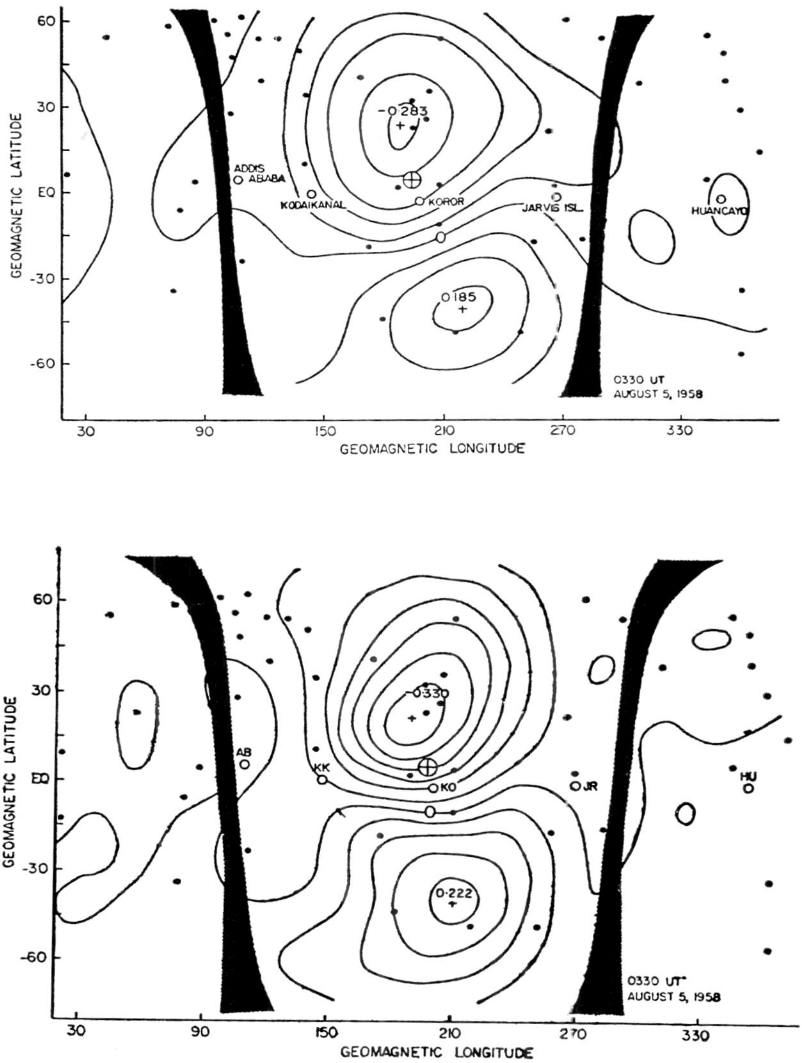

FIG. 5. 10. Geomagnetic S_q variation observed (top) as magnetic potential at the ground and that calculated (bottom). The subsolar point ⊕ and the terminator (thick bands) are shown. The potential lines are given with spacing 5×10^{-2} Weber m^{-1} corresponding to 5×10^4 A (SCHIELDGE *et al.*, 1973).

ionosphere structure and wind have advanced mainly along this line. SCHIELDGE *et al.* (1973) introduced, besides realistic conductivity and geomagnetic field, a wind system which consists of four modes of the solar atmospheric tide, the (1, -2), (2, 2), (2, 3), and (2, 4)-modes. They considered only the symmetric part of $[\sigma](\boldsymbol{W}\times\boldsymbol{B}_0)$ with respect to the equator for producing electric field and driving symmetric current. Only the anti-symmetric part of $[\sigma](\boldsymbol{W}\times\boldsymbol{B}_0)$ was assumed to drive field-aligned current and anti-symmetric ionospheric current. They tried to simulate the equatorial electrojet as well as the global S_q current system by the four modes, varying the amplitudes and phases from day to day to explain the observed S_q variation. The semidiurnal tide was found important to explain the behavior during a solar minimum period. The simulation, as they called it, seems fairly successful in explaining the observations, although, in general, a set of the orthogonal functions can give almost any desired fit. Figure 5. 10 illustrates some of their results. A further simulation for the symmetrical part of S_q has been attempted by RICHMOND *et al.* (1976) who considered a steady wind as well as tidal ones. The used steady eastward wind has been found ineffective to produce electric current as is discussed in the previous section and is illustrated in Fig. 5. 7. The drift which is produced by the resulting electrostatic field i.e. $(\boldsymbol{E}\times\boldsymbol{B}_0)$-drift is constant with local time, while ground-based incoherent scatter measurements show a local time variation. However, the results of simulation are very dependent on the height distribution of $\boldsymbol{W}_y$. GROVES (1972) pointed out that the prevailing wind is possibly westward below 130 km, thereby producing a current opposite to the observation.

As to the tidal wind mode, RICHMOND *et al.* (1976) have used the first negative (symmetric) diurnal mode, the first (symmetric) and second (symmetric) semidiurnal modes into account, separately, and also in combination. They have considered not the classical tidal modes but modified modes due to ion drag in the ionosphere. Their basic equations of motion were those which had been used by VOLLAND (1974). The ion drag effect was approximated as $\nu\boldsymbol{W}$ where ν is constant. Though the treatment cannot escape a criticism against using a constant ν and an equation such as

$$\frac{1}{\sin\theta}\left[\frac{\partial}{\partial\theta}(W_\theta \sin\theta)+imW_\phi\right]-i\frac{m\Omega a}{\rho_0 gh}p_1=0 \tag{5.36}$$

which can prove true only with constant ion drag, the deduced wind seems useful for the purpose. They present the following conclusion:

1) Winds in the ionospheric dynamo region (80–200 km) which are consistent both with observation and with theory are capable of explaining S_q currents.

Note that this wind model was very similar in amplitude to that by SALAH and EVANS (1977; see Fig. 3. 34) who used a seasonally averaged observation of the St. Santin incoherent scatter radar by AMAYENC (1974).

2) Ionospheric winds can also explain observed middle and low latitude electric field, if we take into account the uncertainties still present concerning the global ionospheric wind pattern and also the uncertainties concerning the observed ($\boldsymbol{E}\times\boldsymbol{B}$)-drifts. Specifically, the simulation has made it clear that a good agreement between theory and observation is obtained by the modified (1, −2) plus (2, 4)-modes. Of these two modes, the modified (1, −2)-mode is the main generator of the S_q current. The (2, 4)-mode is important, however, to the electric field production, especially its variability as observed in midlatitudes.

3) The primary sources of S_q currents can be attributed to diurnal winds

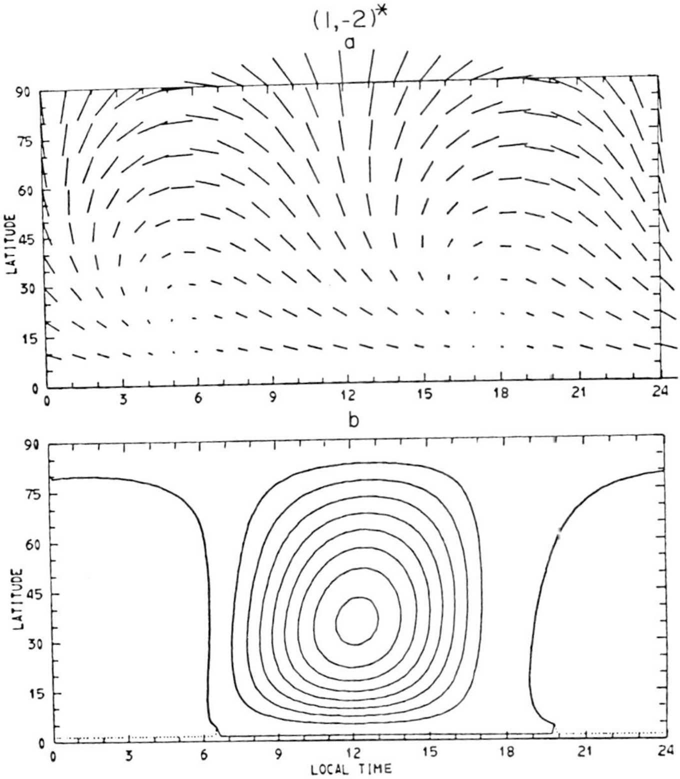

FIG. 5. 11. Wind pattern of the modified (1, −2)-mode, the (1, −2)*-mode, at 150 km as modified by the ion drag. A vector length equal to 5° of latitude represents a wind speed 50 m sec^{-1} (top); and the corresponding current system (bottom). Contour spacings are 10^4 A and current flows counterclockwise during the day (RICHMOND *et al.*, 1976).

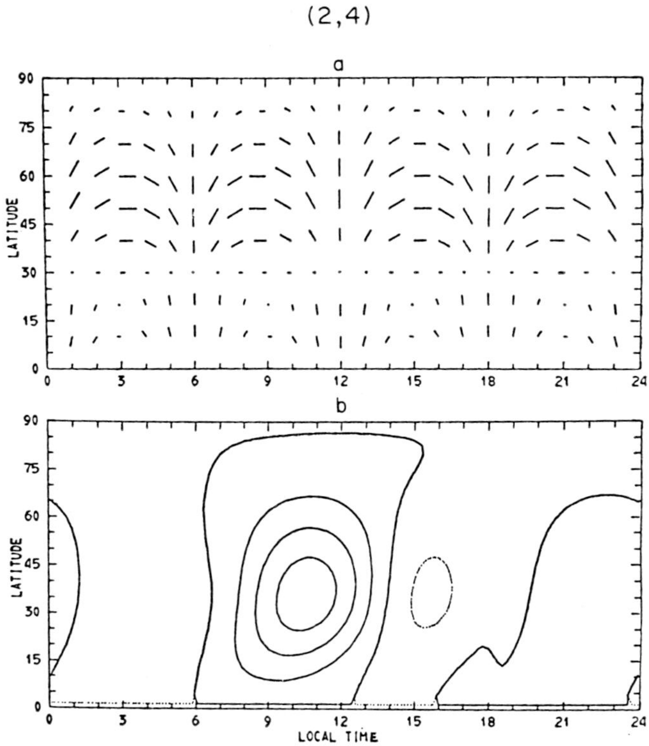

FIG. 5. 12. Similar to Fig. 5. 11 except for the (2, 4)-modes at 115 km.

of the (1, −2)-mode in the upper E and F1 ionosphere regions, which are driven by solar heating in situ.

They considered also the effects of magnetospheric eastward electric field which penetrates into the ionosphere at $\theta=\theta_0=30°$

$$E_\phi = \{0.67 \cos[(t+\phi)-54°]+0.38\cos[2(t+\phi)-240°] \\ +0.40\cos[3(t+\phi)-192°]+0.17\cos[4(t+\phi)-15°]\} \qquad (mV\,m^{-1}) \tag{5.37}$$

where t is measured by degrees. However, the magnetospheric sources are found to be only of secondary importance on quiet days. Their main results are illustrated in Figs. 5. 11–5. 13.

Although the simulation by Richmond *et al.* seems successful in reproducing observations, other possibilities cannot be ruled out. The old idea that the relevant wind be only semidiurnal can still survive (STENING, 1977).

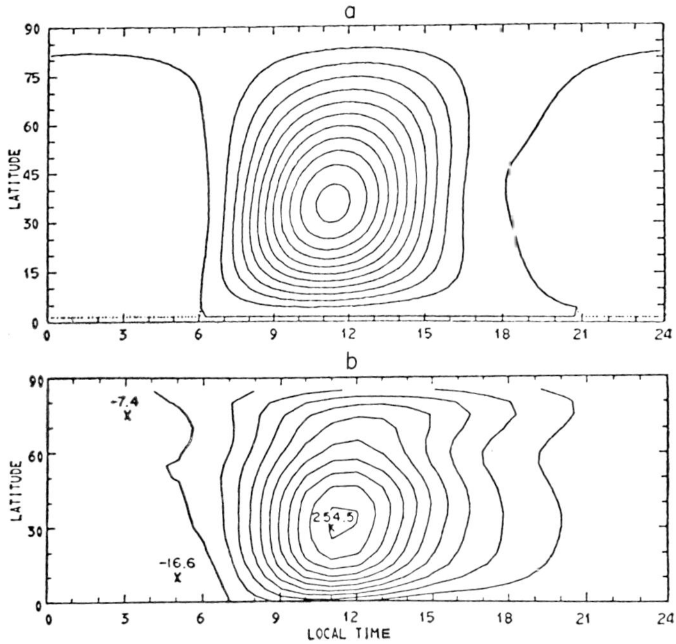

FIG. 5. 13. Current systems produced by the wind of the modified (1, −2)-mode plus the (2, 4)-mode as separately given in Fig. 5. 11 and 5. 12; contour spacing is 10^4 A (top). S_q current system deduced by using IGY magnetic data, contour spacing being 2.5×10^4 A (bottom). Counterclockwise current in both diagrams (RICHMOND *et al.*, 1976).

KIRCHHOFF and CARPENTER (1976) have shown that the geomagnetic variation can be explained by a model (2, 4)-mode wind and the electric field observed by incoherent scatter radar for $\Sigma K_p < 24$ which may approximate a S_q condition (Fig. 5. 14). Also the semidiurnal wind, though not dominant in producing the S_q current vortices, may be necessary to explain some low latitude phenomena, especially the counter equatorial-electrojet (STENING, 1977).

The dynamo theory, discussed so far, assumes that the ionosphere does not modify the tidal winds. This condition is approximately true in the E region but increasingly erroneous with increasing altitude. In a self-consistent approach, the wind and the electric current or field should be derived simultaneously for a given heat source at each height and a set of boundary conditions both at the ground and at the top of the atmosphere. This was discussed in Section 3. 2 of the previous chapter. In this approach, $(\boldsymbol{J} \times \boldsymbol{B}_0)$, the Ampere force, appears in the equation of motion (3. 1)–(3. 3). Also ad-

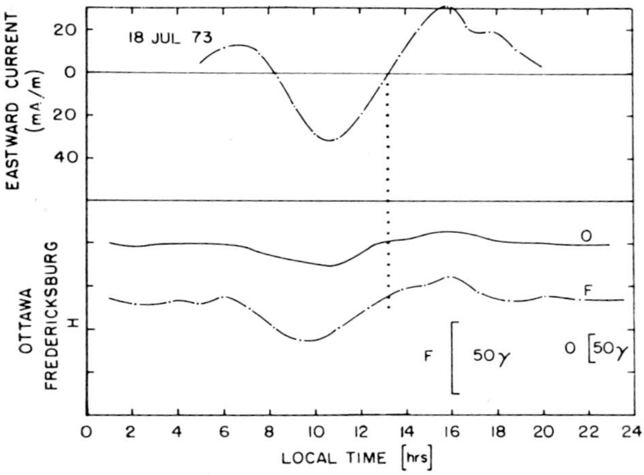

FIG. 5. 14. Height integrated E region current which was calculated by using a (2, 4)-mode tidal wind and the radar-observed electric field at Millstone Hill (top) and the observed magnetogram *H* (horizontal intensity) traces at Ottawa and Fredericksburg. Note that the zero crossing of the calculated current and that of *H* traces (assuming the zero level to be that of nighttime) are in an agreement (KIRCHHOFF and CARPENTER, 1976).

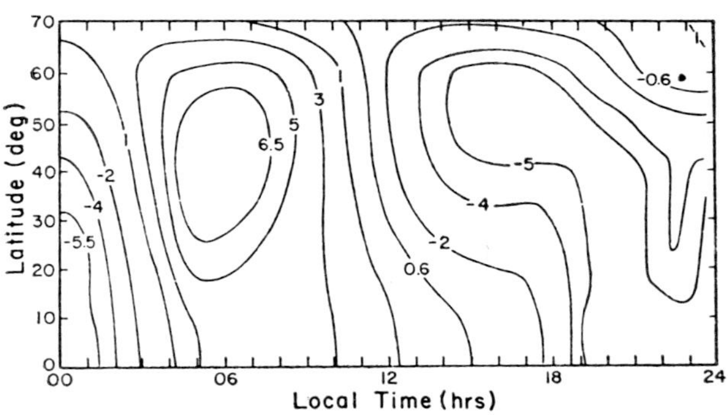

FIG. 5. 15. E region electrostatic potential generated by the tidal wind and the F region wind. Contour levels are in units of 1 kV (HEELIS *et al.*, 1974). Tarpley's model of tidal winds (1970) was used in the calculation.

ditional term in the energy equation (3. 5) is for joule heating $\boldsymbol{J} \cdot \boldsymbol{E}$. Of course, the equation for electric field and current (5. 4)–(5. 6) must be included simultaneously. The approach is prohibitively complex, and a major simplification is necessary in order to make the problem tractable. The emphasis in the simplification should vary with the problem of concern. In the E and F1 region, $\boldsymbol{W}$ is *usually* larger than $(\boldsymbol{E}\times\boldsymbol{B}_0)$-drift and $\boldsymbol{E}$ can be neglected without any serious error in the dynamics of the neutral gas (e.g. RICHMOND *et al.*, 1976). This simplification (e.g. KATO and MATSUSHITA, 1969) divides the original set of equation into two independent parts, one for the neutral gas and the other for the dynamo theory. Joule heating can be neglected as a second order quantity in the first order perturbation method. The above-mentioned works all belong to this category. The simplest case is that in which $(\boldsymbol{J}\times\boldsymbol{B}_0)$, or the ion drag force, as is more properly called now, is entirely neglected, as in the works of TARPLEY (1970), STENING (1969) and SCHIELDGE *et al.* (1973). Further, the wind in the dynamo theory has been considered to be of a particular tidal mode or a combination of a few modes. Even when ion drag is effective, the original single mode has been considered to exist with a modification. Note that if ion drag exists, there is no mathematical justification for the completeness of the eigenfunction of each mode, and no way of finding the excitation of each mode. Nevertheless, as in many other field of physics, the results even with the mathematical ambiguity might approximate reality.

Another simplification was done by TAFFE (1969) who attempted to obtain a self-consistent solution for a uniform anisotropic conductivity and plane stratified atmosphere. However, as is discussed later in Section 5. 4, the boundary which he entirely neglected is essential in the production of $\boldsymbol{E}$ in the ionosphere. His solution is therefore unrelevant; he failed to consider the mode conversion by the boundary.

In the F region the reaction of the hydromagnetic force becomes considerable, invalidating the current and electric field distribution which are obtained without considering the interaction between the wind and electric currents. The F region dynamo has been discussed along this line by RISHBETH (1971), MATUURA (1974) and HEELIS *et al.* (1974). The driving force of the F region motion is given through the horizontal pressure and density gradient in the F region as obtained from an atmosphere model. Simplifications to the dynamical equations, such as neglecting the Coriolis force (HEELIS, 1974) or the inertia force (MATUURA, 1974), were necessary. The produced electric field is transmitted down to the E region where an electric current flows to dissipate the electric field. The resultant current systems in E and F regions resemble, in shape and intensity, the S_q current system observed, but the current directions are almost opposite, cancelling the integrated intensity which is effective on the sea-level magnetic variation. However, the resultant electric field is so large that combination of this electric field and that by the tidal dynamo in the E region may explain the semidiurnal variation of the observed electric field at Millstone Hill (Fig. 5. 15).

In the development of the experimental study on the upper atmosphere dynamics, incoherent scatter radars have been proved to be very powerful.

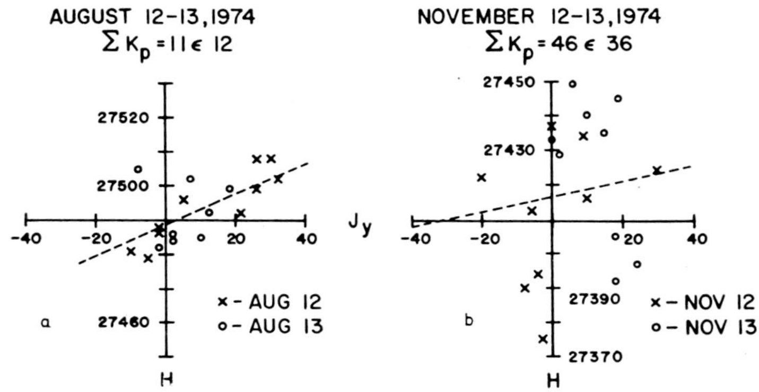

FIG. 5. 16. J_y versus H, the geomagnetic horizontal component, for (a) a slightly disturbed period, August 12 and 13, 1974, and (b) a very disturbed period, November 12 and 13, 1974. The linear regression lines of H on J_y are also shown. The November line is not statistically significant (HARPER, 1977a).

Although the measured ionosphere is limited to the region above the facilities, we are now able to know about $\boldsymbol{E}$ and $\boldsymbol{W}$ both simultaneously and continuously. Other parameters as $T_{e,i}$ are also available by the same facilities. An important test for the tidal dynamo theory has been attempted by HARPER (1977a). He calculated $\boldsymbol{J}$ over Arecibo from (5. 12) and (5. 12′) by using $\boldsymbol{E}$ which was measured in the F region, and $\boldsymbol{W}$ which was based on measurement between 100 and 130 km, and on extrapolation below 100 km. Above 130 km, W_θ was based on measurement, but W_ϕ was on a model. The conductivity used in calculation was based on the measured electron density. The total current after integration of the calculated $\boldsymbol{J}$ was compared to the magnetic field observation on the same day at San Juan close to Arecibo (Fig. 5. 16). It was found that, on slightly disturbed days, the calculated current was responsible to the observed magnetic variation. This was a test favorable for the tidal dynamo. On very disturbed days some cause other than the ionospheric current was found necessary. Note that both diurnal and semidiurnal winds were necessary to well explain the observation. Such simultaneous observations at other stations and on many occasions would be very useful for the development of the theory.

5.4 **Validity of the Conventional Theory**

The conventional approach in the tidal dynamo theory is based on several assumptions. One is div $\boldsymbol{J}=0$ which is valid (hydromagnetic approximation) as far as the displacement current is negligible. This assumption is clearly allowed.

The second assumption is $J_z=0$ at $z=z_{1,2}$, the boundary condition, which implies that the current flows only between z_1 and z_2. If the physical state of both hemispheres is symmetric with respect to the equator, which approximates the equinoctial state, this boundary condition may be valid. This assumption is clearly not valid in general, however.

The third assumption is $J_z=0$ in $z_1<z<z_2 (z=z_{1,2}$ being excluded). This is an approximation chosen for its computational simplifications. By (5. 5), $J_z \neq 0$ gives

$$\hat{E}_z = \frac{J_z - \sigma_{zx}\hat{E}_\theta - \sigma_{zy}\hat{E}_\phi}{\sigma_{zz}} \tag{5. 38}$$

which is substituted into $J_{\theta,\phi}$ in (5.5) to give

$$\boldsymbol{J} = [\bar{\sigma}]\hat{\boldsymbol{E}} + \frac{1}{\sigma_{zz}}\begin{bmatrix}\sigma_{xz}\\ \sigma_{yz}\end{bmatrix} J_z \,. \tag{5. 39}$$

Since $(\sigma_{xz}/\sigma_{zz})$ and $(\sigma_{yz}/\sigma_{zz})$ are small, amounting to 20 at the most near the equator (Fig. 4.3 and (5.5″)) and $|J_z| \approx (L_z/L_{\theta,\phi})|J_{\theta,\phi}|$ by (5.4) where $L_{\theta,\phi,z}$ are the scale sizes of $\boldsymbol{J}$ along θ, ϕ, z and $L_z \simeq 10^{-2} L_{\theta,\phi}$, we conclude

$$\boldsymbol{J} = [\bar{\sigma}]\hat{\boldsymbol{E}} + O(10^{-1}\boldsymbol{J})\,. \tag{5. 40}$$

Thus, the third assumption may be a reasonable one.

The fourth condition is that $\partial \boldsymbol{E}_{x,y}/\partial z$ can be neglected. More exactly, this implies $|(\partial E_{x,y}/\partial z)\Delta z| \ll |E_{x,y}|$ for $|\Delta z|<|z_2-z_1|$. Consider the difference of $\boldsymbol{E}_{x,y}$ between two points A (x, y, z) and C $(x, y, z+\Delta z)$, where $|\Delta z| \ll a$. In Fig. 5. 17 the left diagram indicates $\overline{\text{ABCD}}$ in a magnetic meridian plane; $\overline{\text{AB}}$ and $\overline{\text{CD}}$ are parallel to $\boldsymbol{B}_0$; $\overline{\text{AB}}=\overline{\text{CD}}=\Delta z$ cosec$I$ and $\overline{\text{AD}}=\overline{\text{BC}}=\Delta z$ cot$I$; $I \neq 0$. Then, contour integration of $\boldsymbol{E}$ along the parallelogram vanishes because of curl $\boldsymbol{E}=0$. This gives $[E(\overline{\text{DA}})\,\Delta z \cot I + E(\overline{\text{AB}})\,\Delta z \operatorname{cosec} I + E(\overline{\text{BC}})\,\Delta z \cot I + E(\overline{\text{CD}})\,\Delta z \operatorname{cosec} I]=0$, where $E(\overline{\text{DA}})$ etc. denote the average electric field parallel to $\overline{\text{DA}}$ etc. Neglecting a very small electric field $E(\overline{\text{AB}})$ and $E(\overline{\text{CD}})$ along the highly conductive magnetic field lines,

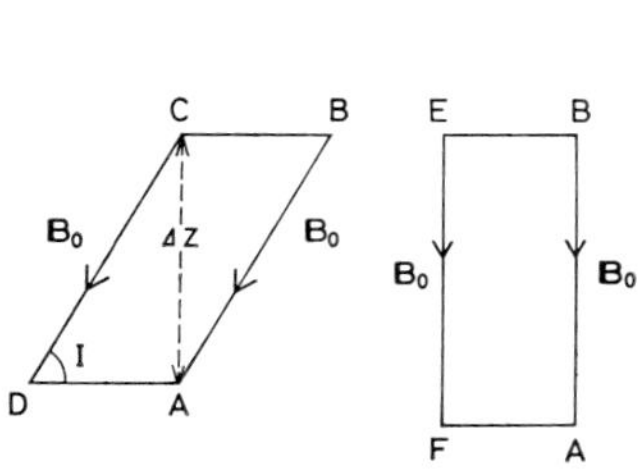

FIG. 5. 17. The paralleloid (left) is in a magnetic meridian, and the rectangle (right) is orthogonal to the meridian and parallel to $\boldsymbol{B}_0$. The length along $\boldsymbol{B}_0$ is the same for both quadrangles.

$E(\overline{\mathrm{DA}})=E(\overline{\mathrm{CB}})$.

$$|E_x(\mathrm{A})-E_x(\mathrm{C})|=|E_x(\mathrm{A})-E_x(\overline{\mathrm{AD}})+E_x(\overline{\mathrm{AD}})-E_x(\overline{\mathrm{BC}})+E_x(\overline{\mathrm{BC}})-E_x(\mathrm{C})|$$

$$\leq|E_x(\mathrm{A})-E_x(\overline{\mathrm{AD}})|+|E_x(\overline{\mathrm{AD}})-E_x(\overline{\mathrm{BC}})|+|E_x(\overline{\mathrm{BC}})-E_x(\mathrm{C})|=\left\{\left|\frac{\partial E_x(\mathrm{A})}{\partial x}\right|\right.$$

$$\left.+\left|\frac{\partial E_x(\mathrm{C})}{\partial x}\right|\right\}|\Delta z|\cot I\simeq\{|E_x(\mathrm{A})|+|E_x(\mathrm{C})|\}\frac{|\Delta z|}{a}\cot I. \qquad (5.40')$$

The right-hand side diagram in Fig. 5.17 is a quadrangle $\overline{\mathrm{ABEF}}$ parallel to $\boldsymbol{B}_0$ and $\overline{\mathrm{AB}}$ is common to the left parallelogram; $\overline{\mathrm{AF}}=\overline{\mathrm{BE}}<\Delta z$. Similar operation on $\overline{\mathrm{ABEF}}$ as that on $\overline{\mathrm{ABCD}}$ gives

$$|E_y(\mathrm{A})-E_y(\mathrm{B})|\leq\{|E_y(\mathrm{A})|+|E_y(\mathrm{B})|\}\frac{|\Delta z|}{a}\cot I$$

if we choose $\overline{\mathrm{BE}}=\overline{\mathrm{AF}}<|\Delta z|$, and

$$|E_y(\mathrm{C})-E_y(\mathrm{B})|\leq\{|E_y(\mathrm{C})|+|E_y(\mathrm{B})|\}\frac{|\Delta z|}{a}\cot I.$$

So, we have

$$|E_y(\mathrm{A})-E_y(\mathrm{C})\leq\{|E_y(\mathrm{A})|+2|E_y(\mathrm{B})|+|E_y(\mathrm{C})|\}\frac{|\Delta z|}{a}\cot I. \qquad (5.40'')$$

Therefore, by (5.40′) and (5.40″) and $|\Delta z|\ll a$

$$|\boldsymbol{E}(\mathrm{A})-\boldsymbol{E}(\mathrm{C})|\simeq\left|\frac{\partial \boldsymbol{E}_{x,y}}{\partial z}\cdot\Delta z\right|\simeq\left(|\boldsymbol{E}_{x,y}|\cdot\frac{|\Delta z|}{a}\right)\ll|\boldsymbol{E}_{x,y}|.$$

Note, however, that even a small field along $\boldsymbol{B}_0$, as is neglected here, can produce a considerable field-aligned current.

The above consideration leads us to conclude that the conventional approach should be fairly correct so far as $J_z=0$ at $z=z_{1,2}$ is the case.

Besides confirmation of the validity of the conventional dynamo theory one important motive in the present discussion (KATO and TSUDA, 1980) is to know precisely how the electric field is produced. In the conventional approach one might confuse the boundary condition and the approximation, both being $J_z=0$. While one could infer the importance of the vertical current for the production of electric field by polarization, the use of $J_z=0$ for $z_1<z<z_2$ on the basic equation should obscure any possible insight into the problem. What we shall follow will be a first step for solving very complex problems of the three-dimensional dynamo theory to solve (5. 7). The ionosphere model is made extremely simple: The plane atmosphere with a constant dip, I. The wind velocity, $\boldsymbol{W}$, is of a sinusoidal wave in θ, ϕ and z; along ϕ the wavelength corresponding to diurnal; along θ the Hough function can be approximated by a combination of the two waves traveling oppositely. This simple ionospheric model will be found to be still informative on the present rudimental stage of our knowledge.

The first case (Case I) assumes $\sigma_{0,1,2}$ to be all constant vertically as well as horizontally. We can get the rigorous analytic solution in this case and an insight on the problem will be possible. However, any quantitative discussion will be possible only in considering realistic conductivities which vary vastly in the ionosphere. The solution will be numerically obtained in this second case (Case II). In these two cases, $J_z=0$ at $z=z_{1,2}$ as in the convention. In the third case (Case III) the ionosphere model is equal to that of Case I, but the boundary condition is $J_{/\!/}=0$ at $z=z_2$ and $J_z=0$ at $z=z_1$, a condition which is very new. The rigorous analytic solution will be found. The last case will consider height-dependent conductivities and a numerical result is obtained for this new boundary condition.

Case I. $d\sigma_{0,1,2}/dz=0$. Now, (5.7) is

$$\varepsilon^2\left(\sin^2 I\frac{\partial^2 S}{\partial x^2}+\frac{\partial^2 S}{\partial y^2}-2\sin I\cos I\frac{\partial^2 S}{\partial x\partial z}+\cos^2 I\frac{\partial^2 S}{\partial z^2}\right)$$

$$+\cos^2 I\frac{\partial^2 S}{\partial x^2}+2\sin I\cos I\frac{\partial^2 S}{\partial x\partial z}+\sin^2 I\frac{\partial^2 S}{\partial z^2}=\frac{\operatorname{div}\boldsymbol{J}_0}{\sigma_0} \tag{5.41}$$

where

$$\boldsymbol{J}_0=[\sigma](\boldsymbol{W}\times\boldsymbol{B}_0)$$

$$\varepsilon=\sqrt{\frac{\sigma_1}{\sigma_0}}\ll 1$$

$$\boldsymbol{W}=\begin{bmatrix}\tilde{W}_x\\ \tilde{W}_y\\ 0\end{bmatrix}\exp[-i(\alpha x+\beta y+kz)] \tag{5.42}$$

$\tilde{W}_{x,y}=$const

$\alpha, \beta, k=$wave number $\alpha=\beta=10^{-7}\mathrm{m}^{-1}$, $k=5\times10^{-4}$–$5\times10^{-5}\mathrm{m}^{-1}$

$\boldsymbol{B}_0=$centered dipole.

Since we are interested in S such as $S\propto\exp[-i(\alpha x+\beta y)]$, (5.41) reduces to an ordinary second order linear differential equation. The solution consists of a particular solution, S_p, and the general solution, S_g, of the homogeneous equation which is obtained by putting the right-hand side of (5.41) zero. It is straightforward to get S_p as

$$S_p=\tilde{S}_p\exp[-i(\alpha x+\beta y+kz)] \tag{5.43}$$

where

$$\tilde{S}_p=\frac{i\varepsilon^2 B_0\{(-\alpha\sin I+k\cos I)[\tilde{W}_y-(\sigma_2/\sigma_1)\sin I\cdot\tilde{W}_x]+\beta[\sin I\cdot\tilde{W}_x+(\sigma_2/\sigma_1)\tilde{W}_y]\}}{(\alpha\cos I-k\sin I)^2}$$

$$\times(1-\varepsilon^2\cot^2 I)\simeq i\varepsilon^2\left(\frac{B_0}{k}\right)\left(\frac{\cos I}{\sin^2 I}\right)\left(\tilde{W}_y-\frac{\sigma_2}{\sigma_1}\tilde{W}_x\right). \tag{5.44}$$

The electric field E_p by S_p is of the order of $\varepsilon^2[|W_yB_0|+(\sigma_2/\sigma_1)|W_xB_0|]$ and mainly along $\boldsymbol{k}$ by $|k|\gg|\alpha|, |\beta|$. Note, however, that the small E_p is very effective to produce a field-aligned current J_p'' as

$$J_p'' = \sigma_0 E_p'' = -\sigma_1 B_0$$

$$\times \frac{\{(-\alpha \sin I + k \cos I)[W_y - (\sigma_2/\sigma_1)\sin I \cdot W_x] + \beta[\sin I \cdot W_x + (\sigma_2/\sigma_1) W_y]\}}{(\alpha \cos I + k \sin I)}$$

$$\times (1 - \varepsilon^2 \cot^2 I) \tag{5.45}$$

$$|J_p''| \simeq |J_0|. \tag{5.45'}$$

The vertical component of $\boldsymbol{J}_p''$ and $\boldsymbol{J}_0$ tends to cancel each other and produces a small vertical current f_0 as

$$f_0 = J''_{pz} + J_{0z} = \left(\frac{\alpha}{k}\right)\left(\frac{\sigma_1 B_0}{\sin I}\right)\Bigg[\Bigg(1 - \frac{\sigma_2}{\sigma_1}\frac{\beta}{\alpha}\sin I$$

$$+ \varepsilon^2 \frac{\cos^4 I - \cos^2 I \sin I - (\sigma_2/\sigma_1)(\beta/\alpha)\cos^2 I)}{\sin^2 I}\Bigg) W_y - \sin I$$

$$\times \left(\frac{\sigma_2}{\sigma_1} + \frac{\beta}{\alpha}\sin I + \frac{\sigma_2}{\sigma_1}\varepsilon^2 \frac{\cos^4 I - \cos^2 I \sin I - (\sigma_2/\sigma_1)(\beta/\alpha)\cos I}{\sin^2 I}\right) W_x\Bigg]. \tag{5.46}$$

Although $|f_0|$ is of the order of $(\alpha/k)|\boldsymbol{J}_0|$, it is f_0 which accumulates charges on the boundary at $z = z_{1,2}$ to produce $\boldsymbol{E}_g$ which is obtained as the general solution of the homogeneous equation S_g where

$$S_g = [\tilde{S}_1 \exp(\varepsilon\hat{\gamma} z) - \tilde{S}_2 \exp(-\varepsilon\hat{\gamma} z)] \exp[-i(\alpha x + \beta y - \alpha z \cot I)] \tag{5.47}$$

where

$$\hat{\gamma} = \frac{\sqrt{\alpha^2 + \beta^2 \sin^2 I}}{\sin^2 I}. \tag{5.48}$$

In (5.47), $\tilde{S}_{1,2}$ are arbitrary constants to be determined by the boundary conditions at $z = z_{1,2}$.

Consider $J_z = 0$ at $z = z_{1,2}$. Then, by tedious but elementary algebra, we get

$$\tilde{S}_1 = \frac{1}{2}(\tilde{\tilde{S}}_1 + \tilde{\tilde{S}}_2), \quad \tilde{S}_2 = \frac{1}{2}(\tilde{\tilde{S}}_1 - \tilde{\tilde{S}}_2)$$

where

$$\tilde{\tilde{S}}_1 = \frac{\varepsilon\alpha B_0}{k\hat{\gamma}\sin^2 I}\Bigg\{\frac{\alpha \cos I[1 + (\sigma_2/\sigma_1)(\beta/\alpha)\sin I]}{\hat{\gamma}^2 \sin^4 I}$$

$$\times \frac{[\exp(-ikz_2) - \exp(-ikz_1)]}{(z_2 - z_1)} - \frac{1}{\sin I}\frac{[z_1 \exp(-ikz_2 - z_2 \exp(-ikz_1)]}{(z_2 - z_1)}\Bigg\}$$

$$\times \left[\left(1 - \frac{\sigma_2}{\sigma_1}\frac{\beta}{\alpha}\right)\tilde{W}_y - \sin I\left(\frac{\sigma_2}{\sigma_1} + \frac{\beta}{\alpha}\sin I\right)\tilde{W}_x\right] \tag{5.49}$$

$$\tilde{\tilde{S}}_2 = i\frac{\alpha B_0}{\hat{\gamma}^2 \sin^3 I}\left[\sin I\left(\frac{\beta}{\alpha}\cdot\sin I + \frac{\sigma_2}{\sigma_1}\right)\langle W_x\rangle - \left(1 - \frac{\sigma_2}{\sigma_1}\cdot\frac{\beta}{\alpha}\sin I\right)\langle W_y\rangle\right] \tag{5.49'}$$

$$\langle W_{x,y}\rangle = \frac{1}{(z_2-z_1)}\int_{z_1}^{z_2} \tilde{W}_{x,y} \exp(-ikz)\, dz$$

$$= \frac{[\exp(-ikz_2)-\exp(-ikz_1)]\,\tilde{W}_{x,y}}{-ik(z_2-z_1)}$$

$$\propto \text{mean of } W_{x,y} \text{ over } (z_2-z_1).$$

The horizontal electric field is given as

$$E_{gx} = -\frac{\partial S_g}{\partial x} = i\alpha\tilde{\tilde{S}}_2 \exp\{-i[\beta y+\alpha(x-z\cot I)]\}$$

$$E_{gy} = -\frac{\partial S_g}{\partial y} = i\beta\tilde{\tilde{S}}_2 \exp\{-i[\beta y+\alpha(x-z\cot I)]\}\,. \qquad (5.50)$$

The wave number of S_g in (5. 47) is orthogonal to the field lines which are given by ($x-z\cot I=$const) except for a very small variation due to ($\hat{\gamma}z$) in the exponent of the terms in the bracket. Therefore, $\boldsymbol{E}_g$ is almost orthogonal to the field lines as is expected. The field aligned electric field is

$$E_g'' = \varepsilon\hat{\gamma}\sin I\,\tilde{\tilde{S}}_1 \exp[-i(\alpha x+\beta y-\cot I\cdot\alpha z)] \qquad (5.51)$$

and $|E_g''|\simeq\varepsilon^2|E_g|$.

Note that S_g is originally produced by $\boldsymbol{W}$, the wave number of which is *not* orthogonal to the field lines. It is the boundary which has converted the mode of $\boldsymbol{W}$ to that of S_g. Thus, the boundary is essential to the production of the electric field, which must always be orthogonal to $\boldsymbol{B}_0$ in magnetoactive plasma. Of course, this cannot be true to $\boldsymbol{E}_p$ by S_p in (5. 43); $\boldsymbol{E}_p$ is the same in mode i.e. of the same wave number as $\boldsymbol{W}$ as in (5. 42), but $\boldsymbol{E}_p$ is extremely small.

The overall picture of the electric field production by tidal dynamo is as

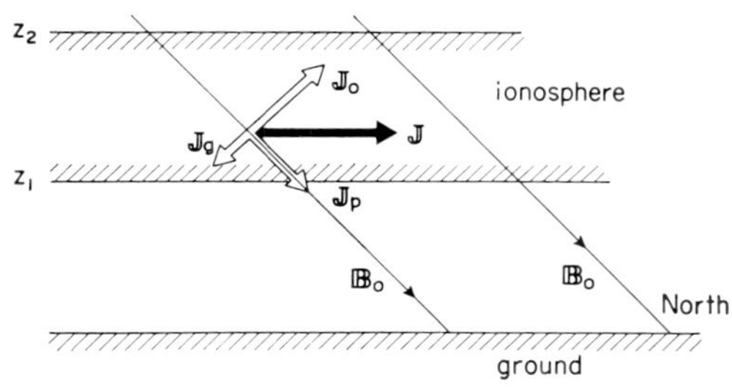

FIG. 5. 18. Dynamo mechanism. $\boldsymbol{J}_0$ is the applied current due to only winds orthogonal to $\boldsymbol{B}_0$. $\boldsymbol{J}_p$ is the field-aligned current due to a small polarization field in the absence of boundary. $\boldsymbol{J}_g$ is the current due to the polarization field by ($\boldsymbol{J}_p+\boldsymbol{J}_0$) in the presence of boundary which completely impedes the vertical current. $\boldsymbol{J}$ is the resultant horizontal current.

follows (Fig. 5. 18): At first an initial current $\boldsymbol{J}_0$ in (5. 41) tends to flow, and a small field $\boldsymbol{E}_{\rm p}$ is set up due to divergence of $\boldsymbol{J}_0$ i.e. div $\boldsymbol{J}_0 \neq 0$. Then, $\boldsymbol{E}_{\rm p}$ drives a field-aligned current $\boldsymbol{J}_{\rm p}''$ (in (5. 45)) which tends to cancel $\boldsymbol{J}_{0z}$ as in (5. 46) where $|J_{0z}+J_{pz}''| \ll |J_{0z}|$. However, the small resulting vertical current accumulates charge on the boundary at $z_{1,2}$. Eventually, the electric current due to the polarization field inhibits the total vertical current to flow, resulting in setting-up $\boldsymbol{E}_{\rm g}$ which can be as large as that really observed.

Next, the above result will be compared to that of the conventional approach in which the total horizontal current only is considered. By (5. 9), (5. 12) and (5. 12′) we have, generally, for *any distribution of* $\sigma_{0,1,2}$

$$\frac{1}{\sin^2 I}\frac{\partial^2 S}{\partial x^2}+\frac{\partial^2 S}{\partial y^2} = -\frac{i}{\boldsymbol{\Sigma}_1}\left(\frac{\alpha}{\sin^2 I}\int_{z_1}^{z_2}\sigma_1 \bar{E}_{0x}\mathrm{d}z+\frac{\alpha}{\sin I}\int_{z_1}^{z_2}\sigma_2\bar{E}_{0y}\mathrm{d}z\right.$$

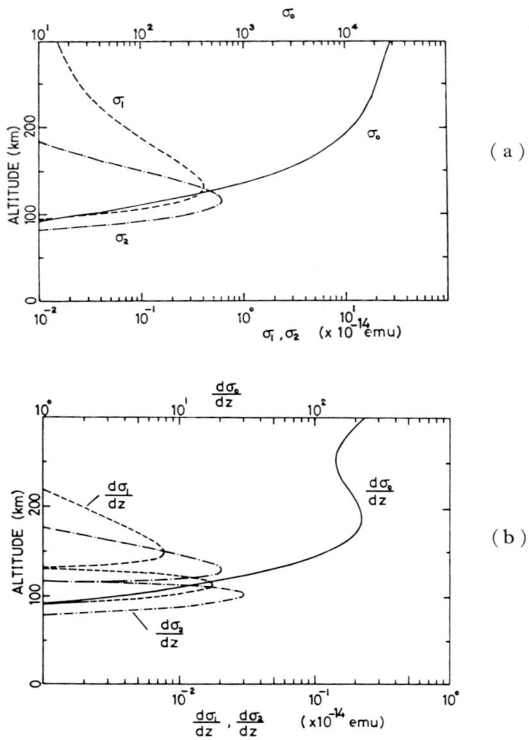

FIG. 5. 19. (a) $\sigma_{0,1,2}$ versus z, the solid, broken and chain curves, respectively. Multiply by 10^{11} to obtain mho m^{-1}. (b) $(\mathrm{d}\sigma_{0,1,2}/\mathrm{d}z)$ versus z; $\sigma_{0,1,2}$ varies with z as in (a). The broken and chain curves denote the magnitude of $(\mathrm{d}\sigma_{1,2}/\mathrm{d}z)$ which become negative above the maximum of $\sigma_{1,2}$.

$$+\beta\int_{z_1}^{z_2}\sigma_1\bar{E}_{0y}\mathrm{d}z-\frac{\beta}{\sin I}\int_{z_1}^{z_2}\sigma_2\bar{E}_{0x}\mathrm{d}z\Bigg) \tag{5.52}$$

which gives the following solution :

$$S = \tilde{S}\exp\left[-i(\alpha x+\beta y)\right]$$

$$\begin{aligned}\tilde{S} = \frac{i\alpha}{\boldsymbol{\Sigma}_1(\alpha^2+\beta^2\sin^2 I)}\Bigg[\int_{z_1}^{z_2}\sigma_1\bar{E}_{0x}\,\mathrm{d}z\\ -\sin I(\beta/\alpha)\int_{z_1}^{z_2}\sigma_2\bar{E}_{0x}\mathrm{d}z+\sin^2 I(\beta/\alpha)\int_{z_1}^{z_2}\sigma_1\bar{E}_{0y}\,\mathrm{d}z\\ -\sin I\int_{z_1}^{z_2}\sigma_2\bar{E}_{0y}\,\mathrm{d}z\Bigg]\end{aligned} \tag{5.53}$$

where

$$\bar{\boldsymbol{E}}_0 = (\tilde{\boldsymbol{W}}\times\boldsymbol{B}_0)\exp(-ikz) = (-\tilde{W}_y,\tilde{W}_x)B_0\sin I\,\mathrm{e}^{-ikz}.$$

When $\sigma_{1,2}$ are constant as in the present case, we can take $\sigma_{1,2}$ out of the integration. And by $\boldsymbol{\Sigma}_1=(z_2-z_1)\sigma_1$, it is found that S in (5. 53) and $\tilde{\tilde{S}}_2$ in (5. 49) are *identical*. Thus, the resulting horizontal fields are also identical except for a factor as $\mathrm{e}^{-\alpha\cot I z}$ which amounts to almost unity (0. 99 for $I=45°$ and $\alpha z=0.01$). Even if we consider height-dependent $[\sigma]$, the situation remains unchanged in essence, as will be shown below.

Case II. $\mathrm{d}\sigma_{0,1,2}/\mathrm{d}z\neq 0$. In this case (Fig. 5. 19) the basic equation corresponding to (5. 41) contains additional terms in the left-hand side due to $(\mathrm{d}\sigma_{0,1,2}/\mathrm{d}z)\neq 0$ as follows:

$$-\left(\frac{\mathrm{d}\sigma_2}{\mathrm{d}z}\right)\cos I\frac{\partial S}{\partial y}-\left(\frac{\mathrm{d}\sigma_1}{\mathrm{d}z}-\frac{\mathrm{d}\sigma_0}{\mathrm{d}z}\right)\sin I\cos I\frac{\partial S}{\partial x}$$
$$+\left(\frac{\mathrm{d}\sigma_0}{\mathrm{d}z}\sin^2 I+\frac{\mathrm{d}\sigma_1}{\mathrm{d}z}\cos^2 I\right)\frac{\partial S}{\partial z}. \tag{5.54}$$

div $\boldsymbol{J}_0$ in the right-hand side also suffers a change corresponding to the above terms replacing (grad S) by $(\boldsymbol{W}\times\boldsymbol{B}_0)$. Since (5. 52) is valid to any distribution of conductivities which are all integrated with respect to z as (5. 14), the result by the conventional method is readily obtained by calculating (5. 53) for given $\sigma_{0,1,2}$ as in Fig. 5. 19. It is found that, inspite of the difference between the two approaches, the resulting electric fields are nearly equal as illustrated in Fig. 5. 20. Of course, only numerical integration of the basic equation is now possible. In the numerical approach we have followed the steps of the analytical approach; at first a particular solution has been numerically obtained for arbitrary boundary condition, and then two solutions for two independent initial conditions at $z=z_2$ have been obtained for the homogeneous equation. Combine the particular solution and the two independent solutions multiplied by an arbitrary constants on each of them. The boundary condition $J_z=0$ at $z=z_{1,2}$ determines the constants. Probably because of wide variation of parameters especially σ_0, the relevant solution has required a very high accuracy in computation.

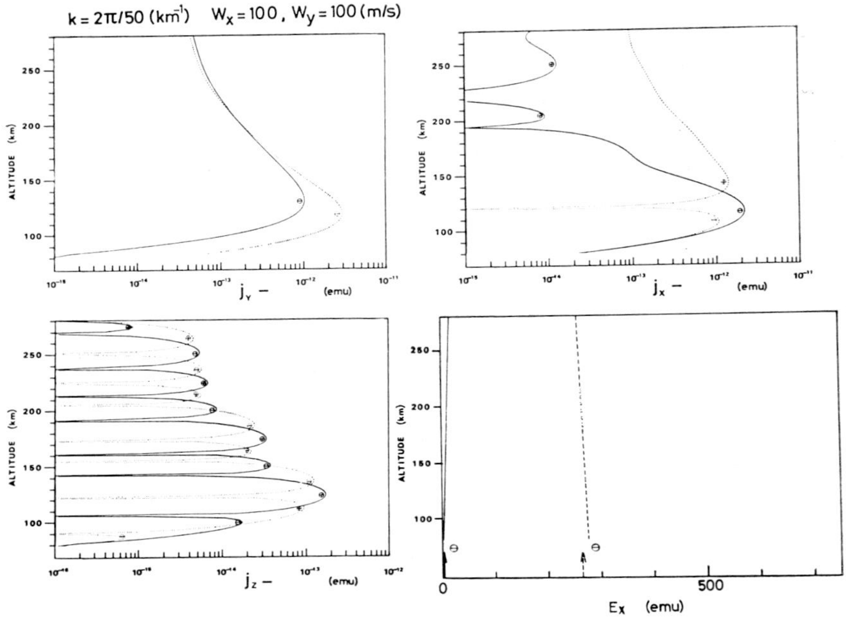
k = 2π/50 (km⁻¹)   Wx = 100 , Wy = 100 (m/s)
ALTITUDE (km)
jY −
(emu)
jX −
jZ −
EX (emu)
0
500

(a)

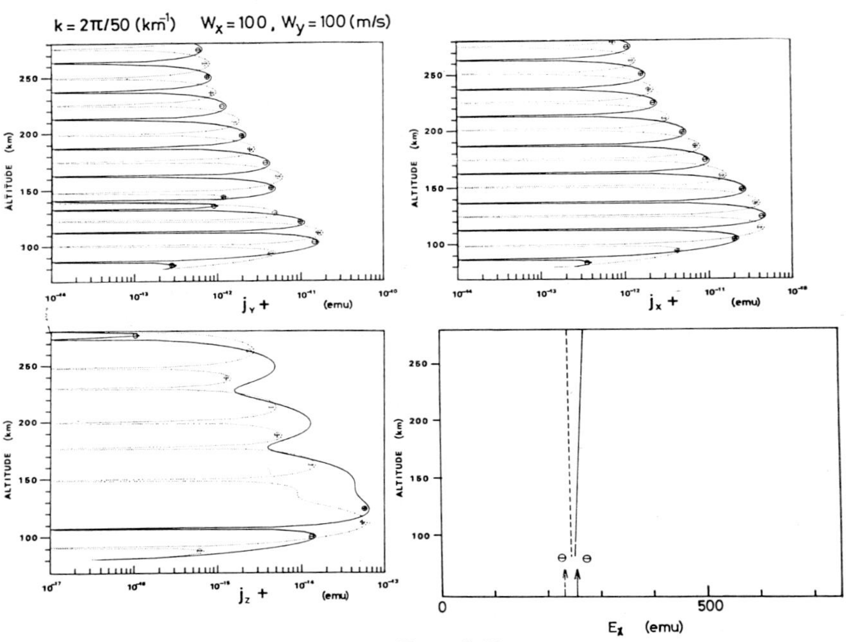
k = 2π/50 (km⁻¹)   Wx = 100 , Wy = 100 (m/s)
ALTITUDE (km)
jY +
(emu)
jX +
jZ +
EX (emu)
0
500

(b)

FIG. 5. 20

The boundary condition $J_z=0$ at $z_{1,2}$ has been used implicitly in the conventional approach of the dynamo theory, but the real meaning of the condition seems ambiguous. Mathematically, any quantity, as $\boldsymbol{J}$ in the present case, consists of both symmetric and anti-symmetric parts with respect to the equator. It is expected that the symmetric part of $\boldsymbol{J}$ is larger than the anti-symmetric part in the equinoctial time, and perhaps even in the solstitial period. For the symmetric part of $\boldsymbol{J}$, the above boundary condition may be relevant, because otherwise some electric current leaking from the upper boundary at one point might enter the ionosphere through the boundary at the conjugate point on the other hemisphere, and this current flow should contradict the symmetry condition. However, the condition is sufficient and not necessary. It is possible to consider those situations in which $J_z \neq 0$ at z_2, and yet no current flows between the symmetric points across the equator (Fig. 5. 21). In *Cases* III and IV we shall adopt $J_{/\!/}=0$ at z_2, and $J_z=0$ at z_1 using the same ionosphere model in *Cases* I and II, respectively. This situation can really occur.

Case III. $\mathrm{d}(\sigma_{0,1,2})/\mathrm{d}z=0$. Analytic solution is obtained as in *Case* I as

$$E_x = -\frac{\left[B_0 \sin I\left(\frac{\sigma_2}{\sigma_1}\tilde{W}_x-\tilde{W}_y\right)\exp\left(-ikz_2\right)\right]}{\left[1+\left(\frac{\sigma_2}{\sigma_1}\right)\left(\frac{\beta}{\alpha}\right)\sin I\right]}\exp\left[-i(\alpha x+\beta y-\alpha z\cdot\cot I)\right]$$

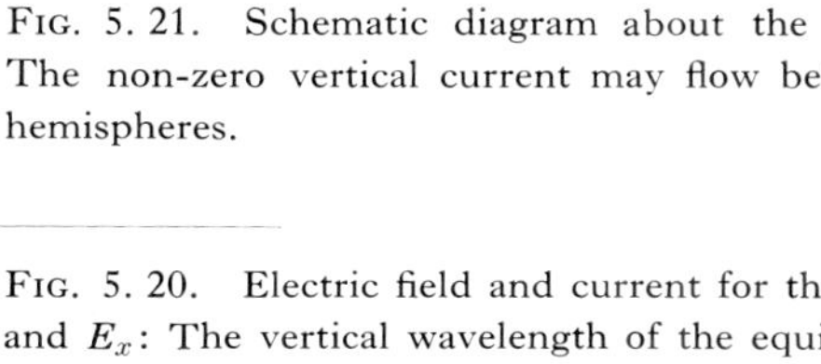

FIG. 5. 21. Schematic diagram about the new upper boundary condition. The non-zero vertical current may flow between the nighttime and daytime hemispheres.

FIG. 5. 20. Electric field and current for the condition J_z $(z_2)=0$. (a) $J_{x,y,z}$ and E_x: The vertical wavelength of the equivalent tidal wave is 50 km and the amplitude of $W_{x,y}$ is 100 m sec^{-1}. Along the north-south direction, the wave is a standing wave proportional to (sin αx). The solid lines are the result for sin $(\omega t-\beta y-kz)$. The dotted lines, except in the right-hand side on the bottom where the broken line is used, are the result for cos $(\omega t-\beta y-kz)$. ⊕ and ⊖ denote positive and negative. The arrows in E_x denote the value obtained by the conventional approach (see text). (b) Similar to (a) for cos (αx) (KATO and TSUDA, 1980). Multiply by 10^5 to obtain, in m.k.s. units, $\boldsymbol{J}$ in A m^{-2} and by 10^{-6} to obtain $\boldsymbol{E}$ in V m^{-1}.

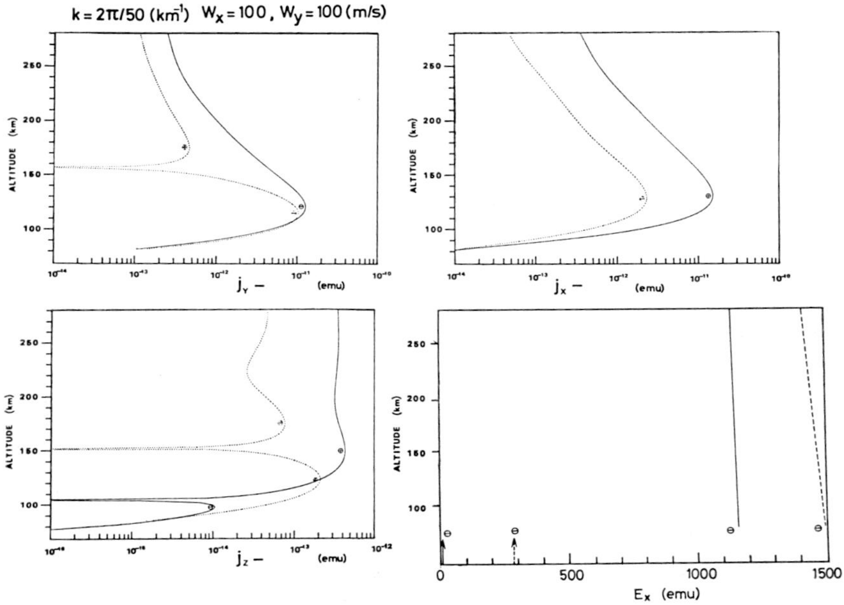
k = 2π/50 (km⁻¹) Wx = 100 , Wy = 100 (m/s)
ALTITUDE (km)
jY − (emu)
jX − (emu)
jZ − (emu)
Ex (emu)

(a)

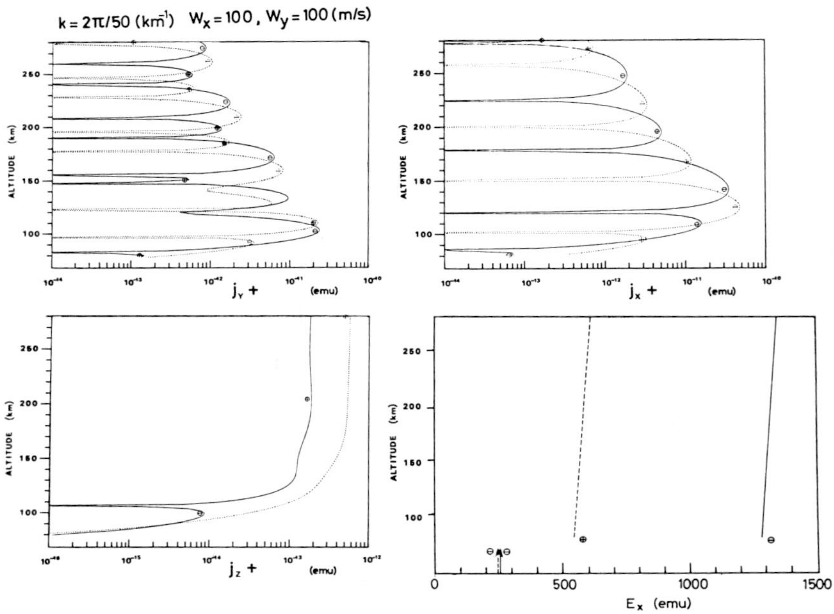
k = 2π/50 (km⁻¹) Wx = 100 , Wy = 100 (m/s)
ALTITUDE (km)
jY + (emu)
jX + (emu)
jZ + (emu)
Ex (emu)

(b)

Fig. 5.22

$$E_y = -\left(\frac{\beta}{\alpha}\right)\frac{\left[B_0 \sin I\left(\frac{\sigma_2}{\sigma_1}\tilde{W}_x - \tilde{W}_y\right)\exp(-ik\,z_2)\right]}{\left[1+\left(\frac{\sigma_2}{\sigma_1}\right)\left(\frac{\beta}{\alpha}\right)\sin I\right]}\exp\left[-i(\alpha x+\beta y-\alpha z\cdot\cot I)\right]. \tag{5.55}$$

$\boldsymbol{E}$ in (5. 55) can be larger or smaller than that in (5. 50) depending on (σ_2/σ_1) and I.

Case IV. $d(\sigma_{0,1,2})/dz \neq 0$ (Fig. 5. 19). The same numerical integration as in *Case* II leads us to the result illustrated in Fig. 5. 22.

It is found that the electric field in *Cases* III and IV is very different from that in *Cases* I and II, and also in the conventional approach. The present ionosphere model is so simplified that the difference which is found here may not be exact in more realistic model, in which the earth's sphericity, the curved field lines, the horizontal variation of conductivities are considered. Nevertheless, it must be kept in mind that $\boldsymbol{E}$ in Section 5. 3 is based on $J_z=0$ at z_2 and cannot be necessarily true, no matter how realistic ionospheric models might be.

For the anti-symmetric part the relevant boundary condition remains unknown. MAEDA and MURATA (1965) have attempted to treat the case to find field-aligned current between both hemispheres. The upper boundary condition is that J_z just inside $z=z_2$ equals $\{2[S_a(\theta, \phi, z_2)]/R\}$, where S_a is the anti-symmetric part of the potential S which is a solution of the Laplace equation (5. 1), because they assumed an *isotropic* and *constant* conductivity in (5. 7). R represents the total magnetospheric resistance between the conjugate points. The fact that $(2S_a)$ appears in the boundary condition above comes from the difference of S_a between the conjugate points at each (θ, ϕ, z_2) and $(180° - \theta, \phi, z_2)$, between which S_a has opposite sign and equal magnitude. In this approach it is hard to evaluate a proper R. In the magnetosphere R would be so small that S_a should vanish for any (θ, ϕ, z_2), the case, which was, as is in Section 5. 2 and 5. 3, discussed by MAEDA (1974) and SCHIELDGE *et al.* (1973).

The conclusion of the present section is as follows: The conventional dynamo theory, in which the height-integrated equation as (5. 9) is used, is able to give correct electric field, if the upper boundary condition is $J_z=0$ for those currents which are driven by the symmetric part of $\boldsymbol{J}_0=[\sigma](\boldsymbol{W}\times\boldsymbol{B}_0)$. It is open to question whether other boundary condition really exists. In the process of refining the theory to get better agreement to observation the question would have to be answered.

FIG. 5. 22. Solution of the new upper boundary condition as $J_{/\!/}=0$ $(J_z \neq 0)$ at z_2 (KATO and TSUDA, 1980). (a) Similar to Fig. 5. 20 (a) except for the boundary condition. (b) Similar to Fig. 5. 20 (b) except for the boundary condition.

APPENDIX

Appendix I (*Chapter* 2)

i) *Derivation of* (2.38), (2.41), (2.43), *and* (2.45)

The integrations come from the following I:

$$I = \iint_{-\infty}^{\infty} \frac{e^{-i(\alpha x+\gamma z)}}{\alpha^2 - A\gamma^2 - B} \, d\alpha d\gamma. \tag{A.2.1}$$

I takes three forms of which, for $\omega > \omega_a$, $I = I_1$ as, with $A = -a^2$ and $B = -b^2$

$$I_1 = \iint_{-\infty}^{\infty} \frac{e^{-i(\alpha x+\gamma z)}}{\alpha^2 + a^2\gamma^2 - b^2} \, d\alpha d\gamma. \tag{A.2.2}$$

First, consider integration with respect to γ. The pole of the integrand is

$$\begin{aligned} &= \pm \sqrt{b^2 - a^2\gamma^2} \quad \text{for} \quad b > a\gamma \\ &= \pm i\sqrt{a^2\gamma^2 - b^2} \quad \text{for} \quad b < a\gamma. \end{aligned} \tag{A.2.3}$$

Introduction of a small negative imaginary part in ω makes the poles on the real axes in the complex γ-plane displace vertically as in Figs. A. 1 and 2.

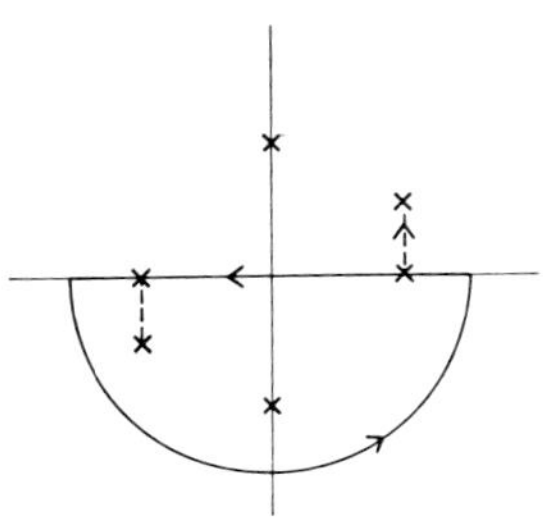

FIG. A. 1

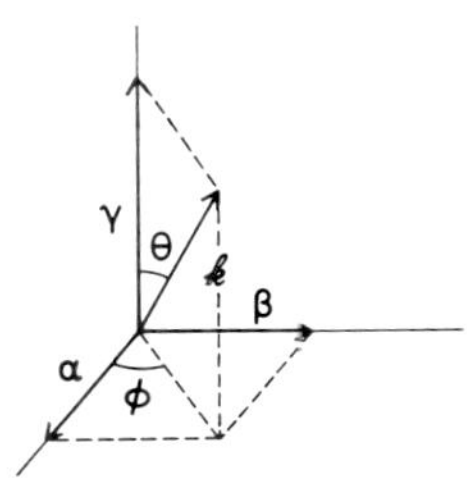

FIG. A. 2

FIG. A. 1. Poles and integration path in the complex γ plane.
FIG. A. 2. Wave number $\boldsymbol{k}$ (α, β, γ).

Solution which we are concerned with is that which vanishes with $t\to-\infty$ i.e. the outgoing wave from the source. The integration about γ between $-\infty$ and ∞ can be changed to a path integral around an infinite lower semi-circle. Then, the integral equals the total sum of the residue at the poles inside times $2\pi i$ as

$$-\pi i\frac{\exp[-ix\sqrt{b^2-a^2\gamma^2}]}{\sqrt{b^2-a^2\gamma^2}}\quad \text{for}\quad b>a$$

$$\pi\frac{\exp[-x\sqrt{a^2\gamma^2-b^2}]}{\sqrt{a^2\gamma^2-b^2}}\quad \text{for}\quad b<a.$$

Finally, integration with respect to α gives

$$I=\pi\left[i\int_{b>a|\gamma|}\frac{\exp(-ix\sqrt{b^2-a^2\gamma^2}-iz\gamma)}{\sqrt{b^2-a^2\gamma^2}}\mathrm{d}\gamma\right.$$
$$\left.-\int_{a|\gamma|<b}\frac{\exp(-ix\sqrt{a^2\gamma^2-b^2}-iz\gamma)}{\sqrt{a^2\gamma^2-b^2}}\mathrm{d}\gamma\right]$$
$$=\frac{\pi}{a}\left[i\int_{b>|s|}\frac{\cos(x\sqrt{b^2-s^2})\,\mathrm{e}^{-izs/a}}{\sqrt{b^2-s^2}}\mathrm{d}s\right.$$
$$+\left\{\int_{|s|<b}\frac{\sin(x\sqrt{b^2-s^2})\,\mathrm{e}^{-i(z/a)s}}{\sqrt{b^2-s^2}}\mathrm{d}s\right.$$
$$\left.\left.-\int_{|s|>b}\frac{\exp(-x\sqrt{s^2-b^2})}{\sqrt{s^2-b^2}}\,\mathrm{e}^{-i(z/a)s}\mathrm{d}s\right\}\right].$$

Thus (OBERHETTINGER, 1957, p. 30), (2.38) is obtained through

$$I=\frac{\pi^2}{a}i\left[J_0\left(b\sqrt{x^2+\frac{z^2}{a^2}}\right)-iN_0\left(b\sqrt{x^2+\frac{z^2}{a^2}}\right)\right]$$
$$=\frac{\pi^2}{a}iH_0^{(2)}\left(b\sqrt{x^2+\frac{z^2}{a^2}}\right).\qquad (\text{A}.2.4)$$

For $\omega_a>\omega>\omega_g$ and $\omega_g>\omega$, $I=I_{2,3}$; we go along similar lines to lead to (2.41), (2.43), and (2.45), respectively.

ii) Derivation of (2.51), (2.53), (2.54), *and* (2.55)

The integration required is I with the following form:

$$I=\int_{-\infty}^{\infty}\int_{-\infty}^{\infty}\int_{-\infty}^{\infty}\frac{\mathrm{e}^{-i(\alpha x+\beta y+\gamma z)}}{\alpha^2+\beta^2-A\gamma^2-B}\,\mathrm{d}\alpha\mathrm{d}\beta\mathrm{d}\gamma\qquad (\text{A}.2.5)$$

which consists of three cases as $I=I_{1,2,3}$. As to $I=I_1$ for $\omega>\omega_a$ we shall introduce a modified spherical coordinate system for (α,β,γ) as $(\mathscr{K},\theta,\phi)$ where θ is measured from (x,y,z) (see Fig. A.2).

$$\mathscr{K}=\sqrt{a^2+\beta^2+a^2\gamma^2}$$
$$\alpha=\mathscr{K}\sin\theta\cos\phi\qquad (\text{A}.2.6)$$

$$\beta = \mathscr{K} \sin\theta \sin\phi$$

$$\gamma = \frac{\mathscr{K}}{a} \cos\theta$$

$$I_1 = \int_{\phi=0}^{2\pi} \int_{\theta=0}^{\pi} \int_{k=0}^{\infty} \frac{\exp(-i\mathscr{K} R\cos\theta)}{a(\mathscr{K}^2 - b^2)} \mathscr{K}^2 \sin\theta \mathrm{d}\mathscr{K} \mathrm{d}\theta \mathrm{d}\phi \quad \text{(A.2.7)}$$

where $R = \sqrt{x^2+y^2+\dfrac{z^2}{a^2}}$.

Integration with respect to ϕ and θ is elementary as

$$I_1 = \frac{2\pi i}{aR}\int_0^{\infty} \frac{\exp(iR\mathscr{K})}{\mathscr{K}^2 - b^2} \mathrm{d}\mathscr{K}. \quad \text{(A.2.8)}$$

As in *i*) the method of path integral is used to get with a small negative imaginary part of ω and

$$I_1 = -\frac{2\pi^2 \mathrm{e}^{-ibR}}{aR}. \quad \text{(A.2.8')}$$

For $\omega_a > \omega > \omega_g$

$$I_2 = \iiint \frac{\mathrm{e}^{-i(\alpha x + \beta y + \gamma z)}}{\alpha^2 + \beta^2 + a^2\gamma^2 + b^2} \mathrm{d}\alpha \mathrm{d}\beta \mathrm{d}\gamma. \quad \text{(A.2.9)}$$

A polar coordinate in (α, β)-plane is introduced as $(\mathscr{K}', \theta')$ where $\mathscr{K}'$ is measured from (x, y). Integration with respect to θ' gives (WATSON, 1944, p. 176) with $R' = \sqrt{x^2+y^2}$

$$I_2 = \int_{\gamma=-\infty}^{\infty} \int_{k=0}^{\infty} \frac{\mathscr{K}' J_0(\mathscr{K}' R') \mathrm{e}^{-iz\gamma}}{\mathscr{K}'^2 + a^2\gamma^2 + b^2} \mathrm{d}\mathscr{K}' \mathrm{d}\gamma. \quad \text{(A.2.10)}$$

Further (WATSON, 1944, p. 425; OBERHETTINGER, 1957, p. 93)

$$I_2 = \frac{a\exp[-b\sqrt{x^2+y^2+(z^2/a^2)}]}{\pi\sqrt{x^2+y^2+(z^2/a^2)}}.$$

For $\omega_g > \omega > 0$ $I = I_3$ as

$$I_3 = \iiint \frac{\mathrm{e}^{-i(\alpha x + \beta y + \gamma z)}}{\alpha^2 + \beta^2 - a^2\gamma^2 - b^2} \mathrm{d}\alpha \mathrm{d}\beta \mathrm{d}\gamma. \quad \text{(A.2.11)}$$

A polar coordinate system as in (A.2.9) is introduced to obtain

$$I_3 = 2\pi \int_{\gamma=-\infty}^{\infty} \int_{k'=0}^{\infty} \frac{\mathscr{K}' J_0(R'\mathscr{K}') \mathrm{e}^{-iz}}{\mathscr{K}'^2 - a^2\gamma^2 - b^2} \mathrm{d}\mathscr{K}' \mathrm{d}\gamma \quad \text{(A.2.12)}$$

which leads by the path integral method to

$$I_3 = \frac{\pi i}{a}\int \frac{\exp[i(|z|/a)\sqrt{\mathscr{K}'^2 - b^2}]}{\sqrt{\mathscr{K}'^2 - b^2}} \mathrm{d}\mathscr{K}' \qquad \text{for } \mathscr{K}' > b$$

$$= \frac{\pi}{a}\int \frac{\exp[-(|z|/a)\sqrt{b^2 - \mathscr{K}'^2}]}{\sqrt{b^2 - \mathscr{K}'^2}} \mathrm{d}\mathscr{K}' \qquad \text{for } \mathscr{K}' < b. \quad \text{(A.2.13)}$$

Then, the integrand is divided into real and imaginary parts for integration with respect to $\mathscr{K}'$. Finally (OBERHETTINGER, 1957, pp. 77, 78; WATSON, 1944, pp. 405, 417, 435)

$$I_3 = 2\pi^2 \frac{\exp\left(-b\sqrt{(z^2/a^2)-(x^2+y^2)}\right)}{a\sqrt{(z^2/a^2)-(x^2+y^2)}} \quad \text{for } x^2+y^2<\frac{z^2}{a^2}$$

$$= -2\pi^2 i \frac{\exp\left(-ib\sqrt{x^2+y^2-(z^2/a^2)}\right)}{a\sqrt{x^2+y^2-(z^2/a^2)}} \quad \text{for } x^2+y^2>\frac{z^2}{a^2}. \quad \text{(A. 2. 14)}$$

Appendix II (Chapter 3)

i) Definition of $P_{n,m}(\theta)$ and $P_n{}^m(\theta)$, Associated Legendre function of order m, degree n

$$P_{n,m} = (\sin\theta)^m \frac{\mathrm{d}^m P_n(\theta)}{\mathrm{d}(\cos\theta)^m} = (1-\mu^2)^{(m/2)} \frac{\mathrm{d}^m P_n(\mu)}{\mathrm{d}\mu^m} \quad \text{(A. 3. 1)}$$

for $n \geq m$ where $P_n(\theta)$ is the Legendre function of degree n. $P_{n,m}$ is called to be Neumann-type. And

$$P_n{}^m = \sqrt{2\frac{(n-m)\,!}{(n+m)\,!}}\, P_{n,m}. \quad \text{(A. 3. 2)}$$

Note that $P_n{}^m$ is called to be Schmidt-type and is normalized as

$$\int_0^{2\pi}\int_0^{\pi}\left[P_n{}^m \begin{Bmatrix}\text{or} & \cos m\phi \\ & \sin m\phi\end{Bmatrix}\right]^2 \sin\theta \,\mathrm{d}\theta\,\mathrm{d}\phi = \frac{4\pi}{2n+1}. \quad \text{(A. 3. 3)}$$

Both $P_{n,m}$ and $P_n{}^m$ constitutes A) orthogonal and B) complete set about n for each m:

$$\text{A)} \int_0^{\pi} P_{n,m}P_{n',m} \sin\theta\,\mathrm{d}\theta = \int_0^{\pi} P_n{}^m P_{n'}{}^{m'} \sin\theta\,\mathrm{d}\theta = 0 \quad \text{for } n \neq n'. \quad \text{(A. 3. 4)}$$

B) Any function which is finite for $0 \leq \theta \leq \pi$ can be expanded by a series of $P_n{}^m$ or $P_{n,m}$.

ii) Derivation of the eigenvalue equation as (3. 25) (e.g. KATO, 1966a) Define $D_{j,n}{}^{(\omega,m)}$ as

$$D_{j,m}{}^{(\omega,m)} = \frac{1}{[j(j+1)-(m/f)]}\left[\frac{(j+1)^2(j-m)}{2j-1}\Gamma_{j-1,n}{}^{(\omega,m)} + \frac{j^2(j+m+1)}{2j+3}\Gamma_{j+1,n}{}^{(\omega,m)}\right] \quad \text{(A. 3. 5)}$$

or

$$\frac{(j+1)^2(j-m)}{(2j-1)}\Gamma_{j-1,n}{}^{(\omega,m)} - N_{j,m}D_{j,m}{}^{(\omega,m)} + \frac{j^2(j+m+1)}{(2j+3)}\Gamma_{j+1,n}{}^{(\omega,m)} = 0 \quad \text{(A. 3. 6)}$$

which is the correct expression of KATO's formula (3. 10) (1966a) which contained no equal sign by error.

Eliminating $\alpha_{j\pm1,n}{}^{(\omega,m)}$ and $\beta_{j,n}{}^{(\omega,m)}$ in (3.23) by (3.22) and (3.24), we obtain

$$\frac{(j+1)^2(j-m)}{(2j-1)} D_{j-1,n}{}^{(\omega,m)} - M_{j,n}{}^{(\omega,m)} \Gamma_{j,n}{}^{(\omega,m)} + \frac{j^2(j+m+1)}{(2j+3)} D_{j+1,n}{}^{(\omega,m)} = 0 \tag{A.3.7}$$

where

$$M_{j,n}{}^{(\omega,m)} = f^2\left[j(j+1) - \frac{m}{f}\right] - \frac{ghj^2(j+1)^2}{4a^2\Omega^2}. \tag{A.3.8}$$

By putting $j=m$, (A.3.7) gives

$$M_{m,n}{}^{(\omega,m)} - \frac{m^2(2m+1)}{(2m+3)} \frac{D_{m+1,n}{}^{(\omega,m)}}{\Gamma_{m,n}{}^{(\omega,m)}} = 0. \tag{A.3.9}$$

(A.3.6) gives by putting $j=m+1$

$$\frac{D_{m+1,n}{}^{(\omega,m)}}{\Gamma_{m,n}{}^{(\omega,m)}} = \frac{m^2(m+2)^2/(2m+1)}{N_{m+1,m} - [(m+1)^2(2m+2)/(2m+5)][D_{m+1,n}{}^{(\omega,m)}/\Gamma_{m+2,n}{}^{(\omega,m)}]}. \tag{A.3.10}$$

Then, substituting (A.3.10) into (A.3.9)

$$M_{m,n}{}^{(\omega,m)} - \frac{m^2(m+2)^2}{(2m+3)} \frac{1}{N_{m+1,m} - \dfrac{(m+1)^2(2m+2)}{(2m+5)} \dfrac{D_{m+1,n}{}^{(\omega,m)}}{\Gamma_{m+2,n}{}^{(\omega,m)}}}. \tag{A.3.11}$$

Then, (A.3.7) gives for $j=m+2$

$$\frac{D_{m+1,n}{}^{(\omega,m)}}{\Gamma_{m+2,n}{}^{(\omega,m)}} = \frac{2m+3}{2(m+3)^2}\left[M_{m+2,n}{}^{(\omega,m)} - \frac{(m+2)^2(2m+3)}{(2m+7)} \frac{D_{m+3,n}{}^{(\omega,m)}}{\Gamma_{m+2,n}{}^{(\omega,m)}}\right]. \tag{A.3.12}$$

Also, (A.3.6) gives for $j=m+3$

$$\frac{D_{m+3,n}{}^{(\omega,m)}}{\Gamma_{m+2,n}{}^{(\omega,m)}} = \frac{(m+4)^2(m+2)/(2m+5)}{N_{m+3,m} - [(m+3)^2(2m+4)/(2m+9)][D_{m+3,n}{}^{(\omega,m)}/\Gamma_{m+4,n}{}^{(\omega,m)}]}. \tag{A.3.13}$$

Substitution of (A.3.13) into (A.3.12) and, further, another substitution of (A.3.12) into (A.3.11) give

$$M_{m,n}{}^{(\omega,m)} - \frac{a_{m,n}}{N_{m+1,n} - \dfrac{a_{m+1,m}}{M_{m+2,n}{}^{(\omega,m)} - \dfrac{a_{m+2,m}}{[N_{m+3,m} - \{(m+3)^2(2m+2)/(2m+9)\}\{D_{m+3,n}{}^{(\omega,m)}/\Gamma_{m+4,n}{}^{(\omega,m)}\}]}}} \tag{A.3.14}$$

where $a_{j,n}$ is given in the text by (3.26). Repetition of this process gives (3.25).

iii) Derivation of $W_{\theta,\phi}{}^{(\omega,m)}$

By (3.8), (3.13), (3.30) and the definition of μ and $\mathscr{D}$

$$W_{\phi,\omega}=\frac{\omega}{4a\Omega^2(f^2-\mu^2)}\sum_{m,n}\left(\frac{1}{m}\right)\left(\frac{m}{f}\mu\mathscr{D}-m^2\right)\left(\frac{p_{1,\omega}}{\rho_0}\right)$$

$$=\frac{\omega}{4a\Omega^2(f^2-\mu^2)\sin\theta}\sum_{m,n}\left(\frac{p_{\omega,m,n}}{\rho_0}\right)\left(\frac{1}{m}\right)$$

$$\times\left(\frac{m}{f}\mu\mathscr{D}-m^2\right)\Theta_{\omega,m,n}\mathrm{e}^{\mathrm{i}m\phi} \qquad \text{(A.3.15)}$$

in which we difine G as

$$G=\left(\frac{m}{f}\mu\mathscr{D}-m^2\right)\Theta_{\omega,m,n}.$$

G reduces by (3.20) as follows:

$$G\equiv\left(\frac{m\mu}{f}\mathscr{D}-m^2\right)\left[\left(\mathscr{D}+\frac{m\mu}{f}\right)\Theta_1+\frac{m^2}{f^2}(f^2-\mu^2)\Theta_2\right]$$

where subscripts n and superscripts (ω, m) are omitted from $\Theta_{1,2}$ in (3.20). Further manipulation by (3.21) results in

$$G\equiv\frac{m\mu}{f}(1-\mu^2)\left[\left(\mathscr{L}+\frac{m}{f}\right)\Theta_1-2\frac{m^2}{f^2}\mu^2\Theta_2\right]$$

$$-(f^2-\mu^2)\left[\frac{m^2}{f^2}\mathscr{D}\Theta_1+\frac{m^3}{f^2}\left(-\frac{\mu}{f}\mathscr{D}+m\right)\Theta_2\right]$$

$$=-(f^2-\mu^2)\left(\frac{m}{f}\right)^2\left[\mathscr{D}\Theta_1-m\left(\frac{\mu}{f}\mathscr{D}-m\right)\Theta_2\right]$$

which is substituted into (A.3.15), yielding (3.37) as

$$W_{\phi,\omega}=\frac{-\omega}{4a\Omega^2\sin\theta}\sum_{m,n}\left(\frac{m}{f^2}\right)\left(\frac{p_{\omega,m,n}}{\rho_0}\right)\left[\mathscr{D}\Theta_1-m\left(\frac{\mu}{f}\mathscr{D}-m\right)\Theta_2\right]$$

$$\times\mathrm{e}^{i(\omega t+m\phi)}. \qquad \text{(A.3.16)}$$

Note that (A.3.16) contains no factor as $(f^2-\mu^2)^{-1}$. The recurrence relation of $P_{j,m}$ gives

$$\mathscr{D}\Theta_1=\sum_j\alpha_{j,n}\mathscr{D}P_{j,m}=\sum_j\left\{-\frac{(j-1)(j-m)}{(2j-1)}\alpha_{j-1,n}\right.$$

$$\left.+\frac{(j+2)(j+m+1)}{2j+3}\alpha_{j+1,n}\right\}P_{j,m} \qquad \text{(A.3.17)}$$

and

$$\left(\frac{\mu}{f}\mathscr{D}-m\right)\Theta_2=\sum_j\left[\frac{\mu}{f}\left\{-\frac{(j-1)(j-m)}{2j-1}\beta_{j-1,n}+\frac{(j+2)(j+m+1)}{2j+3}\right.\right.$$

$$\left.\left.\beta_{j+1,n}\right\}-m\beta_{j,n}\right]P_{j,m}. \qquad \text{(A.3.18)}$$

By (3.22), $\alpha_{j-1,n}$ and $\alpha_{j+1,n}$ in (A.3.17) are given in terms of $\beta_{j-2,n}$, $\beta_{j,n}$, and $\beta_{j+2,n}$ which in turn are given by $\Gamma_{j-2,n}$, $\Gamma_{j,n}$, and $\Gamma_{j+2,n}$. Also (A.3.18) is given in similar terms. (3.38) is thus obtained. As to $W_{\theta,\omega}$, a similar manipulation gives

$$W_{\theta,\omega} = -\frac{i\omega}{4a\Omega^2(f^2-\mu^2)\sin\theta}\left(\mathscr{D}-\frac{m}{f}\mu\right)\left(\frac{p_{1,\omega}}{\rho_0}\right)e^{i(\omega t+m\phi)}$$

$$= \frac{i\omega}{4a\Omega^2\sin\theta}\sum_{m,n}\left(\frac{m}{f}\right)^2\left(\frac{p_{\omega,m,n}}{\rho_0}\right)\left[\Theta_1+\left(\mathscr{D}-\frac{m}{f}\mu\right)\Theta_2\right]e^{i(\omega t+m\phi)} \quad \text{(A.3.19)}$$

which results in (3.35) through similar manipulation as for (A.3.16).

Note that in (3.37) and (3.39) $A_j^{(\omega,m)}$, $B_j^{(\omega,m)}$, $J_j^{(\omega,m)}$, $L_j^{(\omega,m)}$, and $S_j^{(\omega,m)}$ equal those by KATO (1966) except for factors $1/[(j-2)(j-1)+(m/f)]$ etc.; $\Phi_v^{(-1)}$ in (3.23) in KATO's paper in 1966 was in error in the term proportional to $J_n^{(-1)}$ as to replace $(n-1)(n-1)$ by $(n-1)(n-2)$.

iv) Equation of motion in spherical coordinates

Expression of the equation of motion (1.1) in spherical coordinate is fairly complex and, to some readers, is unfamiliar. What follows is a tedious but sound way to the result. The method is a transformation from the (x, y, z)-system in rotation around the z-axis to the spherical coordinate (r, θ, ϕ)-system in the same rotation; θ is measured from the z-axis. Generally, vector $\boldsymbol{A}$ is given between the two systems as

$$\begin{bmatrix} A_r \\ A_\theta \\ A_\phi \end{bmatrix} = \boldsymbol{T}\begin{bmatrix} A_x \\ A_y \\ A_z \end{bmatrix} \quad \text{(A.3.20)}$$

where

$$\boldsymbol{T} = \begin{bmatrix} \sin\theta\ \cos\phi, & \sin\theta\ \sin\phi, & \cos\theta \\ \cos\theta\ \cos\phi, & \cos\theta\ \sin\phi, & -\sin\theta \\ -\sin\phi, & \cos\phi, & 0 \end{bmatrix}. \quad \text{(A.3.21)}$$

When $\boldsymbol{A} = \mathscr{G}\boldsymbol{W}$ (A.3.22)

where

$$\mathscr{G} = \boldsymbol{W}\cdot\nabla = W_r\frac{\partial}{\partial r}+\frac{W_r}{r}\frac{\partial}{\partial\theta}+\frac{W_\phi}{r\sin\theta}\frac{\partial}{\partial\phi}, \quad \text{(A.3.23)}$$

and

$$\begin{bmatrix} W_x \\ W_y \\ W_z \end{bmatrix} = \boldsymbol{T}^*\begin{bmatrix} W_r \\ W_\theta \\ W_\phi \end{bmatrix} \quad \text{(A.3.24)}$$

we get by (A.3.22)

$$\begin{bmatrix} A_r \\ A_\theta \\ A_\phi \end{bmatrix} = \boldsymbol{T}\mathscr{G}\left(\boldsymbol{T}^*\begin{bmatrix} W_r \\ W_\theta \\ W_\phi \end{bmatrix}\right)$$

where $\boldsymbol{T}^*$ is inverse to $\boldsymbol{T}$ and is obtained by interchanging the row and column of $\boldsymbol{T}$.
Thus

$$A_r = \left(W_r \frac{\partial}{\partial r} + \frac{W_\theta}{r}\frac{\partial}{\partial \theta} + \frac{W_\phi}{r\sin\theta}\frac{\partial}{\partial \phi}\right)W_r - \frac{W_\theta{}^2 + W_\phi{}^2}{r}$$

$$A_\theta = \left(W_r \frac{\partial}{\partial r} + \frac{W_\theta}{r}\frac{\partial}{\partial \theta} + \frac{W_\phi}{r\sin\theta}\frac{\partial}{\partial \phi}\right)W_\theta + \frac{W_\theta W_r}{r} - \frac{\cot\theta}{r}W_\phi{}^2$$

$$A_\phi = \left(W_r \frac{\partial}{\partial r} + \frac{W_\theta}{r}\frac{\partial}{\partial \theta} + \frac{W_\phi}{r\sin\theta}\frac{\partial}{\partial \phi}\right)W_\phi + \frac{W_\phi W_r}{r} + \frac{\cot\theta}{r}W_\phi W_\theta \tag{A.3.25}$$

$$\frac{\partial}{\partial t}\begin{bmatrix} W_r \\ W_\theta \\ W_\phi \end{bmatrix} + \begin{bmatrix} A_r \\ A_\theta \\ A_\phi \end{bmatrix} + 2\Omega \begin{bmatrix} -\sin\theta W_\phi \\ -\cos\theta W_\phi \\ \cos\theta W_\theta + \sin\theta W_r \end{bmatrix}$$

$$= -\frac{1}{\rho}\begin{bmatrix} \frac{\partial}{\partial r} \\ \frac{1}{r}\frac{\partial}{\partial \theta} \\ \frac{1}{r\sin\theta}\frac{1}{\partial \phi} \end{bmatrix} p + \begin{bmatrix} g_r \\ g_\theta \\ 0 \end{bmatrix} + \begin{bmatrix} F_r \\ F_\theta \\ F_\phi \end{bmatrix} \tag{A.3.26}$$

where $g_\theta = 0$ if the centrifugal force is neglected. Linearization of (A.3.26) results in (3.63)–(3.65).

Appendix III (Chapter 4)

i) Mobilities

The rigorous deduction is unknown due to the fact that the collision integral in the Boltzmann equation in the gas of three kinds of particles is too complicated. But, phenomenologically, the Lengevin-type equation can be used to give

$$\mp e(\boldsymbol{E} + \boldsymbol{V}_{e,i} \times \boldsymbol{B}_0) - m_{e,i}\nu_{e,i}(\boldsymbol{V}_{e,i} - \boldsymbol{W}) = 0 \tag{A.4.1}$$

where $\mp$ is for electrons $(-)$ and ions $(+)$, and the inertia is neglected. Then,

$$\boldsymbol{V}_{e,i} = \mu_{e,i}{}^{(1)}\boldsymbol{E}_\perp + \mu_{e,i}{}^{(2)}\frac{(\boldsymbol{E}\times\boldsymbol{B}_0)}{B_0} + \mu_{e,i}{}^{(0)}\boldsymbol{E}_{//} \tag{A.4.1'}$$

where

$$\mu_{e,i}{}^{(1)} = \frac{\mp\Omega_{e,i}\nu_{e,i}}{\Omega_{e,i}{}^2+\nu_{e,i}{}^2}\cdot\frac{1}{B_0} \quad \text{(Pedersen mobility)}$$

$$\mu_{e,i}{}^{(2)} = \frac{\Omega_{e,i}{}^2}{\Omega_{e,i}{}^2+\nu_{e,i}{}^2}\cdot\frac{1}{B_0} \quad \text{(Hall mobility)} \qquad \text{(A. 4. 2)}$$

$$\mu_{e,i}{}^{(0)} = \frac{\mp\Omega_{e,i}}{\nu_{e,i}}\cdot\frac{1}{B_0} \quad \text{(Parallel mobility)}$$

and

$$\sigma_1 = n_e e(\mu_i{}^{(1)}-\mu_e{}^{(1)}) = \frac{n_e e(\Omega_e\Omega_i+\nu_e\nu_i)(\Omega_e\nu_i+\Omega_i\nu_e)}{B_0(\Omega_e{}^2+\nu_e{}^2)(\Omega_i{}^2+\nu_i{}^2)}$$

$$\sigma_2 = n_e e(\mu_e{}^{(2)}-\mu_i{}^{(2)}) = \frac{n_e e(\Omega_e\nu_i-\Omega_i\nu_e)(\Omega_e\nu_{\;}+\Omega_i\nu_e)}{B_0(\Omega_e{}^2+\nu_e{}^2)(\Omega_i{}^2+\nu_i{}^2)}$$

$$\sigma_0 = n_e e(\mu_i{}^{(0)}-\mu_e{}^{(0)}) = \frac{n_e e(\Omega_e\nu_i+\Omega_i\nu_e)}{B_0\nu_e\nu_i} \qquad \text{(A. 4. 3)}$$

$$\sigma_3 = \sigma_1+\frac{\sigma_2{}^2}{\sigma_1} = \frac{n_e e(\nu_e\Omega_i+\nu_i\Omega_e)}{B_0(\Omega_e\Omega_i+\nu_e\nu_i)}$$

where $\Omega_{e,i}$=angular-gyrofrequency, $eB_0/m_{e,i}$. For more species of ions no concise form as (A. 4. 3) exists and

$$\sigma_1 = \frac{1}{B_0}\sum_j \frac{n_j\nu_j\Omega_j}{\Omega_j{}^2+\nu_j{}^2}e_j$$

$$\sigma_2 = \frac{-1}{B_0}\sum_j \frac{n_j\Omega_j{}^2}{\Omega_j{}^2+\nu_j{}^2}e_j$$

$$\sigma_0 = \frac{1}{B_0}\sum_j \frac{n_j\Omega_j}{\nu_j}e_j \qquad \text{(A. 4. 4)}$$

where j denotes j-th ions; for electrons $\Omega_j=-\Omega_e$, $e_j=-e$, and for ions Ω_j is defined similarly as that of one species.

ii) Derivation of (4. 8)

Addition of two equations in (A. 4. 1), one for electrons and the other for ions, gives by (4. 1) and (4. 2)

$$-(m_e\nu_e+m_i\nu_i)\,\boldsymbol{W}+m_e\nu_e\boldsymbol{V}_e+m_i\nu_i\boldsymbol{V}_i = (\boldsymbol{J}\times\boldsymbol{B}_0)/n_e \qquad \text{(A. 4. 5)}$$

which further with (4. 2) gives (4. 8).

iii) Ambipolar diffusion

If $p_{e,i}$ is considered, both ions and electrons tend to move with additional differential velocity $\Delta\boldsymbol{V}_{e,i}$ to that given by (4. 1). However, the electrostatic force $\boldsymbol{E}_p$ is immediately set up, preventing both ions and electrons from moving differentially. Namely (e.g. KATO, 1963b) the process goes as follows:

$$\Delta \boldsymbol{V}_{e,i} = \mp \frac{e}{m_{e,i}} \frac{\nu_{e,i}}{\nu_{e,i}^2+\Omega_{e,i}^2}\left(\boldsymbol{E}_{p\perp} \pm \frac{\nabla_\perp p_{e,i}}{n_e e}\right)$$
$$\mp \frac{e}{m_{e,i}}\left(\boldsymbol{E}_{p,//} \pm \frac{\nabla_{//} p_{e,i}}{n_e e}\right) \qquad \text{(A.4.6)}$$

where, in the four double signs, the upper sign is for electrons and the lower sign is for ions. By putting $\Delta \boldsymbol{V}_e = \Delta \boldsymbol{V}_i = \Delta \boldsymbol{V}$,

$$\boldsymbol{E}_{p,\perp} = \frac{[\nu_e\Omega_e/(\nu_e^2+\Omega_e^2)]\nabla_\perp p_e - [\nu_i\Omega_i/(\nu_i^2+\Omega_i^2)]\nabla p_{i,\perp}}{en_e[\nu_e\Omega_e/(\Omega_e^2+\nu_e^2)+\nu_i\Omega_i/(\nu_i^2+\Omega_i^2)]}$$
$$\boldsymbol{E}_{p,//} = -\frac{(\Omega_e/\nu_e)\nabla_{//} p_e - (\Omega_i/\nu_i)\nabla_{//} p_i}{en_e[(\Omega_e/\nu_e)+(\Omega_i/\nu_i)]}. \qquad \text{(A.4.7)}$$

Finally, substitution of (A.4.7) into (A.4.6) gives

$$\Delta \boldsymbol{V}_\perp = -\frac{\Omega_e\Omega_i\nu_e\nu_i}{(\Omega_e\nu_i+\Omega_i\nu_e)(\nu_e\nu_i+\Omega_e\Omega_i)} \frac{\nabla_\perp(p_e+p_i)}{en_eB_0}$$
$$\Delta \boldsymbol{V}_{//} = -\frac{\Omega_e\Omega_i}{\Omega_e\nu_i+\Omega_i\nu_e} \frac{\nabla_{//}(p_e+p_i)}{en_eB_0}. \qquad \text{(A.4.8)}$$

Here the diffusion only in the (magnetic) meridional plane is considered, because the vertical wavelength of $\boldsymbol{W}$ is assumed to be much shorter than the horizontal one.

Appendix IV (Chapter 5)

i) Solution of $\nabla^2\mathcal{S}=0$

The equation is given in terms of spherical coordinate (r, θ, ϕ) as

$$\frac{1}{r^2}\frac{\partial}{\partial r}\left(r^2\frac{\partial \mathcal{S}}{\partial r}\right) + \frac{1}{r^2\sin\theta}\frac{\partial}{\partial\theta}\left(\sin\theta\frac{\partial\mathcal{S}}{\partial\theta}\right) + \frac{1}{r^2\sin^2\theta}\frac{\partial^2\mathcal{S}}{\partial\phi^2} = 0$$

which is separated into the following three equations:

$$\frac{1}{\mathcal{S}_r}\frac{d}{d}\left(r^2\frac{d\mathcal{S}_r}{dr}\right) = C_1$$
$$\frac{d}{d\theta}\left(\sin\theta\frac{d\mathcal{S}_\theta}{d\theta}\right) + C_1\sin\theta\, S_\theta - \frac{C_2\mathcal{S}_\theta}{\sin\theta} = 0 \qquad \text{(A.5.1)}$$
$$\frac{d^2\mathcal{S}_\phi}{d\phi^2} + C_2\mathcal{S}_\phi = 0$$

where $\mathcal{S} = \mathcal{S}_r(r)\mathcal{S}_\theta(\theta)\mathcal{S}_\phi(\phi)$ and $C_{1,2}$ are constants. For stable solution $\mathcal{S}_\phi$ should be periodic in $0<\phi<2\pi$

$$\mathcal{S} \propto e^{im\phi} \quad \text{for} \quad m = 0, 1, 2, \ldots \qquad \text{(A.5.2)}$$

which requires $C_2=m^2$ as the eigenvalues in the last equation in (A. 5. 1). Upon $\mathscr{S}_\theta$ in the second equation in (A. 5. 1) we impose the boundary condition that $\mathscr{S}_\theta=0$ at $\theta=0$ and π which give the eigenvalue $C_1=n(n+1)$ where $n=m, m+1, m+2, \ldots$ Then

$$\mathscr{S}_\theta \propto P_{n,m}(\theta) \qquad \text{(A. 5. 3)}$$

where $P_{n,m}(\theta)$ is *Associated Legender function* of order m and degree n. Finally the first equation gives

$$\mathscr{S}_{\mathrm{r}} \propto r^n \text{ or } r^{-(n+1)}.$$

Thus

$$\mathscr{S} = \sum_{m,n} [a_{m,n} r^n + b_{m,n} r^{-(n+1)}] P_{n,m}(\theta) e^{im\phi}. \qquad \text{(A. 5. 4)}$$

Appendix V (*General*)

Distribution of the atmospheric parameters

Figure A. 3 (p. 212) shows the distribution of the atmosphere scale height and the normalized height by the scale height. The standard atmospheric parameters are also shown in Table A. 1 and 2 (pp 213–215).

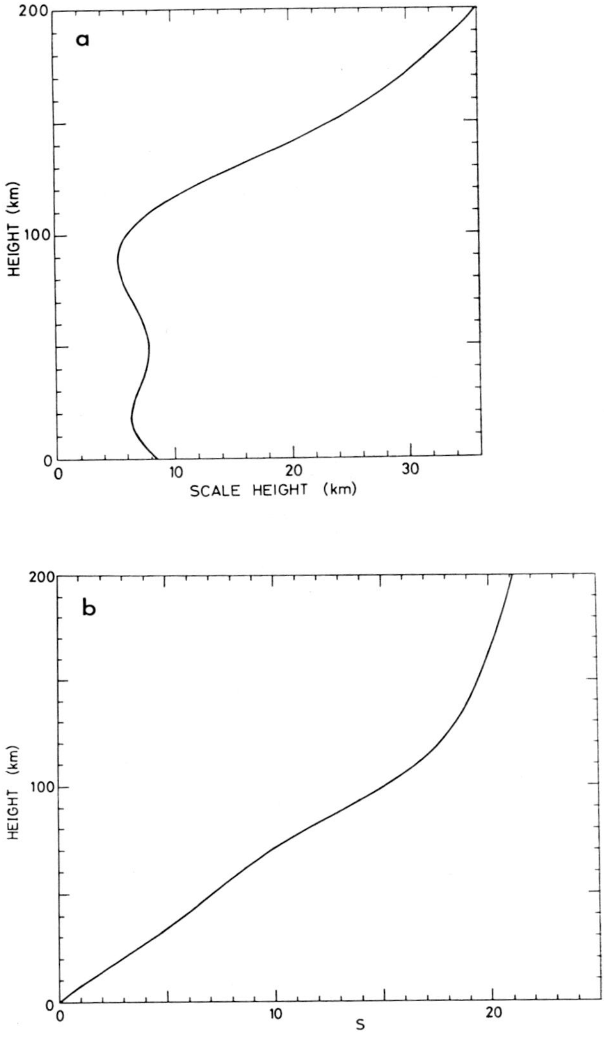

FIG. A. 3. (a) Scale height of the atmosphere, H. (b) Heights measured by the scale height, $s = \int_0^z (\mathrm{d}z/H)$. The parameters used are based on CIRA model, 1972.

TABLE A.1*

ATMOSPHERIC PROPERTIES 30°N LATITUDE

Altitude (km)		Temperature (°K)		Pressure (mb)		Density, ρ (kgm^{-3})		Speed of sound (m/s)	Viscosity** $\mu_0 \times 10^5$	Thermal Conductivity** $\kappa_t \times 10^6$
$\mathscr{Z}$	z	January	July	January	July	January	July	July	(kg/m·sec)	(kcal/m·sec·°K)
0	0	288.52	304.58	1.021×10^{3}	1.014×10^{3}	1.233	1.159	349.9	1.868	6.359
1	1.002	285.24	295.58	9.064×10^{2}	9.044×10^{2}	1.107	1.066	344.7	1.825	6.192
2	2.003	281.86	289.54	8.035	8.048	9.931×10^{-1}	9.683×10^{-1}	341.1	1.796	6.079
3	3.006	275.10	283.72	7.107	7.143	9.000	8.771	337.7	1.768	5.970
4	4.008	268.39	277.82	6.268	6.325	8.135	7.931	334.1	1.739	5.858
5	5.011	261.81	272.13	5.510	5.586	7.331	7.151	330.7	1.711	5.750
6	6.014	255.24	266.44	4.828	4.920	6.589	6.433	327.2	1.683	5.641
7	7.017	248.71	259.36	4.216	4.321	5.905	5.804	322.8	1.647	5.504
8	8.021	242.18	252.27	3.668	3.781	5.276	5.221	318.4	1.611	5.367
9	9.025	235.67	245.22	3.179	3.295	4.699	4.681	313.9	1.574	5.229
10	10.030	229.16	238.18	2.744	2.861	4.172	4.185	309.4	1.537	5.091
15	15.056	208.35	203.15	1.246	1.317	2.083	2.258	285.7	1.346	4.388
20	20.091	208.15	211.95	5.418×10^{1}	5.762×10^{1}	9.068×10^{-2}	9.471×10^{-2}	291.9	1.396	4.567
25	25.133	219.15	222.15	2.438	2.623	3.876	4.114	298.8	1.452	4.772
30	30.184	229.15	232.15	1.138	1.237	1.730	1.856	305.4	1.506	4.971
35	35.243	240.35	243.35	5.491×10^{0}	6.023×10^{0}	7.958×10^{-3}	8.622×10^{-3}	312.7	1.565	5.193
40	40.309	252.35	255.35	2.744	3.036	3.788	4.141	320.3	1.627	5.427
45	45.384	264.35	267.35	1.417	1.579	1.867	2.057	327.8	1.687	5.658
50	50.467	269.15	272.15	7.492×10^{-1}	8.410×10^{-1}	9.697×10^{-4}	1.076	330.7	1.711	5.750
55	55.558	261.15	264.15	3.941	4.456	5.258	5.876×10^{-4}	325.8	1.671	5.597
60	60.656	250.05	252.35	2.022	2.304	2.818	3.180	318.5	1.611	5.368
65	65.764	234.55	233.35	9.990×10^{-2}	1.140	1.484	1.701	306.2	1.512	4.995
70	70.879	219.05	214.35	4.703	5.311×10^{-2}	7.479×10^{-5}	8.632×10^{-5}	293.5	1.409	4.615
75	76.002	203.55	195.35	2.095	2.306	3.585	4.112	280.2	1.302	4.229
80	81.134	191.14	176.34	8.764×10^{-3}	9.189×10^{-3}	1.597	1.815	266.2	1.191	3.836
85	86.274	191.07	172.47	3.586	3.422	6.535×10^{-6}	6.909×10^{-6}	263.3	1.168	3.756
90	91.422	197.72	174.50	1.477	1.273	2.596	2.535	265.2	1.182	3.807
96	97.610	216.04	186.16	5.524×10^{-4}	4.130×10^{-4}	8.781×10^{-7}	7.598×10^{-7}	275.9	1.267	4.105
100	101.742	227.07	192.85	3.018	2.043	4.510	3.577	282.8	1.323	4.302
110	112.096	276.22	266.40	8.150×10^{-5}	4.880×10^{-5}	9.761×10^{-8}	6.002×10^{-8}	337.4	1.766	5.961

* After US Standard Atmosphere Supplements (1966).
** in July.

TABLE A. 2*

SPRING/FALL MODEL, EXOSPHERIC TEMPERATURE=1,000°K

Altitude z, km	Temp. T, °K	Number density m^{-3}				Molecular weight M	Scale height H, km	Pressure p, mb	Log pressure	Density ρ, kg m^{-3}	Log density
		log $n(O_2)$	log $n(O)$	log $n(N_2)$	log $n(He)$						
120	355.0	16.875	16.881	17.602	13.531	26.90	11.62	2.700—5	−4.569	2.461—8	−7.6089
125	439.5	16.583	16.689	17.335	13.449	26.58	14.58	1.842	−4.735	1.340	−7.8730
130	513.0	16.351	16.539	17.123	13.387	26.29	17.23	1.345	−4.871	8.289—9	−8.0815
135	576.9	16.156	16.416	16.946	13.337	26.02	19.60	1.025	−4.989	5.560	−8.2549
140	632.3	15.986	16.311	16.792	13.296	25.76	21.74	8.046—6	−5.094	3.943	−8.4042
145	680.5	15.835	16.219	16.656	13.261	25.51	23.66	6.456	−5.190	2.910	−8.5361
150	722.4	15.697	16.138	16.532	13.231	25.26	25.41	5.266	−5.279	2.214	−8.6547
155	758.8	15.570	16.064	16.419	13.205	25.01	26.99	4.351	−5.361	1.725	−8.7632
160	790.4	15.452	15.995	16.313	13.181	24.77	28.43	3.633	−5.440	1.369	−8.8635
165	817.9	15.340	15.932	16.213	13.160	24.53	29.75	3.059	−5.514	1.104	−8.9572
170	841.8	15.234	15.873	16.119	13.141	24.29	30.97	2.595	−5.586	9.006—10	−9.0455
175	862.5	15.132	15.817	16.028	13.123	24.05	32.10	2.214	−5.655	7.427	−9.1292
180	880.5	15.034	15.763	15.941	13.106	23.81	33.15	1.900	−5.721	6.180	−9.2090
185	896.2	14.939	15.712	15.857	13.090	23.58	34.13	1.637	−5.786	5.181	−9.2856
190	909.8	14.847	15.663	15.776	13.076	23.34	35.05	1.417	−5.849	4.373	−9.3592
195	921.6	14.757	15.615	15.697	13.062	23.11	35.92	1.231	−5.910	3.712	−9.4304
200	931.9	14.669	15.568	15.619	13.048	22.87	36.75	1.073	−5.970	3.166	−9.4994
205	940.8	14.583	15.523	15.543	13.035	22.64	37.54	9.376—7	−6.028	2.713	−9.5665
210	948.6	14.498	15.479	15.468	13.023	22.41	38.30	8.218	−6.085	2.335	−9.6318
215	955.3	14.414	15.436	15.394	13.011	22.18	39.04	7.221	−6.141	2.016	−9.6955
220	961.2	14.332	15.393	15.322	12.999	21.95	39.74	6.361	−6.196	1.747	−9.7577
225	966.3	14.250	15.351	15.250	12.988	21.72	40.43	5.615	−6.251	1.518	−9.8187
230	970.7	14.169	15.310	15.179	12.977	21.50	41.10	4.967	−6.304	1.323	−9.8784
235	974.5	14.089	15.269	15.109	12.966	21.28	41.75	4.402	−6.356	1.156	−9.9370

240	977.9	14.010	15.228	15.039	12.955	21.06	42.39	3.909—7	−6.408	1.013—10	−9.9945
245	980.8	13.931	15.188	14.970	12.945	20.85	43.01	3.477	−6.459	8.890—11	−10.0511
250	983.3	13.852	15.148	14.901	12.934	20.64	43.63	3.098	−6.509	7.821	−10.1067
255	985.5	13.774	15.109	14.832	12.924	20.44	44.23	2.765	−6.558	6.895	−10.1614
260	987.4	13.697	15.070	14.764	12.914	20.24	44.82	2.471	−6.607	6.091	−10.2153
265	989.0	13.620	15.031	14.697	12.904	20.04	45.41	2.212	−6.655	5.390	−10.2684
270	990.5	13.543	14.992	14.629	12.894	19.85	45.98	1.982	−6.703	4.778	−10.3208
275	991.7	13.466	14.953	14.562	12.884	19.66	46.55	1.779	−6.750	4.243	−10.3723
280	992.8	13.390	14.915	14.495	12.874	19.48	47.10	1.599	−6.796	3.774	−10.4232
285	993.8	13.314	14.877	14.429	12.865	19.30	47.65	1.439	−6.842	3.362	−10.4734
290	994.6	13.238	14.839	14.362	12.855	19.13	48.19	1.296	−6.887	2.999	−10.5230
295	995.3	13.162	14.801	14.296	12.845	18.96	48.72	1.169	−6.932	2.680	−10.5719
300	995.9	13.087	14.763	14.230	12.836	18.80	49.25	1.056	−6.976	2.398	−10.6202
310	996.9	12.937	14.687	14.098	12.817	18.49	50.27	8.636—8	−7.064	1.927	−10.7151
320	997.7	12.787	14.613	13.967	12.798	18.21	51.25	7.092	−7.149	1.557	−10.8078
330	998.2	12.638	14.538	13.837	12.779	17.94	52.21	5.846	−7.233	1.263	−10.8985
340	998.7	12.490	14.464	13.707	12.761	17.69	53.13	4.835	−7.316	1.030	−10.9872
350	999.0	12.342	14.390	13.578	12.742	17.45	54.02	4.011	−7.397	8.429—12	−11.0742
360	999.2	12.195	14.316	13.449	12.724	17.23	54.88	3.338	−7.476	6.925	−11.1596
370	999.4	12.048	14.243	13.320	12.705	17.03	55.72	2.786	−7.555	5.710	−11.2434
380	999.6	11.902	14.169	13.192	12.687	16.83	56.54	2.332	−7.632	4.723	−11.3258
390	999.7	11.756	14.096	13.065	12.669	16.65	57.34	1.956	−7.709	3.918	−11.4069
400	999.8	11.611	14.024	12.937	12.650	16.47	58.13	1.645	−7.784	3.260	−11.4868

* After U.S. standard atmosphere supplement (1966).

REFERENCES

Abreu Neto, A. L. and C. H. Liu, Generation of acoustic-gravity waves in isothermal atmosphere by moving sources and ground movements, Ionosphere Radio Laboratory, University of Illinois, Technical Report No. 46, May, 1972.

Amayenc, P., Tidal oscillations of the meridional neutral wind at mid-latitudes, *Radio Sci.*, **9**, 281–293, 1974.

Antoniadis, D. A., Thermospheric winds and exospheric temperatures from incoherent scatter radar measurements in four seasons, *J. Atmos. Terr. Phys.*, **38**, 187–195, 1976.

Aso, T., T. Tsuda, and S. Kato, Meteor radar observations at Kyoto University, *J. Atmos. Terr. Phys.*, **41**, 517–525, 1979.

Axford, W. I., The formation and vertical movement of dense ionized layers in the ionosphere due to neutral wind shears, *J. Geophys. Res.*, **68**, 769–779, 1963.

Beer, T., *Atmospheric Waves*, Adam Hilger Ltd., London, 1974.

Bernard, R., Tides in the E region observed by incoherent scatter over St. Santin, *J. Atmos. Terr. Phys.*, **36**, 1105–1120, 1974.

Bertin, F., K. A. Hughes, and L. Kersley, Atmospheric waves induced by the solar eclipse of 30 June, 1973, *J. Atmos. Terr. Phys.*, **39**, 457–461, 1977.

Blanc, M., P. Amayenc, P. Bauer, and C. Taieb, Electric field induced drifts from the French incoherent scatter facility, *J. Geophys. Res.*, **82**, 87–97, 1977.

Blumen, W. and R. G. Hendl, On the role of joule heating as a source of gravity wave energy above 100 km, *J. Atmos. Sci.*, **26**, 210–217, 1969.

Booker, J. R. and F. P. Bretherton, The critical layer for internal gravity waves in a shear flow, *J. Fluid Mech.*, **27**, 513–539, 1967.

Broche, P., M. Crochet, and J. C. De Maistre, Gravity waves generated by the 30 June 1973 solar eclipse in Africa, *J. Atmos. Terr. Phys.*, **38**, 1361–1364, 1976.

Butler, S. T. and K. A. Small, The excitation of atmospheric oscillations,

Proc. Roy. Soc., Ser. A, **274**, 91–121, 1963.

CHAPMAN, S. and J. BARTELS, *Geomagnetism*, Vol. II, pp. 620–637, 750–794, Oxford University Press, London, 1940.

CHAPMAN, S. and R. S. LINDZEN, *Atmospheric Tides*, D. Reidel Publ. Co., Dordrecht-Holland, 1970.

CHARNEY, J. G. and P. G. DRAZIN, Propagation of planetary-scale disturbances from the lower into the upper atmosphere, *J. Geophys. Res.*, **66**, 83–109, 1961.

CHIMONAS, G., The equatorial electrojet as a source of long period travelling ionospheric disturbances, *Planet. Space Sci.*, **18**, 583–589, 1970.

CHIMONAS, G. and W. I. AXFORD, Vertical movement of temperate-zone sporadic E layers, *J. Geophys. Res.*, **73**, 111–117, 1968.

CHIMONAS, G. and C. O. HINES, Atmospheric gravity waves launched by auroral currents, *Planet. Space Sci.*, **18**, 565–582, 1970.

CLARK, R. R., Meteor wind measurements at Durham, New Hampshire (43°N, 71°W), *J. Atmos. Sci.*, **32**, 1689–1693, 1975.

COLE, K. D., Joule heating of the upper atmosphere, *Aust. J. Phys.*, **15**, 223–235, 1962.

DICKINSON, R. E., Theory of planetary wave-zonal flow interaction, *J. Atmos. Sci.*, **26**, 73–81, 1969.

DICKINSON, R. E. and J. E. GEISLER, Vertical motion field in the middle thermosphere from satellite drag densities, *Mon. Weath. Rev.*, **96**, 606–616, 1969.

DICKINSON, R. E., E. C. RIDLEY, and R. G. ROBLE, Meridional circulation in the thermosphere, 1. Equinox conditions, *J. Atmos. Sci.*, **32**, 1737–1754, 1975.

DUNGEY, J. W., *Cosmic Electrodynamics*, pp. 167–168, Cambridge University Press, London, New York, 1958.

DÜTSCH, H. U., Photochemistry of atmospheric ozone, *Adv. Geophys.*, **15**, 219–322, 1971.

ECKART, C., *Hydrodynamics of Oceans and Atmospheres*, pp. 95–97, 256–279, Pergamon Press, London, New York, Paris, 1960.

ESKINAZI, S., *Fluid Mechanics and Thermodynamics of Our Environment*, pp. 54–131, Academic Press, New York, San Francisco, London, 1975.

EVANS, J. V., Theory and practice of ionospheric study by Thomson scatter radar, *Proc. IEEE*, **57**, 496–530, 1969.

EVANS, J. V., Incoherent scatter contributions to studies of the dynamics of the lower thermosphere, *Rev. Geophys. Space Phys.*, **16**, 195–216, 1978.

FÄLTHAMMAR, C.-G., Problems related to macroscopic electric fields in the magnetosphere, *Rev. Geophys.*, **15**, 457–466, 1977.

FELLOUS, J. L., R. BERNARD, M. GLASS, M. MASSEBEUF, and A. SPIZZICHINO, A study of the variations of atmospheric tides in the meteor zone, *J. Atmos. Terr. Phys.*, **37**, 1511–1524, 1975.

FLATTERY, T. W., Hough functions, Tech. Rep. No. 21 to NSF, The University of Chicago, 1967.

FORBES, J. M. and H. B. GARRETT, Solar diurnal tide in the thermosphere, *J. Atmos. Sci.*, **33**, 2226–2241, 1976.

FORBES, J. M. and H. B. GARRETT, Seasonal-latitudinal structure of the diurnal thermospheric tide, *J. Atmos. Sci.*, **35**, 148–159, 1978.

FRANCIS, S. H., Acoustic-gravity mode and large-scale traveling ionospheric disturbances of a realistic dissipative atmosphere, *J. Geophys. Res.*, **78**, 2278–2301, 1973.

FRANCIS, S. H., A theory of medium-scale ionospheric disturbances, *J. Geophys. Res.*, **79**, 5245–5260, 1974.

FRIEDMAN, J. P., Propagation of internal gravity waves in a thermally stratified atmosphere, *J. Geophys. Res.*, **71**, 1033–1054, 1966.

FUJITAKA, K. and T. TOHMATSU, A tidal theory of the ionospheric intermediate layer, *J. Atmos. Terr. Phys.*, **35**, 425–438, 1973.

FUKAO, S., S. KATO, S. YOKOI, R. M. HARPER, R. F. WOODMAN, and W. E. GORDON, One full-day radar measurement of lower stratospheric winds over Jicamarca, *J. Atmos. Terr. Phys.*, **40**, 1331–1337, 1978.

FUKAO, S., T. SATO, S. KATO, and R. M. HARPER, Mesospheric winds and waves over Jicamarca on 23–24 May, 1974, *J. Geophys. Res.*, **84**, 4379–4386, 1979.

FUKUYAMA, K., Airglow variations and dynamics in the lower thermosphere and upper mesosphere. 1. Diurnal variation and its seasonal dependency, *J. Atmos. Terr. Phys.*, **38**, 1279–1287, 1976.

GAGE, K. S. and J. L. GREEN, Evidence for specular reflection from monostatic VHF radar observations of the stratosphere, *Radio Sci.*, **13**, 991–1001, 1978.

GELLER, M. A., S. A. BOWHILL, and G. C. HESS, A description of the University of Illinois meteor radar system and some first results, *J. Atoms. Terr. Phys.*, **39**, 15–24, 1977.

GEORGES, T. M., HF doppler studies of traveling ionospheric disturbances, *J. Atmos. Terr. Phys.*, **30**, 735–746, 1968.

GIWA, F. B. A., The response curve in atmospheric oscillations, *Quart. J. Roy. Meteorol. Soc.*, **94**, 192–205, 1968.

GROVES, G. V., Annual and semi-annual zonal wind component and corresponding temperature and density variations 60–130 km, *Planet. Space Sci.*, **20**, 2099–2112, 1972.

HARPER, R. M., A comparison of ionospheric currents, magnetic variations and electric field at Arecibo, *J. Geophys. Res.*, **82**, 3233–3242, 1977a.

HARPER, R. M., Tidal winds in the 100–200 km region at Arecibo, *J. Geophys. Res.*, **82**, 3243–3250, 1977b.

HARPER, R. M., A semidiurnal tide in the meridional wind at F region heights at low latitudes, *J. Geophys. Res.*, **84**, 411–415, 1979.

HARPER, R. M., R. H. WAND, and J. D. WHITEHEAD, Comparison of Arecibo

E-region data and sporadic-E theory: A measurement of the diffusion coefficient, *Radio Sci.*, **10**, 357–361, 1975.
HARPER, R. M., R. H. WAND, C. J. ZAMLUTTI, and D. T. FARLEY, E region ion drifts and winds from incoherent scatter measurement at Arecibo, *J. Geophys. Res.*, **81**, 25–35, 1976.
HARRIS, I. and H. G. MAYR, Diurnal variations in the thermosphere 1. Theoretical formulation, *J. Geophys. Res.*, **80**, 3925–3933, 1975.
HASEGAWA, M., On the progressive change of the field of diurnal variation of terrestrial magnetism, *Proc. Imp. Acad. Japan*, **12**, 277–280, 1936.
HAURWITZ, B., The diurnal surface-pressure oscillation, *Arch. Meteorol. Geophys. Bioklimatol, A, Meteorol. Geophys.*, **14**, 361–379, 1965.
HAURWITZ, B. and B. FOGLE, Waves forms in noctilucent clouds, *Deep Sea Res.*, **16**, 85–95, 1969.
HEARN, A. L. and K. C. YEH, Medium scale TID's and their associated internal gravity waves as seen through height-dependent electron density power spectra, *J. Geophys. Res.*, **82**, 4983–4990, 1977.
HEELIS, R. A., P. C. KENDALL, R. J. MOFFETT, D. W. WINDLE, and H. RISHBETH, Electrical coupling of the E-and F-regions and its effects on F-region drifts and winds, *Planet. Space Sci.*, **22**, 743–756, 1974.
HERRON, T. J., Dispersion of ionospheric waves, *Nature*, **248**, 213–215, 1974.
HINES, C. O., Internal atmospheric gravity waves at ionospheric heights, *Can. J. Phys.*, **38**, 1441–1481, 1960.
HINES, C. O., Dynamical heating of the upper atmosphere, *J. Geophys. Res.*, **70**, 177–183, 1965.
HINES, C. O., Diurnal tide in the upper atmosphere, *J. Geophys. Res.*, **71**, 1453–1459, 1966.
HINES, C. O. and C. A. REDDY, On the propagation of atmospheric gravity waves through regions of wind shear, *J. Geophys. Res.*, **72**, 1015–1034, 1967.
HIRONO, M. and T. KITAMURA, A dynamo theory in the ionosphere, *J. Geomag. Geoelectr.*, **8**, 9–23, 1956.
HOLTON, J. R., *The Dynamic Meteorology of the Stratosphere and Mesosphere*, pp. 3–10, Am. Meteorol. Soc., 1975.
HONG, S. and R. S. LINDZEN, Solar semidiurnal tide in the thermosphere, *J. Atmos. Sci.*, **33**, 135–153, 1976.
HOOKE, W. H., Ionospheric irregularities produced by internal atmospheric gravity waves, *J. Atmos. Terr. Phys.*, **30**, 795–823, 1968.
HOUGH, S. S., On the application of harmonic analysis to the dynamical theory of the tides, *Phil. Trans. Roy. Soc. London*, **191**, 139–185, 1898.
HUNG, R. J. and R. E. SMITH, Dynamics of severe storms through the study of thermospheric-tropospheric coupling, *J. Geomag. Geoelectr.*, **31**, 183–194, 1979.
HUNSUCKER, R. D., Estimate of the relative importance of joule heating and

the Lorentz force in generating atmospheric gravity waves from the auroral electrojet, *J. Geophys. Res.*, **82**, 4826–4828, 1977.

Jacchia, L. G., Static diffusion models of the upper atmosphere with empirical temperature profiles, Res. in Space Sci. Smith Astrophys. Obs., No. 170, December, 1964.

Jones, M. N., Atmospheric oscillations (I), *Planet. Space Sci.*, **18**, 1393–1416, 1970.

Jones, M. N., Atmospheric oscillations (II), *Planet. Space Sci.*, **19**, 609–634, 1971.

Jordan, A. R., Atmospheric gravity waves from winds and storms, *J. Atoms. Sci.*, **29**, 445–456, 1972.

Justus, C. G. and A. Woodrum, Upper atmospheric planetary-wave and gravity-wave observations, *J. Atmos. Sci.*, **30**, 1267–1275, 1973.

Kamiide, Y. and J. L. Horwitz, Chatanika radar observations of ionospheric and field-aligned currents, *J. Geophys. Res.*, **83**, 1063–1070, 1978.

Kato, S., Horizontal wind systems in the ionospheric E region deduced from the dynamo theory of the geomagnetic S_q variation. II. Rotating earth, *J. Geomag. Geoelectr.*, **8**, 24–37, 1956.

Kato, S., Horizontal wind systems in the ionospheric E region deduced from the dynamo theory of the geomagnetic S_q variation. IV, *J. Geomag. Geoelectr.*, **9**, 107–115, 1957a.

Kato, S., Prevailing wind in the ionosphere and geomagnetic S_q variations, *J. Geomag. Geoelectr.*, **9**, 215–217, 1957b.

Kato, S., Joule heating and temperature in the upper atmosphere, *Planet. Space Sci.*, **9**, 939–946, 1962.

Kato, S., Theory of movement of irregularities in the upper atmosphere, *Planet. Space Sci.*, **11**, 823–830, 1963a.

Kato, S., Joule heating at the magnetic equator, *Planet. Space Sci.*, **11**, 1297–1302, 1963b.

Kato, S., Diurnal atmospheric oscillation, 1. Eigenvalues and Hough functions, *J. Geophys. Res.*, **71**, 3201–3209, 1966a.

Kato, S., Diurnal atmospheric oscillation, 2. Thermal excitation in the upper atmosphere, *J. Geophys. Res.*, **71**, 3211–3214, 1966b.

Kato, S., Theoretical basis of the formation of sporadic *E* due to wind motion in the ionosphere, *Radio Sci.*, **1**, New Ser., 226–234, 1966c.

Kato, S., Diurnal and semi-diurnal atmospheric tidal oscillation, eigenvalues and Hough functions, *Rep. Ionos. Space Res. Japan*, **20**, 448–463, 1966d.

Kato, S., Cross-field instability for the formation of sporadic E, *Radio Sci.*, **7**, 417–423, 1972.

Kato, S., Electric field and wind motion at the magnetic euqator, *J. Geophys. Res.*, **78**, 757–762, 1973.

Kato, S., Uniqueness on ionospheric wind deduced from geomagnetic S_q variation, *J. Geomag. Geoelectr.*, **26**, 559–561, 1974.

KATO, S. and S. MATSUSHITA, A consideration on the tidal wave transmission through the ionized atmosphere, *J. Geomag. Geoelectr.*, **21**, 471–478, 1969.

KATO, S., T. ASO, and T. HORIUCHI, Sporadic-E formation by wind shear, comparison between observation and theory, *Radio Sci.*, **7**, 359–362, 1972.

KATO, S., K. KAWAKAMI, and D. ST-JOHN, Theory of gravity wave emission from moving sources in the upper atmosphere, *J. Atmos. Terr. Phys.*, **39**, 581–588, 1977.

KATO, S. and T. TSUDA, A consideration on the tidal dynamo theory, to be published in 1980.

KELVIN, W. (Thomson), On the thermodynamic acceleration of the earth's rotation, *Proc. Roy. Soc., Edinb.*, **11**, 396–405, 1882.

KIRCHHOFF, V. W. J. H. and L. A. CARPENTER, The day-to-day variability in ionospheric electric fields and currents, *J. Geophys. Res.*, **81**, 2737–2742, 1976.

KOHL, H. and J. W. KING, Atmospheric winds between 100 and 700 km and their effects on the ionosphere, *J. Atmos. Terr. Phys.*, **29**, 1045–1062, 1967.

LAMB, H., *Hydrodynamics*, pp. 1–6, 330–357, Dover, New York, 1945.

LEOVY, C. B., Simple models of thermally driven mesospheric circulation, *J. Atmos. Sci.*, **21**, 327–341, 1964.

LINDZEN, R. S., Thermally driven diurnal tide in the atmosphere, *Quart. J. Roy. Meteorol. Soc.*, **93**, 18–42, 1967.

LINDZEN, R. S., Internal gravity waves in atmospheres with realistic dissipation and temperature, III. Daily variation in the thermosphere, *Geophys. Fluid Dyn.*, **2**, 89–121, 1971.

LINDZEN, R. S. and H. L. KUO, A reliable method for numerical integration of a large class of ordinary and partial differential equations, *Mon. Weath. Rev.*, **97**, 732–734, 1969.

LINDZEN, R. S. and S. HONG, Effects of mean winds and horizontal temperature gradients on solar and lunar semidiurnal tides in the atmosphere, *J. Atmos. Sci.*, **31**, 1421–1446, 1974.

LONGUET-HIGGINS, M. S., The eigenfunctions of Laplace's tidal equations over a sphere, *Phil. Trans. Roy. Soc., Ser. A*, **262**, 511–607, 1968.

MAEDA, H., Horizontal wind systems in the ionospheric E region deduced from the dynamo theory of the geomagnetics S_q variation. I. Nonrotating earth, *J. Geomag. Geoelectr.*, **7**, 121–132, 1955.

MAEDA, H., Horizontal wind systems in the ionospheric E region deduced from the dynamo theory of the geomagnetic S_q variation. III, *J. Geomag. Geoelectr.*, **9**, 86–93, 1957.

MAEDA, H., Field-aligned current induced by asymmetric dynamo action in the ionosphere, *J. Atmos. Terr. Phys.*, **36**, 1395–1401, 1974.

MAEDA, H. and H. MURATA, Electric currents induced by non-periodic winds in the ionosphere (1), *J. Geophys. Res.*, **73**, 1077–1092, 1968.

MAEDA, K. and H. MURATA, Ionospheric dynamo theory with consideration for magnetospheric current along the geomagnetic lines of force, *Rep. Ionos. Space Res. Japan*, **19**, 272–285, 1965.

MAEDA, K. and S. KATO, Electrodynamics of the ionosphere, *Space Sci. Rev.*, **5**, 57–79, 1966.

MARTYN, D. F., Cellular atmospheric waves in the ionosphere and troposphere, *Proc. Roy. Soc. London, Ser. A*, **201**, 216–234, 1950.

MARTYN, D. F., Electric currents in the ionosphere. III. Ionization drift due to winds and electric fields, *Phil. Trans. Roy. Soc. London, Ser. A*, **246**, 306–320, 1953.

MARTYN, D. F. and W. G. BAKER, Ionization drift due to winds and electric fields, *Phil. Trans. Roy. Soc. London, Ser. A*, **246**, 281–305, 1953.

MATSUSHITA, S., Interrelations of sporadic E and ionospheric currents, in *Ionospheric Sporadic E*, edited by E. K. Smith and S. Matsushita, pp. 276–291, Pergamon Press, Oxford, 1962.

MATSUSHITA, S. and H. MAEDA, On the geomagnetic solar quiet daily variation field during the IGY, *J. Geophys. Res.*, **70**, 2535–2558, 1965.

MATUURA, N., Electric fields deduced from the thermospheric model, *J. Geophys. Res.*, **79**, 4679–4689, 1974.

MIYAHARA, S., Zonal mean winds induced by vertically propagating atmospheric tidal waves in the lower thermosphere, *J. Meteorol. Soc. Japan*, **56**, 86–97, 1978.

MÖHLMAN, D., Non-uniqueness of ionospheric winds and electric fields, deduced from geomagnetic S_q variations, *Gerlands Beitr. Geophysik (Leipzig)*, **83**, 16–18, 1974.

MUNRO, G. H., Travelling ionospheric disturbances in the F region, *Aust. J. Phys.*, **11**, 91–112, 1958.

MURATA, H., Thermal excitation of diurnal atmospheric oscillation in the lower mesosphere, *Rep. Ionos. Space Res. Japan*, **22**, 213–234, 1968.

MURGATROYD, R. J., Winds in the mesosphere and lower thermosphere, *Proc. Roy. Soc. London, Ser. A*, **288**, 575–589, 1965.

NASTROM, G. D. and A. D. BELMONT, Periodic variations in stratospheric-mesospheric temperature from 20–65 km at 80°N to 30°S, *J. Atmos. Sci.*, **32**, 1715–1722, 1975.

NOXON, J. F., Effect of internal gravity waves upon night airglow temperatures, *Geophys. Res. Lett.*, **5**, 25–27, 1978.

OBERHETTINGER, F., *Tabellen zur Fourier Transformation*, Mathematischen Wissenschaften, Band XC, pp. 10–14, 18–32, 69–78, Springer-Verlag, Berlin, Göttingen, Heidelberg, 1957.

OKUZAWA, T., F. OHMI, and O. TSUKIO, Observation of TID by an array of HF-doppler sounders, *Sol. Terr. Environ. Res. Japan*, **2**, 141–144, 1978.

OYA, H. and T. ASO, Electron density profile by gyro-plasma probe, Sounding rocket data in Japan, Vol. 1 (1958–1972), p. 28, Dept. Space Sci.

ISAS, University of Tokyo, 1972.

PEKERIS, C. L., Atmospheric oscillations, *Proc. Roy. Soc., Ser. A*, **158**, 650–671, 1937.

PEKERIS, C. L., The propagation of a pulse in the atmosphere, *Proc. Roy. Soc., Ser. A*, **171**, 434–449, 1939.

PERKINS, F. W. and S. H. FRANCIS, Artificial production of traveling ionospheric disturbances and large-scale atmospheric motion, *J. Geophys. Res.*, **79**, 3879–3881, 1974.

PETITDIDIER, M. and H. TEITELBAUM, Lower thermosphere emission and tides, *Planet. Space Sci.*, **25**, 711–721, 1977.

PITTEWAY, M. L. V. and C. O. HINES, The viscous damping of atmospheric gravity waves, *Can. J. Phys.*, **41**, 1935–1948, 1963.

PRAKASH, S., S. P. GUPTA, and B. H. SUBBARAYA, A study of the irregularities in the night time equatorial E-region using a Langmuir probe and plasma noise probe, *Planet. Space Sci.*, **18**, 1307–1318, 1970.

PRASAD, S. S. and D. R. FURMAN, Atmospheric gravity waves: their effects on ionospheric temperature and composition, *J. Atmos. Terr. Phys.*, **37**, 17–30, 1975.

RASTOGI, P. K. and R. F. WOODMAN, Mesospheric studies using the Jicamarca incoherent-scatter radar, *J. Atmos. Terr. Phys.*, **36**, 1217–1231, 1974.

RASTOGI, P. K. and S. A. BOWHILL, Gravity waves in the equatorial mesosphere, *J. Atmos. Terr. Phys.*, **38**, 51–60, 1976.

REED, R. J., Semidiurnal tidal motions between 30 and 60 km, *J. Atmos. Sci.*, **24**, 315–317, 1967.

REES, D., Rocket measurement of annual mean prevailing, diurnal and semidiurnal winds in the lower thermosphere at mid-latitudes, *J. Geomag. Geoelectr.*, **31**, 253–265, 1979.

RICHMOND, A. D., The nature of gravity wave ducting in the thermosphere, *J. Geophys. Res.*, **83**, 1385–1389, 1978.

RICHMOND, A. D. and S. MATSUSHITA, Thermospheric response to a magnetic substorm, *J. Geophys. Res.*, **80**, 2839–2850, 1975.

RICHMOND, A. D., S. MATSUSHITA, and J. D. TARPLEY, On the production mechanism of electric currents and fields in the ionosphere, *J. Geophys. Res.*, **81**, 547–555, 1976.

RISHBETH, H., The F-layer dynamo, *Planet. Space Sci.*, **19**, 263–267, 1971.

ROBLE, R. G. and R. E. DICKINSON, The effect of displaced geomagnetic and geographic poles on the thermospheric neutral winds, *Planet. Space Sci.*, **22**, 623–631, 1974.

ROBLE, R. G., J. E. SALAH, and B. A. EMERY, The season variation of the diurnal thermospheric winds over Millstone Hill during solar cycle maxium, *J. Atmos. Terr. Phys.*, **39**, 503–511, 1977.

ROBLE, R. G., B. A. EMERY, J. E. SALAH, and P. B. HAYS, Diurnal variation of the neutral thermospheric winds determined from incoherent scatter radar data, *J. Geophys. Res.*, **79**, 2868–2876, 1974.

ROBLE, R. G., A. D. RICHMOND, W. L. OLIVER, and R. M. HARPER, Ionospheric effects of the gravity wave launched by the September 18, 1974, sudden commencement, *J. Geophys. Res.*, **83**, 999–1009, 1978.

RÖTTGER, J., Drifting patches of equatorial spread-F irregularities, experimental support for the spatial resonance mechanism in the ionosphere, *J. Atmos. Terr. Phys.*, **40**, 1103–1112, 1978.

RÖTTGER, J. and C. H. LIU, Partial reflection and scattering of VHF radar signals from the clear atmosphere, *Geophys. Res. Lett.*, **5**, 357–360, 1978.

ROW, R. V., Acoustic-gravity waves in the upper atmosphere due to a nuclear detonation and an earthquake, *J. Geophys. Res.*, **72**, 1599–1610, 1967.

RÜSTER, R., J. RÖTTGER, and R. F. WOODMAN, Radar measurement of waves in the lower stratosphere, *Geophys. Res. Lett.*, **5**, 119–122, 1978.

SALAH, J. E. and R. H. WAND, Tides in the temperature of the lower thermosphere at mid-latitudes, *J. Geophys. Res.*, **79**, 4295–4304, 1974.

SALAN, J. E. and J. V. EVANS, Tests of electrodynamic consistency from daytime ionospheric drift observations, *J. Geophys. Res.*, **82**, 2413–2418, 1977.

SALAH, J. E., R. H. WAND, and J. V. EVANS, Tidal effects in the E region from incoherent scatter radar observations, *Radio Sci.*, **10**, 347–355, 1975.

SALAH, J. E., R. H. WAND, and R. BERNARD, Comparison of simultaneous tidal oscillations by incoherent scatter radars, *Ann. Geophys.* **33**, 95–102, 1977.

SATO, T., Unified theory of type I and II irregularities in the equatorial electrojet, *J. Geophys. Res.*, **78**, 2232–2243, 1973.

SCHIELDGE, J. P., S. V. VENKATESWARAN, and A. D. RICHMOND, The ionospheric dynamo and equatorial magnetic variations, *J. Atmos. Terr. Phys.*, **35**, 1045–1061, 1973.

SCHUSTER, A., The diurnal variation of terrestrial magnetism, *Phil. Trans. Roy. Soc. London, Ser. A*, **180**, 467–518, 1889.

SIEBERT, M., Atmospheric tides, *Adv. Geophys.*, **7**, 105–187, 1961.

STENING, R. J., An assessment of the contributions of various tidal winds to the S_q-current system, *Planet. Space Sci.*, **17**, 889–908, 1969.

STENING, R. J., Ionospheric dynamo calculations with semidiurnal winds, *Planet. Space Sci.*, **25**, 1075–1080, 1977.

STEWART, B., Terrestrial magnetism, *Encyclopedia Britannica*, **16**, 159–184, 9th, 1883.

STRATTON, J. A., *Electromagnetic Theory*, pp. 230–236, McGraw Hill, New York, London, 1941.

SUGIURA, M. and D. J. POROS, An improved model equatorial electrojet with a meridional current system, *J. Geophys. Res.*, **74**, 4025–4034, 1969.

SWIFT, D. W., The generation of infrasonic waves by auroral electrojets, *J. Geophys. Res.*, **78**, 8205–8221, 1973.

TAFFE, W. J., Hydromagnetic effects on atmospheric tides, *J. Geophys. Res.*, **74**, 5775–5791, 1969.

TARPLEY, J. D., The ionospheric wind dynamo, II. Solar tides, *Planet. Space Sci.*, **18**, 1091–1103, 1970.

TAYLOR, G. I., The oscillations of the atmosphere, *Proc. Roy. Soc., Ser. A*, **156**, 318–326, 1936.

TESTUD, J., Gravity waves generated during magnetic substorms, *J. Atmos. Terr. Phys.*, **32**, 1793–1805, 1970.

THOME, G., Long-period waves generated in the polar ionosphere during the onset of magnetic storms, *J. Geophys. Res.*, **73**, 6319–6336, 1968.

TITHERIDGE, J. E., Periodic disturbances in the ionosphere, *J. Geophys. Res.*, **73**, 243–252, 1968.

TITHERIDGE, J. E., Nonperiodic irregularities in the ionosphere, *J. Geophys. Res.*, **76**, 6955–6960, 1971.

TSUTSUI, M. and T. OGAWA, HF doppler observation of ionospheric effects due to typhoons, *Rep. Ionos. Space Res. Japan*, **27**, 121–123, 1973.

VAN SABBEN, D., Ionospheric current systems caused by non-periodic winds, *J. Atmos. Terr. Phys.*, **24**, 959–974, 1962.

VOLLAND, H., A simplified model of the geomagnetic S_q-current system and the electric fields within the ionosphere, *Cosm. Electrodyn.*, **1**, 428–459, 1971.

VOLLAND, H., Solutions of Laplace's tidal equation for complex frequencies, *J. Atmos. Terr. Phys.*, **36**, 445–460, 1974.

VOLLAND, H. and H. G. MAYR, A three-dimensional model of thermosphere dynamics, 2. Tidal waves, *J. Atmos. Terr. Phys.*, **34**, 1769–1799, 1972.

VOLLAND, H. and L. GRELLMANN, A hydromagnetic dynamo of the atmosphere, *J. Geophys. Res.*, **83**, 3699–3708, 1978.

WALLACE, J. M. and F. R. HARTRANFT, Diurnal wind variations, surface to 30 kilometers, *Mon. Weath. Rev.*, **97**, 446–455, 1969.

WALLACE, J. M. and R. F. TADD, Some further results concerning the vertical structure of atmospheric tidal motions within the lowest 30 kilometers, *Mon. Weath. Rev.*, **102**, 795–803, 1974.

WALKER, J. C. G., Analytic representation of upper atmosphere densities based on Jacchia's static diffusion models, *J. Atmos. Sci.*, **22**, 462–463, 1965.

WAND, R. H., Semi-diurnal tide in the E region from incoherent scatter measurement at Arecibo, *Radio Sci.*, **11**, 641–652, 1976.

WATSON, G. N., *A Treatise on the Theory of Bessel Functions*, pp. 176, 405, 417, 435, Cambridge University Press, London, 1944.

WHITEHEAD, J. D., The formation of the sporadic-E layer in the temperate zones, *J. Atmos. Terr. Phys.*, **20**, 49–58, 1961.

WHITEHEAD, J. D., Ionization disturbances caused by gravity waves in the presence of an electrostatic field and background wind, *J. Geophys. Res.*, **76**, 238–241, 1971.

WILKES, M. V., *Oscillation of the Earth's Atmosphere*, Cambridge University Press, London, 1949.

WILSON, C. R., Auroral infrasonic waves, *J. Geophys. Res.*, **74**, 1812–1836, 1969.

WOODMAN, R. F. and A. GULLEN, Radar observations of winds and turbulence in the stratosphere and mesosphere, *J. Atmos. Sci.*, **31**, 493–505, 1974.

YANOWITCH, M., Effect of viscosity on gravity waves and the upper boundary condition, *J. Fluid Mech.*, **29**, 209–231, 1967.

YOUNG, J. M., C. Y. JOHNSON, and J. C. HOLMES, Positive ion composition of a temperate-latitude sporadic E layer as observed during a rocket flight, *J. Geophys. Res.*, **72**, 1473–1479, 1967.

Though not all are specifically cited in this book, the following books are recommended as monographs and textbooks on atmospheric waves:

BEER, T., *Atmospheric Waves*, Adam Hilger Ltd., London, 1974.

CHAPMAN, S. and R. S. LINDZEN, *Atmospheric Tides*, D. Reidel Publ. Co., Dordrecht-Holland, 1970.

HINES, C. O. and Colleagues, *The Upper Atmosphere in Motion, a Selection of Papers with Annotation*, Am. Geophys. Union, 1974.

HOLTON, J. R., *The Dynamic Meteorology of the Stratosphere and Mesosphere*, Am. Meteorol. Soc., 1975.

TOLSTOY, I., *Wave Propagation*, McGraw Hill, New York, London, 1973.

YEH, K. C. and C. H. LIU, *Theory of Ionospheric Waves*, Academic Press, New York, London, 1972.

SUBJECT INDEX